International Conference on Mechanical Engineering Research (ICMER 2015)

MATEC Web of Conferences Volume 74 (2016)

Kuantan, Malaysia
18 - 19 August 2015

Editors:

Saiful Anwar Che Ghani
Azri Alias
Rizalman Mamat
Mustafizur Rahman

ISBN: 978-1-5108-2927-5

Printed from e-media with permission by:

Curran Associates, Inc.
57 Morehouse Lane
Red Hook, NY 12571

proceedings.com

Some format issues inherent in the e-media version may also appear in this print version.

Printed by Curran Associates, Inc. (2016)

For additional information, please contact EDP Sciences – Web of Conferences at the address below.

EDP Sciences – Web of Conferences
17, Avenue du Hoggar
Parc d'Activité de Courtabœuf
BP 112
F-91944 Les Ulis Cedex A
France

Phone: +33 (0) 1 69 18 75 75
Fax: +33 (0) 1 69 28 84 91

contact-edps@webofconferences.org

Additional copies of this publication are available from:

Curran Associates, Inc.
57 Morehouse Lane
Red Hook, NY 12571 USA
Phone: 845-758-0400
Fax: 845-758-2633
Email: curran@proceedings.com
Web: www.proceedings.com

TABLE OF CONTENTS

Preface The 3rd ICMER2015

The 3rd ICMER2015 is the continuity of the NCMER2010. The year 2010 represents a significant milestone in the history for Faculty of Mechanical Engineering, Universiti Malaysia Pahang (UMP) Malaysia with the organization of the first and second national level conferences (1st and 2nd NCMER) at UMP on May 26-27 and Dec 3-4 2010. The Faculty then has changed the name from National Conference on Mechanical Engineering Research (NCMER) to International Conference on Mechanical Engineering Research (ICMER) in 2011 and this year, 2015 is our 3rd ICMER.

These proceedings contain the selected scientific manuscripts submitted to the conference. It is with great pleasure to welcome you to the "International Conference on Mechanical Engineering Research ICMER2015" that is held at Zenith Hotel, Kuantan, Malaysia. The call for papers attracted submissions of over two hundred abstracts from twelve different countries such as Japan, Iran, China, Kuwait, Indonesia, Norway, Philippines, Morocco, Germany, UAE and many more.

The scientific papers published in these proceedings have been revised and approved by the technical committee of the 3rd ICMER2015. All of the papers exhibit clear, concise, and precise expositions that appeal to a broad international readership interested in mechanical engineering, combustion, metallurgy, materials science as well as in manufacturing and biomechanics. The reports present original ideas or results of general significance supported by clear reasoning and compelling evidence, and employ methods, theories and practices relevant to the research. The authors state clearly the questions and the significance of their research to theory and practice, describe how the research contributes to new knowledge, and provide tables and figures that meaningfully add to the narrative.

In this edition of ICMER representatives coming are from academia, industry, governmental and private sectors. The plenary and invited speakers present, discuss, promote and disseminate research in all fields of mechanical engineering in a constructive manner. Numerous social events that provide opportunities to the participants to renew old contacts and establish new ones were scheduled to accompany and follow the scientific activities. They also aim to acquaint the participants with the capital of state Pahang, Kuantan. In addition, booths were also prepared for the industries to showcase their state-of-the-art products. The organizing committee of the conference thanks all the participants for their fruitful work and personal contribution to the development of these conference proceedings.

Professor Dr. Rizalman Mamat

The Chairman of ICMER2015

edp sciences

MATEC Web of Conferences

Statement of Peer review

In submitting conference proceedings to MATEC Web of Conferences, I certify to the Publisher that I adhere to the **Policy on Publishing Integrity** of the journal in order to safeguard good scientific practice in publishing.

1. All articles have been subjected to peer review administered by the proceedings editors.
2. Reviews have been conducted by expert referees, who have been requested to provide unbiased and constructive comments aimed, whenever possible, at improving the work.
3. Proceedings editors have taken all reasonable steps to ensure the quality of the materials they publish and their decision to accept or reject a paper for publication has been based only on the merits of the work and the relevance to the journal.

Title, date and place of the conference

The 3rd International Conference on Mechanical Engineering Research (ICMER 2015)
18 -19 August 2015
Kuantan, Pahang, MALAYSIA

Proceedings editor(s):

Dr. Saiful Anwar Che Ghani
Dr. Azri Alias
Prof. Dr. Rizalman Mamat
Prof. Dr. Md. Mustafizur Rahman

Date and editor's signature

17, avenue du Hoggar - PA de Courtabœuf – BP 112 - 91944 Les Ulis Cedex A (France)
Tél. : 33 (0)1 69 18 75 75 - Fax : 33(0)1 69 07 45 17 – www.edpsciences.org

Comparative Analysis of Torque and Acceleration of Pre- and Post-Transmission Parallel Hybrid Drivetrains

Saiful A. Zulkifli[1,3,a], Nordin Saad[2] and A. Rashid A. Aziz[2,3]

[1]*Electrical and Electronics Engineering Department*
[2]*Mechanical Engineering Department*
[3]*Centre for Automotive Research and Electric Mobility (CAREM)*
Universiti Teknologi PETRONAS, 32610 Seri Iskandar, Perak, Malaysia

Abstract. Parallel hybrid electric vehicles (HEV) can be classified according to the location of the electric motor with respect to the transmission unit for the internal combustion engine (ICE): they can be pre-transmission or post-transmission parallel hybrid. A split-axle parallel HEV – in which the ICE and electric motor provide propulsion power to different axles – is a sub-type of the post-transmission hybrid, since addition of torque and power from the two power sources occurs after the vehicle's transmission. The term 'through-the-road' (TTR) hybrid is also used for the split-parallel HEV, since power coupling between the ICE and electric motor is not through some mechanical device but through the vehicle itself, its wheels and the road on which it moves. The present work presents torque-speed relationship of the split-parallel hybrid and analyses simulation results of torque profiles and acceleration performance of pre-transmission and post-transmission hybrid configurations, using three different sizes of electric motor. Different operating regions of the pre-trans and post-trans motors are observed, leading to different speed and torque profiles. Although ICE average efficiency in the post-trans hybrid is slightly lower than in the pre-trans hybrid, the post-trans hybrid vehicle has better fuel economy and acceleration performance than the pre-trans hybrid vehicle.

1 Introduction

A parallel hybrid electric vehicle typically has the electric motor-generator on board the chassis, as in most production hybrids of today, but in one variant of the parallel hybrid, the electric motor is part of the wheel itself, called a hub motor, or in-wheel motor (IWM). In this configuration (Figure 1), the ICE operates on one drive-shaft (e.g., front) while the in-wheel motors are fitted into the other pair of wheels (e.g., rear), resulting in a split-axle parallel architecture [1, 2].

There is no specific device coupling the two propulsion sources; instead, the power sources are coupled through the road, giving rise to the term TTR hybrid. This configuration enables an existing ICE-driven vehicle to be converted into a hybrid electric vehicle with minimal modification – by replacing the otherwise non-driven wheels with in-wheel motors (IWM) and powering them with a battery bank located in the vehicle's trunk. Along with motor controllers and an energy management system controller (EMS) that intelligently manages the vehicle's energy and power flow, they form a TTR-IWM hybrid vehicle system.

Figure 2 shows three different configurations of the parallel hybrid drive-train: a TTR-IWM split-axle parallel hybrid, a single-axle post-transmission parallel hybrid, and a single-axle pre-transmission parallel hybrid.

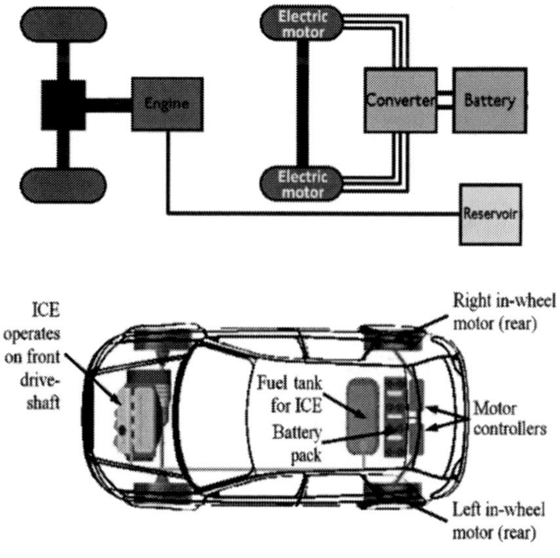

Figure 1. A split-parallel through-the-road hybrid electric vehicle.

[a]saifulazrin_mz@petronas.com.my

 DOI: 10.1051/matecconf/20167400001

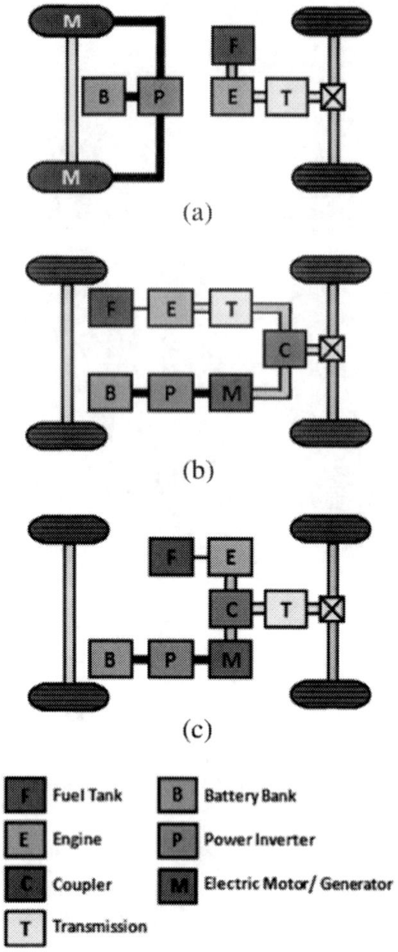

Figure 2. Parallel hybrid drive-train configurations: (a) TTR-IWM split-axle parallel HEV; (b) single-axle, post-transmission parallel HEV; (c) single-axle, pre-transmission parallel HEV [1, 2]

2 Torque-Speed Relationship of Split-Axle Parallel Hybrid

A parallel hybrid vehicle is essentially a torque-coupled system consisting of a mechanically-coupled drive-train. An ordinary parallel hybrid has some form of coupling device such as a pulley, gear or chain assembly, which adds together mechanical powers from the two sources. Figure 3 shows a parallel hybrid drive-train represented by a 3-port, 2-degree-of-freedom mechanical system. Port 1 is a unidirectional input (connected to the ICE), while port 2 (connected to the motor) and port 3 (connected to the wheels) are bidirectional input or output of the system (at least one must be an output at any time) [3, 4].

Ports 1 and 2 can either be connected directly to the power sources or through a mechanical transmission. Ignoring losses, power input to the torque coupler is equal to its power output, as expressed below, Eq. (1), where τ and ω represent torque and speed respectively of the coupling device and individual sources:

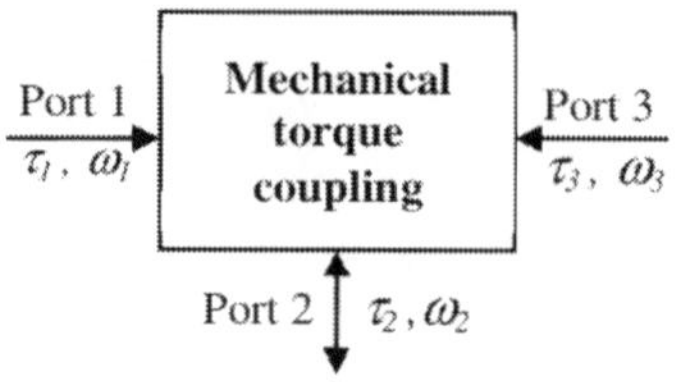

Figure 3. Three-port representation of a torque-coupled parallel hybrid drive-train.

$$\tau_3\omega_3 = \tau_1\omega_1 + \tau_2\omega_2 \quad (1)$$

Torque output of the mechanical coupler τ_3 can be expressed as:

$$\tau_3 = k_1\tau_1 + k_2\tau_2 \quad (2)$$

where k_1 and k_2 are structural parameters of the coupling system, described for example by gear ratios, and are thus fixed parameters of the system. The contributing torques τ_1 and τ_2 are independent of each other and thus, can be individually controlled. However, due to power conservation, Eq. (1) leads to the constraint that speed of the power sources are linked together and governed by the relationship:

(c) $$\omega_3 = \frac{\omega_1}{k_1} = \frac{\omega_2}{k_2} \quad (3)$$

Thus, the sources' speeds cannot be individually controlled. Coupling of the sources can consist of either single-shaft or double-shaft configuration. The former has the motor and engine connected in-line on the same shaft, while the latter has two separate shafts connected via the gear or chain assembly.

The split-axle, TTR parallel hybrid is a torque-coupled configuration - torque from the ICE and electric motor are added up through the vehicle chassis, wheels and road, without a coupling device. It is topologically similar to the dual-shaft parallel configuration, since the motor and ICE operate on separate drive axles. The torques can be individually controlled but their speeds are governed by Eq. (3).

In the case of a retrofit-conversion TTR-IWM vehicle, the ICE is connected to a multi-gear transmission (existing) while there is no transmission for the motor since it is in the wheel hub (similar to a directly-coupled on-board motor). Hence, with port 3 connected to the front wheels and port 1 connected to the ICE, k_1 is determined by the gear position of the ICE's transmission. Port 2 is connected to the motor with $k_2 = 1$ (direct-coupled, in-wheel motor). Thus, Eq. (2) and Eq. (3) become:

$$\tau_3 = k_1\tau_1 + \tau_2 \quad (4)$$

$$\omega_3 = \frac{\omega_1}{k_1} = \omega_2 \quad (5)$$

The in-wheel motor speed ω_2 is thus equal to the speed of the ICE-driven front wheels, ω_3. This is similar to a double-shaft, device-coupled parallel drive-train with the motor shaft coupled directly to the front-wheel axle (post-transmission coupling) with $k = 1$. Figure 4 and Figure 5 show simulated torque profiles of pre-transmission and post-transmission hybrid drive-trains (see discussion below). Torque profiles of a TTR-IWM hybrid are similar to a single-axle post-transmission parallel hybrid.

3 Comparisons of Torque Profile and Acceleration Performance

To investigate difference in torque and acceleration between TTR-IWM (or split-parallel) hybrid and pre-transmission hybrid vehicle, simulation tests of acceleration are carried out using *ADVISOR*, a software based on *Matlab-Simulink* and commonly used for advanced vehicular powertrain studies. A constant vehicle speed request of 300 km/h is given as simulation input, resulting in each drivetrain attempting to provide maximum possible torque. The simulation tests are carried out on a small vehicle, mounted with electric motors of three different sizes (16-kW, 32-kW and 75-kW continuous power). Vehicle and motor specifications are given in Table 1 and Table 2 respectively.

Table 1. Simulation test vehicle specifications.

	Small Car
Base mass:	500 kg
IC Engine:	1.0 L SI
ICE Max Power:	41 kW
Transmission:	4-Speed Automatic
Wheel Radius:	0.225 m
Accessory Load:	Constant 700 W
Powertrain Control:	Conventional (4-Spd AT)

Table 2. Simulation test motor specifications.

	PM16	PM32	AC75
Motor Type:	Permanent Magnet	Permanent Magnet	AC Induction
Cont. Power:	15.8 kW	32 kW	75 kW
Max. Torque:	124 Nm	150 Nm	271 Nm
Peak Efficiency:	92 %	94 %	92 %
Manufac-turer:	Unique Mobility	Unique Mobility	Westing-house
Weight:	20.6 kg	60.8 kg	91 kg

3.1 Torque Comparison

Figures 4 and 5 show that motor torque profile in split-parallel hybrid is continuous (except for drops during gear change); but the torque profile in pre-transmission hybrid shows torque increase after every gear change (most obvious in the PM16 and PM32 motors). The reason for this difference is the effect and location of the ICE's transmission gearbox: in the pre-transmission hybrid, with every upshift of the gear, there is torque increase from any source connected to its input, including the electric motor's torque [5, 6].

In the split-parallel and post-transmission hybrid however, since the motor's torque is added after the gearbox, the torque is not multiplied by the gearbox, resulting in the smooth torque profiles. From Figures 4 and 5, it can also be seen that in both drivetrains, the PM16 (16-kW permanent-magnet) motor has the highest torque profiles, followed by PM32 and AC75 motors. Power delivered by the motor is the product of torque and speed. As can be seen from Figures 6 and 7, which show operation maps of the PM16 and PM32 motors, the operating points do not fall in the same speed region.

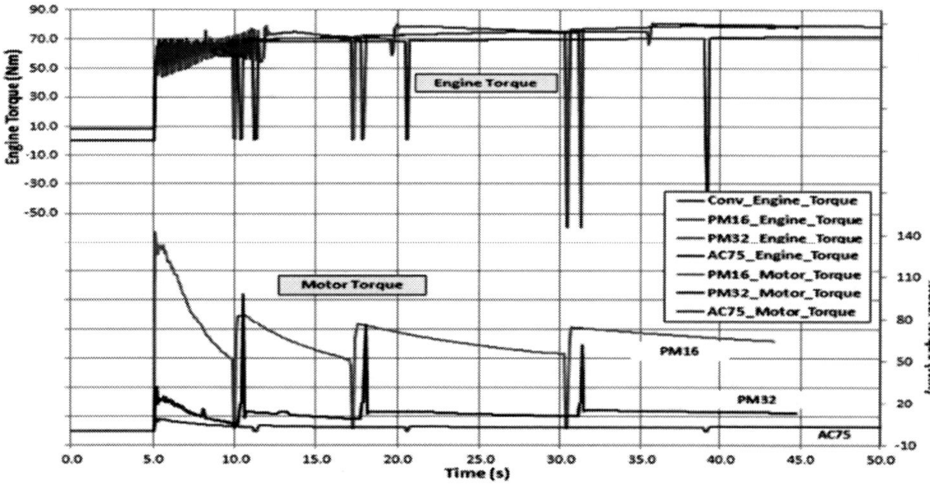

Figure 4. Engine and motor torque profiles in pre-transmission hybrid drive-train.

 DOI: 10.1051/matecconf/20167400001

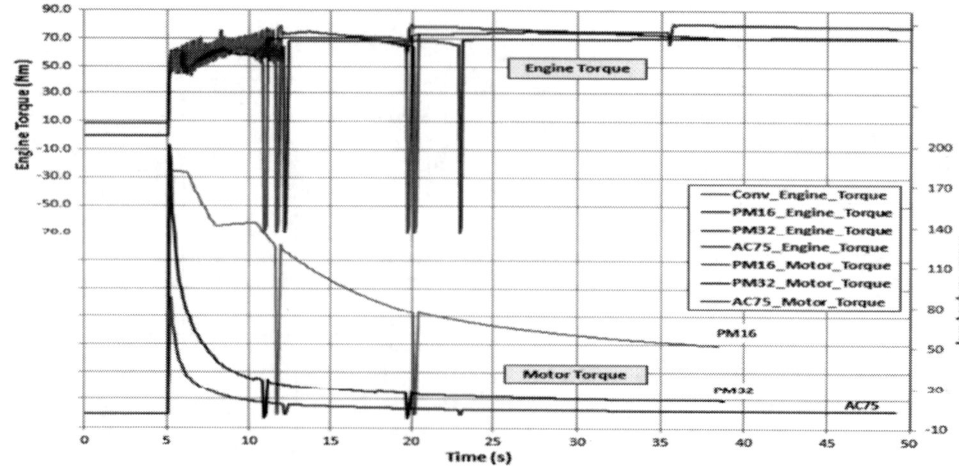

Figure 5. Engine and motor torque profiles in post-transmission hybrid drive-train.

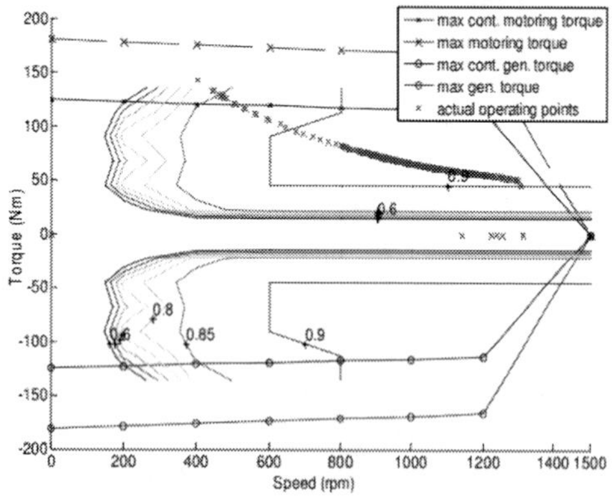

(a)

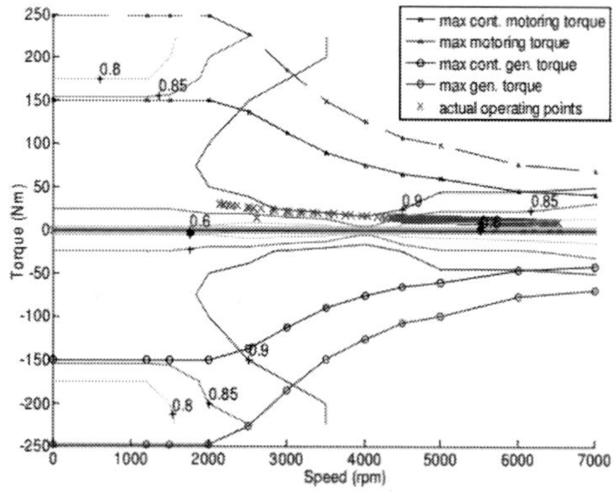

(b)

Figure 6. Operation map of (a) PM16 motor and (b) PM32 motor in pre-transmission hybrid configuration.

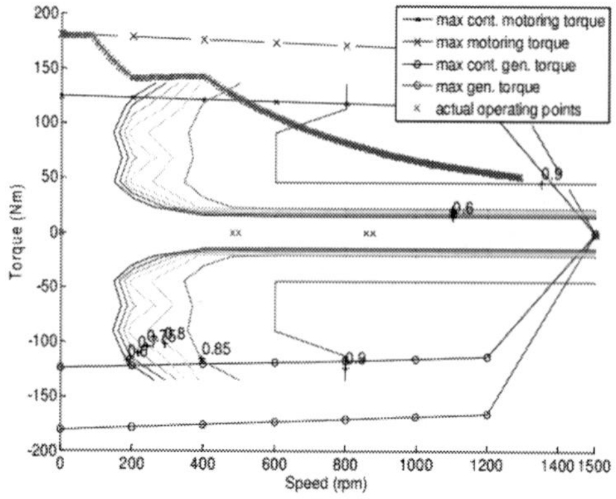

(a)

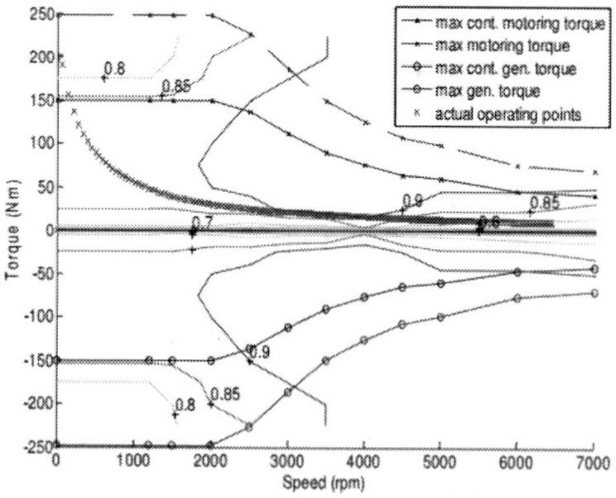

(b)

Figure 7. Operation map of (a) PM16 motor and (b) PM32 motor in post-transmission hybrid configuration.

Figure 8(a) shows time chart of motor speed for both PM 16 and PM 32 motors, in a post-transmission hybrid. If both motor torque curves are plotted on the same graph of torque vs. speed as in Figure 8(b), the torque difference is not as staggering as it seems in the charts of torque vs. time in Figure 5. The different speed regions between PM16 and PM 32 motors can also be clearly seen in Figure 8(b).

Figures 9, 10 and 11 show that motor torque profiles in the split-parallel hybrids are higher than those in the pre-transmission hybrids, for all motors: PM16, PM32 and AC75. This is related to the motors' operating speeds.

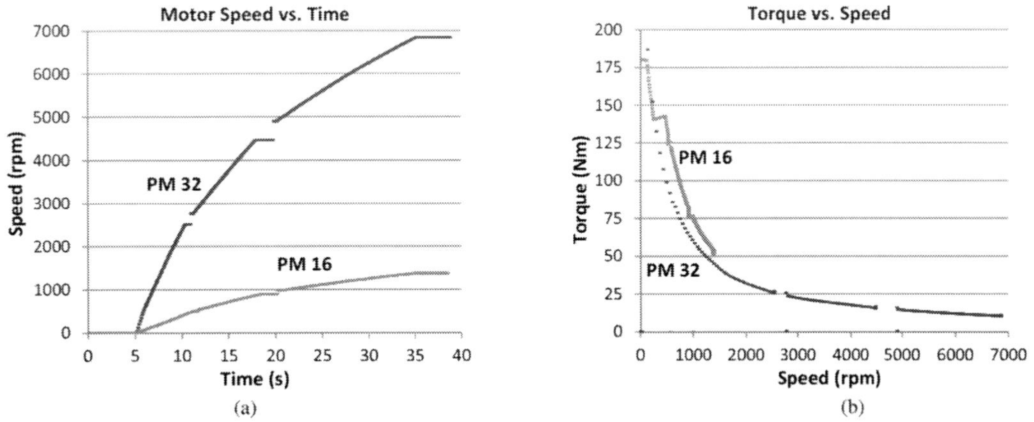

Figure 8. (a) Motor speed vs. time and (b) torque vs. speed for PM16 and PM32 motors in post-transmission hybrid configuration.

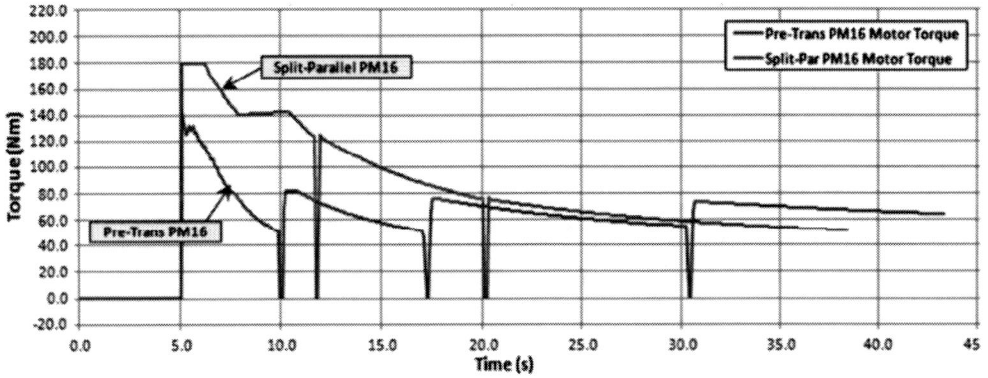

Figure 9. Motor torque profile of pre-transmission & split-parallel hybrids: 16-kW motor.

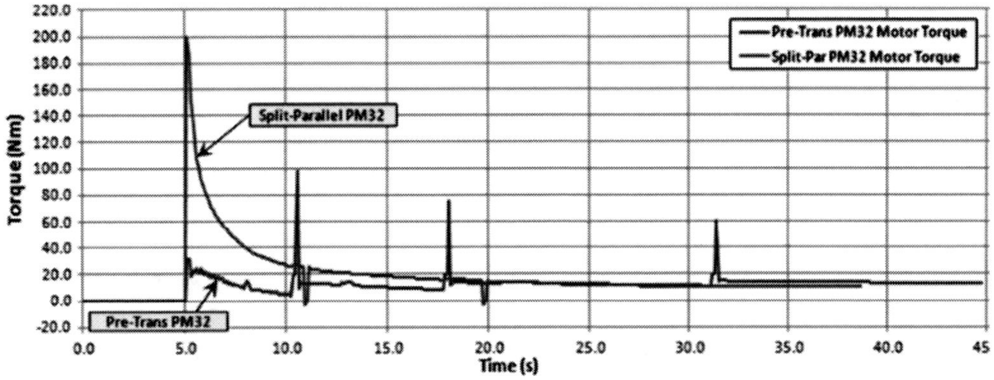

Figure 10. Motor torque profile of pre-transmission & split-parallel hybrids: 32-kW motor.

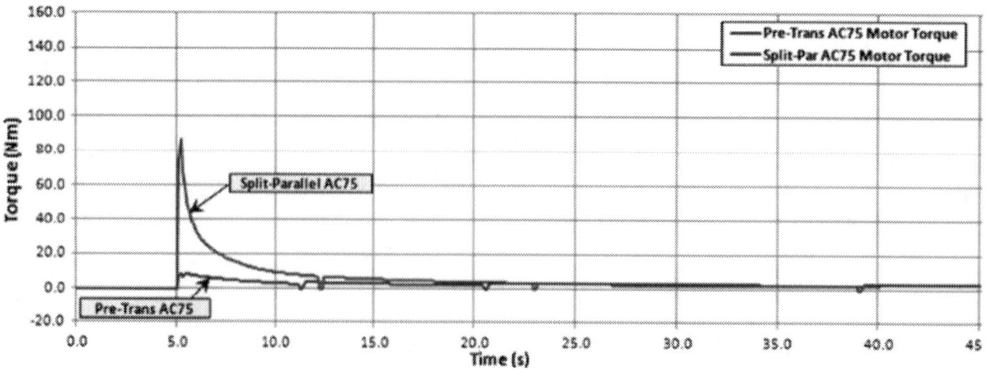

Figure 11. Motor torque profile of pre-transmission & split-parallel hybrids: 75-kW motor.

As shown in Figure 12 and Figure 13, the pre-transmission motors operate almost instantaneously to a certain speed - about 350 rpm for PM16 and 1800 rpm for PM32 - while the post-transmission motors operate from zero speed.

In the first 18 seconds, the post-transmission motor produces higher torque at low speeds, while the pre-transmission motor delivers lower torque due to its lower initial speeds. After about 18 seconds, each motor then operates in the same torque and speed region.

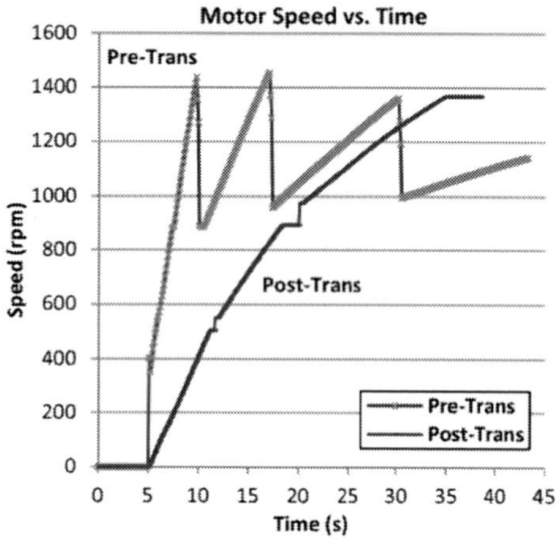

(a)

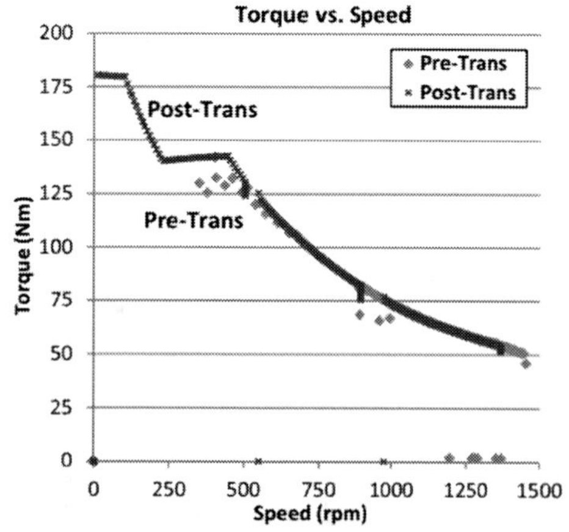

(b)

Figure 12. Comparison of pre-transmission and post-transmission hybrid configurations with PM16 motor: (a) motor speed vs. time and (b) torque vs. speed.

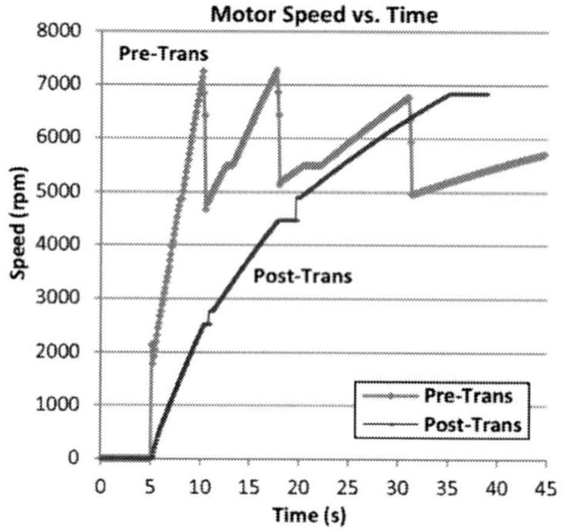

(a)

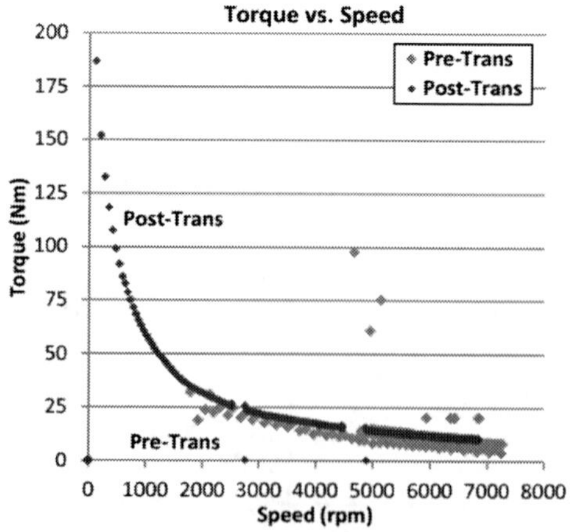

(b)

Figure 13. Comparison of pre-transmission and post-transmission hybrid configurations with PM32 motor: (a) motor speed vs. time and (b) torque vs. speed.

DOI: 10.1051/matecconf/20167400001

3.2 Acceleration Performance

Figure 14 shows that with each motor, acceleration performance improves from the conventional vehicle, and is always better (faster) than the pre-transmission hybrid, with the AC75 hybrid showing the biggest difference and the PM16 hybrid having the fastest response. The same patterns are observed in fuel consumption (Fig. 15).

However, for emissions (Fig. 16), all TTR-IWM hybrids show increased levels from the pre-transmission hybrids, for all emission types (biggest difference in AC75 hybrid).

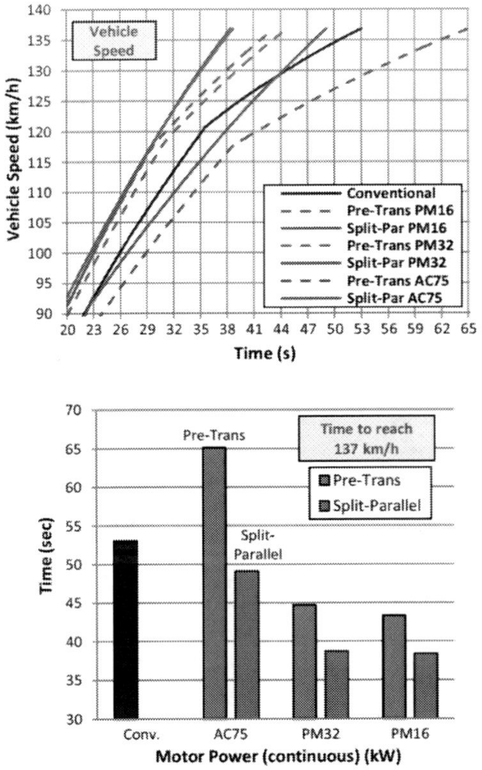

Figure 14. Vehicle speed and acceleration performance comparison.

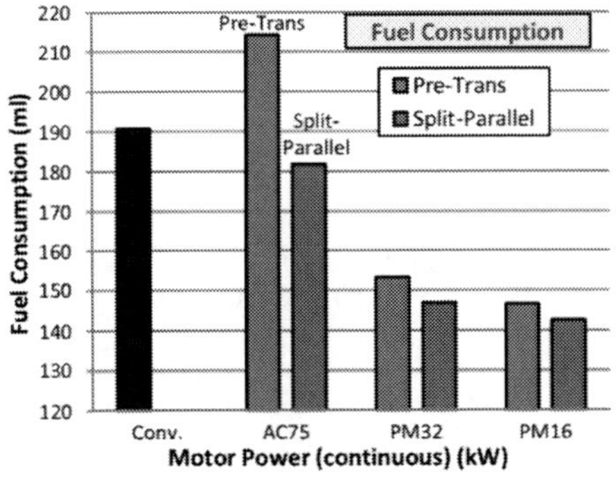

Figure 15. Fuel consumption comparison.

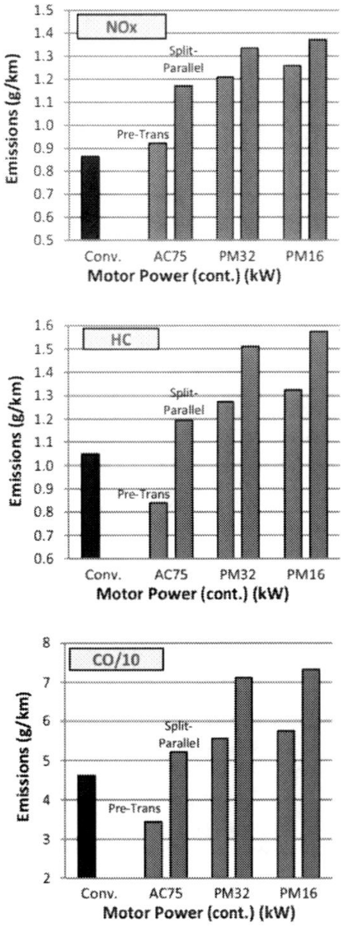

Figure 16. Emissions comparison – NOx, HC and CO.

Figures 17 and 18 show power flow and energy usage of each sub-system within the hybrid powertrain, for both pre-transmission and post-transmission hybrids. From Table 3, it is seen that total energy delivered by the motor in a pre-transmission hybrid vehicle is higher than that in a split-parallel hybrid, for all motor sizes. Similarly, total engine output is always higher in the pre-transmission hybrid, for any motor, resulting in overall more fuel consumption in the pre-transmission hybrid, compared to the split-parallel.

This is due to the higher motor torque and power in the post-transmission hybrid, which results in the vehicle reaching the final desired speed faster than the pre-transmission hybrid. Thus, the split-parallel hybrid has better fuel economy across all motor sizes, with the highest fuel savings of 25.2 % in the PM16 TTR hybrid.

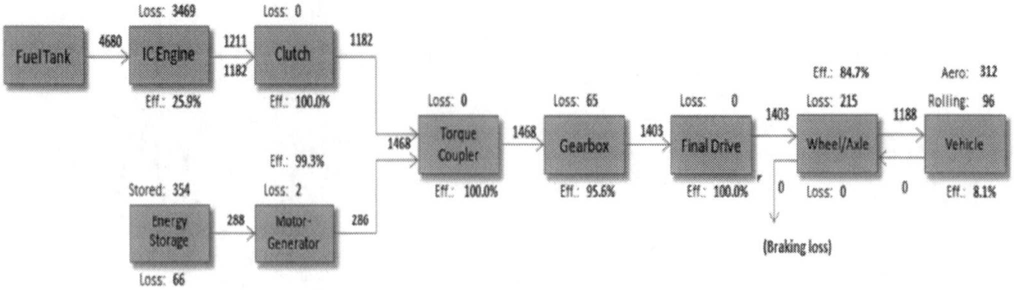

Figure 17. Power flow and energy usage in pre-transmission hybrid drive-train

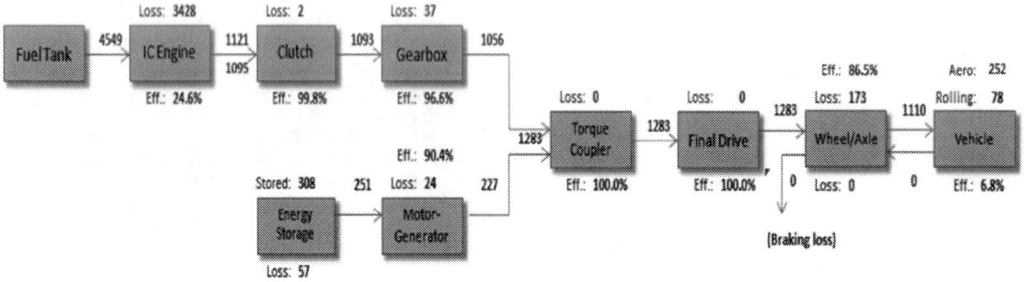

Figure 18. Power flow and energy usage in post-transmission hybrid drive-train

Figure 19 shows graphs of fuel consumption versus the combustion engine's output energy, showing straight-line relationship as expected, for both drivetrains. However, it can be seen that the graph of pre-transmission hybrid lies below the graph of post-transmission (split-axle) hybrid, which means that for every operating point of the ICE, the post-transmission hybrid consumes slightly more fuel, or is less efficient, than the pre-transmission hybrid.

This is corroborated by the ICE average efficiency values in Table 3, which shows post-transmission hybrids having slightly lower ICE efficiencies. Although engine efficiency of the post-transmission hybrid is slightly lower (less than 2.2 percentage points), overall it still consumes less fuel to reach the final target speed in the acceleration simulation, compared to the pre-transmission hybrid, for all motor sizes.

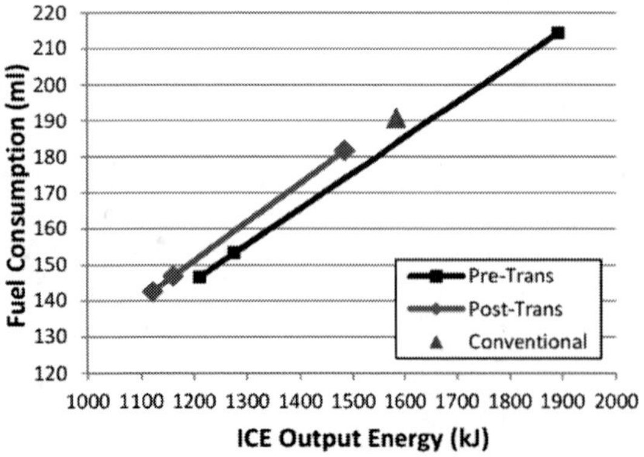

Figure 19. Fuel consumption vs. ICE output energy.

Conclusions

A split-axle parallel hybrid configuration enables conventional vehicles to be converted into a hybrid electric vehicle with minimal physical modification, to be a through-the-road, post-transmission hybrid vehicle with in-wheel motors (TTR-IWM). This paper describes torque-speed relationship for the TTR-IWM hybrid, comparing it to the pre-transmission, parallel hybrid with on-board motor. Simulation results of acceleration test are provided, comparing torque profiles and acceleration performance of the pre-transmission and post-transmission drivetrains, for three different electric motor sizes.

Results show different operating regions of the pre-trans and post-trans motors, leading to different speed and torque profiles. It is found that, although ICE average efficiency in the post-transmission hybrid is slightly lower than in the pre-transmission hybrid, the post-transmission hybrid vehicle has better overall fuel economy and acceleration performance than the pre-transmission hybrid vehicle.

Acknowledgment

This work is partly funded by an *ERGS* grant (*Exploratory Research Grant Scheme*) from the Ministry of Higher Education (MOHE), Malaysia.

[a]saifulazrin_mz@petronas.com.my

Table 3. Performance comparison of split-parallel and pre-transmission parallel hybrid vehicles.

	ICE Output (kJ)	Motor Output (kJ)	Motor Ave. Eff. (%)	ICE Ave. Eff. (%)	Overall Eff. (%)	Fuel Cons. (ml)	Fuel Savings (%)
Conventional	1583	n/a	n/a	26.6	8.7	190.7	N/A
Split-Par AC75	1484	100	89.3	26.1	8.0	181.7	4.7
Split-Par PM32	1160	239	94.0	25.2	7.0	146.9	23.0
Split-Par PM16	1121	227	90.4	25.2	6.8	142.6	25.2
Pre-Trans AC75	1891	135	89.1	28.3	10.3	214.3	-12.4
Pre-Trans PM32	1276	296	99.4	26.7	8.1	153.3	19.6
Pre-Trans PM16	1211	288	99.5	26.5	8.1	146.7	23.1

References

1. S.A. Zulkifli, S. Mohd, N. Saad, A.R.A. Aziz, Appl. Mech. and Materials AMM **663**, 498-503 (2014)
2. S.A. Zulkifli, S. Mohd, N. Saad, A.R.A. Aziz, Int. Journal of Auto. and Mech. Eng. IJAME **11**, 2793-2808 (2015)
3. M. Ehsani, Y. Gao, A. Emadi, *Modern electric, hybrid electric, and fuel cell vehicles: fundamentals, theory and design*, 2nd ed., CRC Press (2010)
4. C.C. Chan, A. Bouscayrol, K. Chen, IEEE Trans. Vehicular Technology **59**, 589-98 (2010)
5. Z. Rahman, M. Ehsani, K.L. Butler, Future Engines SP-1559, 2000-01-3062 (2000)
6. Z. Rahman, K.L. Butler, M. Ehsani, SAE 2000 World Congress, 2000-01-0994 (2000)
7. B. Asaei, M. Habibidoost, *IEEE Vehicle Power and Propulsion Conference*, 1-5 (2010)
8. M. Young, G.M. Molen, D. Oglesby, K. Crawford, K. Walp, R. Lewis, C. Whitt, S. Phillips, *IEEE Vehicle Power and Propulsion Conference*, 511-518 (2007)
9. N. Fallahi, A.H. Niasar, Intelligent Systems in Electrical Engineering **4**, 51-61 (2014)

[a]saifulazrin_mz@petronas.com.my

Statistical Analysis of Deep Drilling Process Conditions Using Vibrations and Force Signals

Hazwan Syafiq[1], M.A.N. Kamarizan[1], M.F. Ghazali[2] and A.R Yusoff[1, a]

[1]*Faculty of Manufacturing Engineering, Universiti Malaysia Pahang, 26600, Pekan, Pahang*
[2]*Faculty of Mechanical Engineering, Universiti Malaysia Pahang, 26600, Pekan, Pahang*

Abstract. Cooling systems is a key point for hot forming process of Ultra High Strength Steels (UHSS). Normally, cooling systems is made using deep drilling technique. Although deep twist drill is better than other drilling techniques in term of higher productivity however its main problem is premature tool breakage, which affects the production quality. In this paper, analysis of deep twist drill process parameters such as cutting speed, feed rate and depth of cut by using statistical analysis to identify the tool condition is presented. The comparisons between different two tool geometries are also studied. Measured data from vibrations and force sensors are being analyzed through several statistical parameters such as root mean square (RMS), mean, kurtosis, standard deviation and skewness. Result found that kurtosis and skewness value are the most appropriate parameters to represent the deep twist drill tool conditions behaviors from vibrations and forces data. The condition of the deep twist drill process been classified according to good, blunt and fracture. It also found that the different tool geometry parameters affect the performance of the tool drill. It believe the results of this study are useful in determining the suitable analysis method to be used for developing online tool condition monitoring system to identify the tertiary tool life stage and helps to avoid mature of tool fracture during drilling process..

1 Introduction

Drilling is the most familiar machine tool operations in manufacturing technology over the time. Almost 75 percent of metal cutting operations are still carried out by conventional twist drills [1]. Since the industry has allocated a plentiful amount of money on drilling tools, the idea to decrease the production cost in drilling operation is realized by introducing of deep drilling process. In this technique, the hole can be drilled at least 10 times of the diameter of drill bit which allow producing a deeper hole [2]. There are several deep drilling techniques that been applied in industry such as boring trepanning association drill, the ejector drilling systems and gun drill. However there are alternative techniques have proposed by researchers to overcome the weakness of deep drill methods called deep twist drilling technique. Deep twist drilling can reduce the production time using high speed and feed rate compared to the other deep drilling techniques. Based on current research of hot forming, it is found that technique of deep twist drilling can improve the cooling rate. By increasing cutting length capability of drill tool to increase the cooling energy efficient that can reduce the number of hot forming die blocks. However, one of primary issues in deep twist drilling technique is premature tool breakage; due to the tool wear, chip clogging and deflection which can affect sustainability of the process. Therefore, there is a need for more clear understanding of deep twist drill phenomena.

Statistical parameters are established method applied in research analysis. Through the implementation of statistical analysis it can improve the process understanding, enhanced level of control and thus contribute in reduces process variability and improved the output quality [3]. There are several numbers of statistical parameters such as root mean square (RMS), arithmetic mean, standard deviation, variance and kurtosis etc. It has been employed widely in manufacturing process analysis activities. Ghoreishi et al. [4] applied variance to analyze the interactions between six controllable variables on the hole taper and circularity in laser percussion drilling. Variance through implementation of ANOVA also use to studies the effect of drilling parameters such as cutting speed, feed rate, point angle and diameter of tool drill [5, 6]. Meanwhile, Gaja [7] implemented RMS to analyze data depth-of-cut detection and tool-workpiece engagement by using acoustic emission monitoring system during milling machining. RMS values also were used to classified tool wear levels in lathe, milling and drilling machines [8, 9, and 10]. Kurtosis and skewness of vibration signal data also be implemented to indicate the different types of drill wear [11].

Jantunen et al. [12] compared seven statistical parameters such as RMS, skewness, kurtosis, mean,

[a] Corresponding author: razlan@ump.edu.my

arithmetic mean, maximum and minimum in tool wear measurement case study. Based on comparison, it found that RMS value is appropriate parameters to represent the case study data. Patrick [13] in his thesis compared six statistical parameters: mean, peak, RMS, variance, crest factor and kurtosis in study the signals pattern of a sharp drill and a worn drill. Results reveal that, four parameters which are RMS, peak, crest factor and the kurtosis are capable to evaluate the patterns and the levels of different measured signals. However, it also found that peak is not suitable to be applied in drilling operation due to stochastic drill behavior and peak level is not an averaging statistical value. It is summarized that better results could be initiated by considering derivative factors such as crest factor and kurtosis. Xu [14] implemented several statistical parameters to find the relations between the drilling process conditions with several machining parameters such as workpiece material, drill diameter, spindle speed, feedrate and the drill corner wear. Statistical parameters applied are mean, RMS, delta and standard deviation features. Based on analysis, in term of material, spindle speed, feed rate and drill diameter comparison, it found that data produce by mean and RMS is almost similar but differs to delta features due to statistical algorithm. Meanwhile others parameters have clear distinguish values especially in feedrate comparison. It is concluded that mean and RMS are very similar in the values and variation forms, thus, some of them can be eliminated for remove of redundancy.

Based on the review, it can summarize that, a lot of study especially in drilling activities has been conducted by using statistical parameters analysis. Even though there is no any statistical analysis of deep drilling case study found, it is believes that the conceptual stages and methodology will be same as standard drilling. This paper discusses the identification of deep twist drill tool condition when different machining parameters, such as cutting speed, feed rate and depth of cut are applied. The identification is done through several statistical parameters; root mean square (RMS), mean, kurtosis, standard deviation and skewness. It is then been compared to find the most suitable statistical parameters that can represent tool condition behavior. In this study, it also leads to the finding of suitable machining parameters and tool geometry that can be used in deep twist drill in order to get maximum deep-hole drilling capability.

2 Experimental setup

The deep twist drilling condition experiments are carried out on a 3-axis CNC vertical machine (HAAS CNC Vertical Machine). The tool employed are high speed steel, HSS long series twist drills with diameter 8 mm, overall length 165 mm and flute length 109mm. Since this is an initial study, it is good to start with a standard geometry of tool drill, as tabulated in Table 1. Based on literature review conducted, there is no study that focus on the performance of this tool even tough manufacturer keep produces and promote this types of geometry. Die material, SKD 61 are used in this study. It is a hard-to-cut material with HRC ranges 50-55 in die and mold field. The work piece is cut into block with size of 90 mm x 70 mm x 80 mm. 25 holes are drilled in a block that held in a vice on top of the dynamometer. Each of holes are drilled by using different cutting conditions such as cutting speed, feed rate and depth of cut from design of experiment, as listed in Table 2. The summaries of cutting condition ranges used in design of experiment, DOE are referred from data used by others researcher and data provided by tool manufacturer. Experiments are divided into 5 regions which represent number of cutting speed. This experiment is repeated to 3 times to get the reliable results. All boreholes were pre-center using pilot drills with diameter 8 mm and 1.5 mm depth. This can minimize the effects associated with tool run-out [15].

Meanwhile, for data acquisition, force and vibration sensors are mounted to the workpiece and spindle machine, as shown in Figure 1. Dynamometer; model Kistler 9257B with built-in 3 axis charge amplifier attached on the work piece. Meanwhile for vibration, the tri-axial accelerometer (Model 356B21) with sensitivity of + 10 mV/g is attached to spindle machine. The analog signals from these sensors were fed into data acquisition board, DAQ with 2000 of sampling rate capability. Meanwhile, for experiments to compare between two tool geometry, holes with optimum results of tool condition from previous experiment are repeated again by other tool geometry such as point angle, helix angle, web thickness and relief angle. Collected data are compared and analyzed. Details for both tool drills specifications are provided in Table 1.

Table 1. Long twist drill series specifications and process parameters.

Tool	Manufacturer tool type	Substrate	Point angle	Helix angle	Flute shape
1	Sherwood SHR-025	HSS	118°	35°	Standard
2	OSG	Solid Carbide	120°	40°	Wide flutes

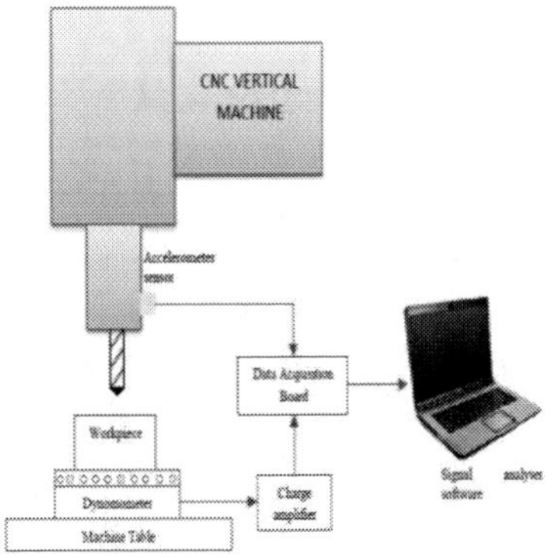

a). Schematic diagram

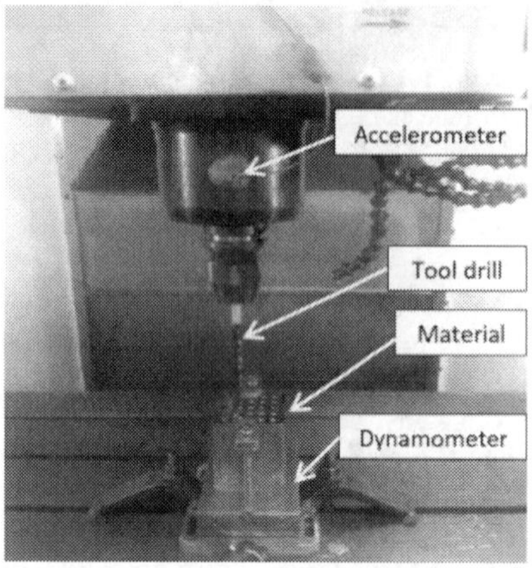

b). Experimental setup

Figure 1. The schematic diagram and actual setup of experiment.

Table 2. Cutting conditions for 25 experiment.

Experiment	Region	Cutting speed (m/min)	Feedrate (mm/rev)	Depth of cut (mm)
1	1	30	0.10	40
2		30	0.15	50
3		30	0.20	60
4		30	0.25	70
5		30	0.30	80
6	2	40	0.10	50
7		40	0.15	60
8		40	0.20	70
9		40	0.25	80
10		40	0.30	40
11	3	50	0.10	60
12		50	0.15	70
13		50	0.20	80
14		50	0.25	40
15		50	0.30	50
16	4	60	0.10	70
17		60	0.15	80
18		60	0.20	40
19		60	0.25	50
20		60	0.30	60
21	5	70	0.10	80
22		70	0.15	40
23		70	0.20	50
24		70	0.25	60
25		70	0.30	70

3 Statistical measurement

In order to analyze data generated by dynamometer and accelerometer sensors, fives statistical parameters are been applied and compared. These parameters are the root mean square (RMS), mean μ, standard deviation σ, kurtosis K and the skewness S. These parameters can be calculated using the following formula as referred in [11].

Mean (μ) calculates the force and vibration signals average for N = 2000 samples:

$$\mu = \frac{1}{N}\sum_{i=1}^{N} x_i \quad (1)$$

Root mean square (RMS) is the most common parameter used in statistical analysis [16]. Its value contains all the energy in the signal and therefore also all the noise and all elements that depend on the cutting process. RMS defined as:

$$RMS = \sqrt{\frac{1}{N}\sum_{i=1}^{N} x_i^{\,2}} \quad (2)$$

where x_i refer to instantaneous amplitude of the signal from force and vibration. Meanwhile for standard deviation(σ), the data was standardized by $(x_i - \mu)$ and it is the mean square value about the mean:

$$\sigma = \sqrt{\frac{1}{N}\sum_{i=1}^{N}(x_i - \mu)^2} \quad (3)$$

Next, the kurtosis (K) is a measure of how outline-prone a distribution [17]. Based on Issam [11] the normal distribution of kurtosis is 3. Meanwhile for distribution that is more or less outliner-prone having kurtosis greater or less than 3. Kurtosis defined as:

$$K = \frac{1}{N}\sum_{i=1}^{N} \frac{(x_i - x)^4}{\sigma^4} \quad (4)$$

Where x is average of force and vibration amplitude or mean, μ. In term of analysis, it found that kurtosis was rather sensitive to the occurrence of spikes or impulses in the time domain of the vibration signal [18]. Skewness (S) is a measure of the asymmetry of the data around the sample mean. Skewness defined as:

$$S = \frac{E(x-\mu)^3}{\sigma^3} \quad (5)$$

Where $E(x)$ is the expected value of x. In the next section, results from experiments are analyzed and discuss.

4 Results and discussion

4.1. Analysis of tool condition behavior

Figure 2 and 3 show the tools condition behavior based on five statistical analyses for vibrations and forces. Tool condition are been classify into three category; good, fracture and blunt. Based on tool condition, only 6 out of 25 holes are capable to finish without failures. Experiments number 1, 2, 3, 4, 6 and 7 are in category good condition meanwhile others are in blunt and fracture categories. The highest cutting speed that can be adapted through standard long series twist drill tool (Sherwood SHR-025) is through experiment number 7 with 40 m/min, feed rate 0.15mm/rev and depth of cut 60mm.

Meanwhile the highest feed rate that can be adapted is 0.25 mm/rev with cutting speed 30 m/min and depth 70 mm through experiment number 4. In term of deep drilling definition, ratio between length to diameter L/D must be at least 10 [1]. Through cutting parameter experiment number 7, it only can achieve 8.75 ratio of L/D, where it not enough to stand as deep drilling theory. However, based on pattern shown by experiment number 6; average value of it vibration is the lowest. It is found that the depth can be increased from 50mm to 80mm to achieve L/D 10. Due to that, cutting condition of experiment number 6 will be used for experiment of tool geometry comparison by changing its depth from 50 to 80 mm.

4.2 Statistical analysis data of vibration and force

Based on data observation, it found that data vibrations recorded by accelerometer more details on the tool deflection meanwhile data captured via dynamometer influenced by chip clogging. In term of selection of appropriate statistical parameter, it is chosen by considering several theories. Such validation with time domain graph pattern generated by sensors and the relationship between cutting condition to the statistical graph generated. In Table 3, it can be seen that all vibrations statistical data except skewness, are capable to differentiate between good and blunt condition according to the statistical ranges for experiments number 1 until 10. However, the phenomena were changed when boreholes number 11 to 25 were analyzed. All statistical analysis except kurtosis cannot differentiate between blunt and fracture condition. This is happen due to small number of difference between blunt and fracture data. So, in order to distinguish between this two data, it requires for sensitive analysis methods. Based on Mahfouz [11], it is found that kurtosis was sensitive to the occurrence of spikes or impulses in the time domain of the vibration signal. It capable to differentiate between blunt and fracture data. In Figure 2 (e), if the value of acceleration is exceeding to 17.13 m/s2 the tool are in 'blunt' region and if lower than that is in 'good' zone. Meanwhile, if the value exceeding to 28.62 and above tool are in 'fracture' region.

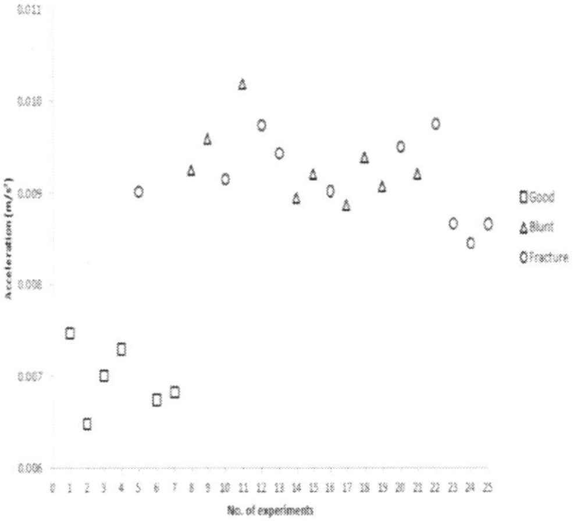

a). Mean analysis

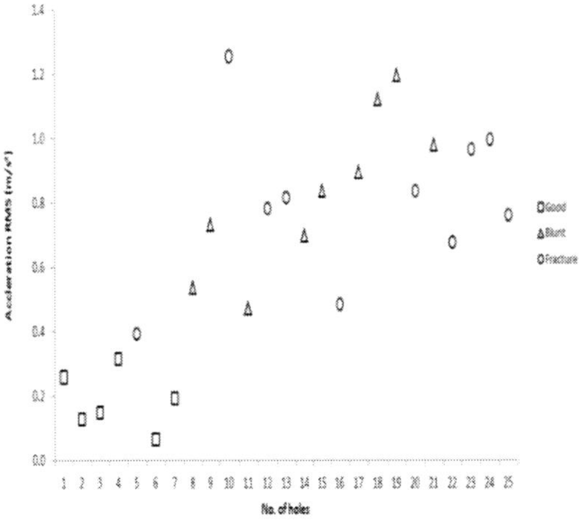

b). RMS analysis

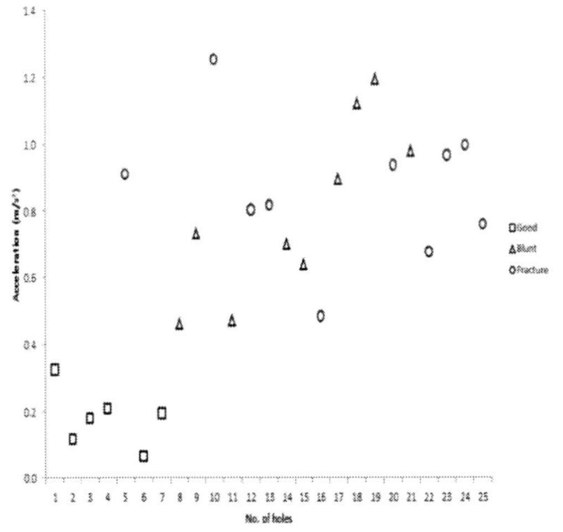

c). Standard deviation analysis

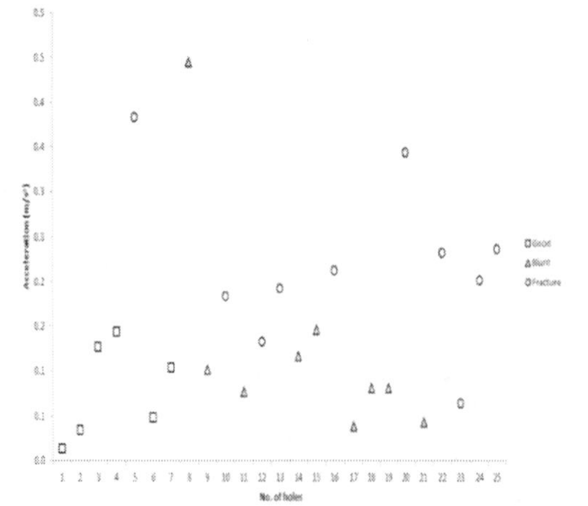

d). Skewness analysis

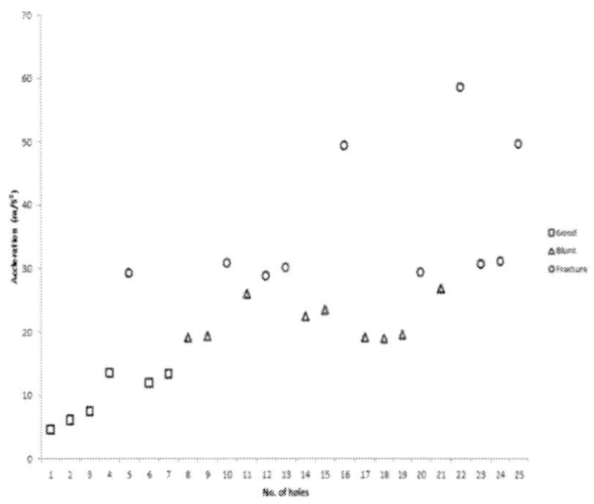

e). Kurtosis analysis

Figure 2(a-e). Statistical analysis of acceleration data

As stated earlier, data captured via dynamometer is influenced by chip clogging. This can be seen in Figure 3 (a-e) at experiments number 1 until 4 where tool are in good condition. It show values are inclining when the depth is increase, due to chip clog. It shows that depth is one of the key players in influencing the forces data. Furthermore, others cutting condition such as cutting speed and feed rate also played role to make chip clogging. For example, when increase the numbers of cutting speeds and feed rate of the machine, its can lead to the growth of amount and size of chips removes and the chances for the tools to break due to chip clogging onto the hole. In Figure 3, it can see that the pattern at experiments number 8 until 25 become blunt and fracture zone. This happens due to the impact from blunt and fracture of the tools. In term of statistical parameter selection, skewness is verified to be the finest gauge of chip clogging, to differentiate between blunt and fracture. It is as indicator there is more closely relation to the changing of cutting conditions. Unlike data produce by acceleration, the range levels of tools conditions for forces are different for every region where it is according to the tools cutting condition as shown in Table 4. For example, in Figure 3 (e) the minimum fracture value for region 3; experiments number 11 until 15 is 4.42, meanwhile the minimum fracture value for region 4; experiments number 16 to 20 is 2.54. This is happen due to the high differences cutting speed and feed rate. The early breakage of tool in region 4 is because of early chip clogging occurred.

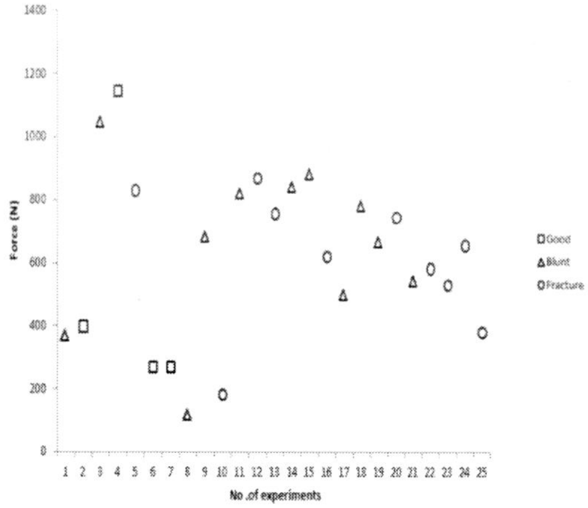

a). Mean analysis

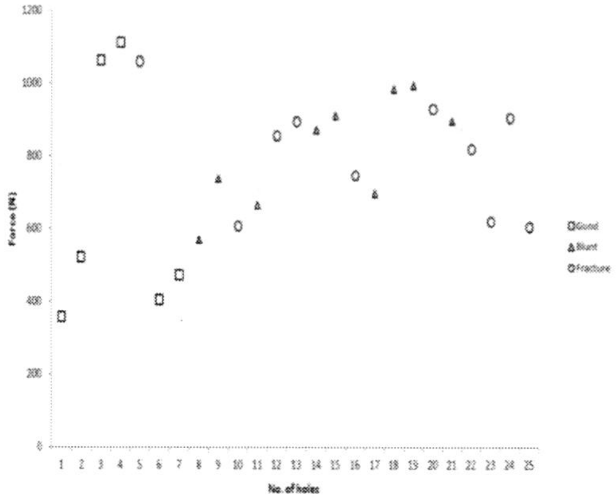

b). RMS analysis

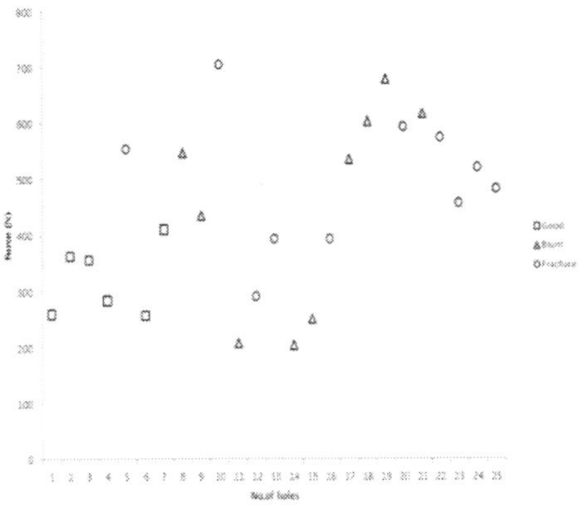

c). Standard deviation analysis

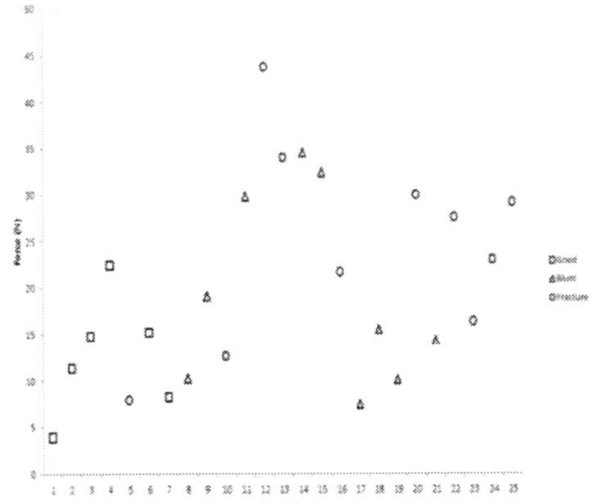

d). Kurtosis analysis

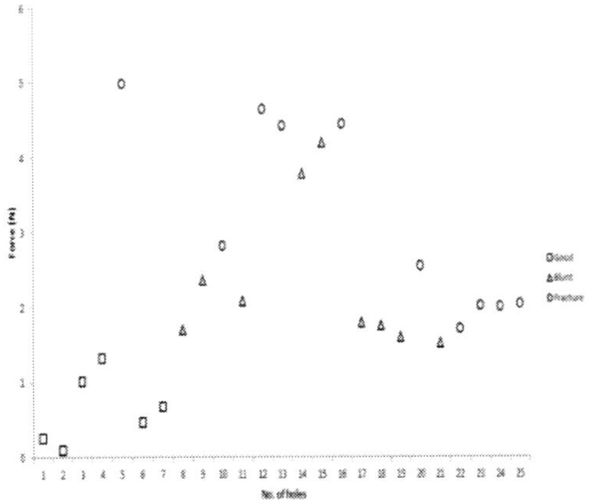

e). Skewness analysis

Figure 3(a-e). Statistical analysis of force data

Table 3. Range of statistical analysis of acceleration data

Tools condition	Statistical analysis range value				
	Mean	RMS	Standard Deviation	Skewness	Kurtosis
Good	0.0065 - 0.0075	0.066 - 0.276	0.056 - 0.265	0.013 - 0.144	4.590 - 13.568
Blunt	0.0089 - 0.0102	0.469 - 1.302	0.432 - 1.297	0.038 - 0.445	18.981 - 26.927
Fracture	0.0085 - 0.0098	0.386 - 1.249	0.474 - 1.249	0.063 - 0.384	28.788 - 58.542

Table 4. Range of statistical analysis of force data

Tool condition	Region	Statistical analysis range value				
		*A	*B	*C	*D	*E
Good	1	369.17 - 1093.71	358.32 - 1113.3	259.22 - 363.17	0.098 - 1.31	3.94 - 22.39
Fracture		852.02	1060.6	552.59	4.98	7.84
Good	2	239.43 - 272.61	404.60 - 474.10	256.72 - 410.60	0.46 - 0.67	8.21 - 15.19
Blunt		150.81 - 626.33	569.65 - 736.38	435.63 - 546.77	1.69 - 2.35	10.23 - 19.09
Fracture		225.511	607.704	702.68	2.80	12.523
Blunt	3	735.13 - 828.89	663.97 - 895.14	204.14 - 250.99	2.08 - 4.19	29.8 - 34.52
Fracture		806.29 - 854.49	857.08 - 895.14	288.81 - 391.73	4.42 - 4.63	33.87 - 43.66
Blunt	4	451 - 880.67	696.77 - 993.79	250.99 - 679.04	1.59 - 4.19	7.45 - 32.36
Fracture		713.14 - 714.75	746.14 - 929.70	391.89 - 591.85	2.54 - 4.44	21.57 - 29.85
Blunt	5	620.19	895.25	616.59	1.52	14.25
Fracture		425.36 - 723.60	606.41 - 929.70	456.87 - 591.85	1.70 - 2.54	16.28 - 29.85

*A= Mean, *B= RMS, *C= Standard deviation, *D= Skewness, *E= Kurtosis

4.3 Data analysis of different tool geometry

As explained before, the modification of cutting condition for experiment number 6 with the new depth 80 mm is used for this experiment. Specifications of tools uses for the comparison are shown in Table 1. Based on the data, as expected the tool is not broken, both tools finish the job smoothly. However, it was found that the value of force for tool 2 is lower than tool 1 up to 34 percent when analyzed through several statistical parameters. This is due to the flutes shape pattern, where flutes shape tool 2 is widen compare to tool 1. Through the widen flutes, the chip flow management is better, and this can allow the tool drill to go deeper. The material of tool 2 also influence the performance where the harden tool material can withstand the impact resulting from the friction between the tool and the surface of the material as well as can prevent from wear. Table 5 shows the values of force for both tools analyze by several statistical parameter.

Table 5. Values of force analyze by statistical parameters

Tool	Tool type	Mean	RMS	Standard deviation	Kurtosis	Skewness
1	Sherwood SHR-025	497.41	589.41	283.68	5.242	1.053
2	OSG	267.84	323.15	180.28	2.758	0.693

4 Conclusion

In this paper, a summary of tool performances and condition behavior of standard deep twist drill has been given. The signal has been analyzed through several statistical parameters techniques. It was found that data generated by vibrations are more focused on tool deflection meanwhile forces data covered chip clogged. Kurtosis and skewness establish to be appropriate statistical parameters to represent deep twist drill tool conditions for vibrations and forces signals. The results also reveal that the geometry influenced the performance of the tools. Tool with wide flutes can afford for more depth of cut, it also recommended for deep drilling technique. It believe the results of this study are useful in determining the suitable analysis method to be used for developing online tool condition monitoring system to identify the tertiary tool life stage and helps to avoid mature of tool fracture during drilling process.

Acknowledgements

The authors would like to be obliged to Ministry of Science, Technology and Innovation (MOSTI) and Universiti Malaysia Pahang for providing laboratory facilities and financial assistance under e-science research fund project no. RDU140506.

References

1. Elzenheimer, J.,Liebeck, T and Tschannerl, M. Noch viel ungenutztes Potenzial beim Bohren. Werkstatt und Betrieb. 2003; 136(11):55-57.
2. Heinemann, R., Hinduja, S., Barrow G .Use of process signals for tool wear progression sensing in drilling small deep holes. International Journal Advance Manufacturing Technology. 2007; 33: 243–250
3. Presentetioneze.com. Statistical methods for manufacturing process improvement. Retrived from: http://www.presentationeze.com/blog/statistical-methods-manufacturing-process improvement/ 15 April 2014
4. Ghoreishi, M., Low, D.K.Y., Li, L. Comparative statistical analysis of hole taper and circularity in laser percussion drilling. International Journal of Machine Tools and Manufacture. 2002; 42: 985–995
5. Kadam, M.S and Pathak, S.S. Experimental Analysis and Comparative Performance of Coated and Uncoated Twist Drill Bit Dry Machining. International Journal of Research in Mechanical Engineering and Technology. 2011; 1: 2249-5762
6. Rao, S. and Dr. Lewlyn L.R. Impact Of Process factors On Twist Drill Wear In Machining gfrp Composites By Applying Taguchi Design Analysis And Anova Technique. International Journal of Advanced Technology & Engineering Research (IJATER). 2014; 4: 2250-3536
7. Gaja, H. Analysis and modeling of depth-of-cut during end milling of deposited material. Masters Theses, Missouri University of Sceince and Technology. 2011.
8. Pontuale, G., Farrelly, F.A., Petri, A., and Pitolli, L. A statistical analysis of acoustic emission signals for tool condition monitoring (TCM). Acoustics Research Letters Online. 2003; 4(1): 1529-7853
9. xiqing, M and Chuangwen, Xu. Tool wear monitoring of acoustic emission signals from milling processes. Proceedings of the 1st International Workshop on Education Technology and Computer Science, ETCS. 2009; 1: 431-435
10. Karali, P. Acoustic Emission based Tool Condition Monitoring System in Drilling. Proceedings of the World Congress on Engineering. 2011; 3: 2078-0958
11. Issam, A.M. Drilling wear detection and classification using vibrations signals and artificial neural network. International Journal of Machine Tools and Manufacture. 2003; 43:707-720.
12. Jantunen, E. and Jokinen, H. Automated On-line Diagnosis of Cutting Tool Condition. International Journal of Flexible Automation and Integrated Manufacturing. 1996; 4:273–287.
13. Sambayi, P.M.K. Drill wear monitoring using instantaneous angular speed: A comparison with conventional technology used in drill monitoring systems. Masters Theses, University of Pretoria. 2012
14. Jie, X. Drill wear prediction and drilling conditionsrecognition with newly generated features. Ph.D Theses, Hiroshima University. 2014
15. Heinemann,R., Hinduja, S,. Barrow, G., Petuelli, G. The Performance of Small Diameter Twist Drills in Deep-Hole Drilling. Journal of Manufacturing Science and Engineering. 2006;128.
16. Jantunen, E. A summary of methods applied to tool condition monitoring in drilling. International Journal of Machine Tools and Manufacture. 2002; 42:997-1010.
17. Signal Processing Toolbox, For use with MATLAB User's guide. The Math Works USA. 1998
18. El-Wardany, T.I., Gao, D. and Elbestawi, M.A. Tool condition monitoring in drilling using vibration signature analysis. International Journal of Machine Tool and Manufacture. 1996; 36(6): 687-711

Stress Distribution of Bolted Joints With Different Lay-Up Types

H. Ahmad
Faculty of Civil and Environmental Engineering, Universiti Tun Hussein Onn Malaysia, 86400 Parit Raja, Batu Pahat, Johor, Malaysia

Abstract. A parametric study of the stress distribution in the composite plates is undertaken to show the effect of different lay-ups. Full 3-D elastic properties are required in the modelling. These properties are calculated using equations taken from the literature and derivation from simple Classical Laminate Plate Theory (CLPT). In previous experimental work, it was shown that tensile failure involved the development of a damage zone at the edge of the hole. In a double-lap joint, it is assumed that uniform stresses are exhibited throughout the plate thickness. Different lay-ups system give different tangential stress distributions and insignificant variation in the radial stress distribution along the hole boundary.

1 Introduction

In early finite element work, most of the researchers were working with 2-D finite element models and plane stress state following CLPT theory. The simplified 2D models ignore the effect from the bolt load. Crews, Hong and Raju [1] completed a parametric study on stress distributions around the hole boundary for a variation of *W/d* values to include bolt properties and contact between bolt and the laminate. Stress distributions are strongly dependent on the anisotropy for both magnitude and location of peak hoop stress on the hole boundary. Stress concentrations for the tested lay-up in the range $2 \leq W/d \leq 10$. The 0° lay-up gave largest stress concentration (about 4.5), followed by cross-ply lay-up (about 3.75) and the lowest stress concentration is with the quasi-isotropic lay-up (about 1.7). This is consistent with the open-hole problem. This work is later extended to include the effects of friction, pin elasticity, clearance and laminate properties in FEA work by Eriksson *et. al* [2] and Hyer and Klang [3]. These parameters change the location and value of ultimate tangential stress. Rowlands *et. al.* [4] compared strain obtained from finite element model to experimental strains using strain gauges on the bearing plane and found that increased friction was able to redistribute the load and correspondingly the position of the main load-carrying fibres away from the bearing plane towards the net-tension plane.

Lay-up stacking also affects both the bearing strength and failure mode in pin joints as reported by Quinn and Matthews [5] and in clamped bolted joints reported later by Park [6] and found that placing 90° layers on the surface inhibits delamination bearing strength in the region $\theta = 0°$ but does not affect the ultimate bearing strength. Stockdale and Matthews [7] reported a 40% increase in pin loaded bearing strength for a finger-tight case and increased as much as 100% at a maximum clamping load (14.7 kN) compared to a pin joint in GFRP system. Eriksson [8] found that normalised strength of clamped to pin joint in CFRP system rose by up to 2.4 times. This is expected as lower stiffness in GFRP gives instability effects which reduced its strength performance. Smith [9] found that increasing bolt load changed the initial failure mechanism from local bearing (splitting and interlaminar cracking) to local shear and compression damage under the bolt head resulting in increased bearing stress. Remote bearing failure (more significant in low torque conditions) was shown to be a mixture of compressive damage and delamination failures.

Kontolatis [10] used a similar GFRP woven fabric system as used by Belmonte *et. al.* [11] in double-lap bolted joints with a clamping torque of 5 N m using protruding bolt. Kontolatis [10] found that net-tension failures occurred with $W/d \leq 4$. These initiated from the stress concentration at the hole edge perpendicular to the loading axis. Mixed mode of bearing and net-tension failures occurred with *W/d*=5 and sufficiently large *e/d* =4. Kontolatis [10] also conducted interrupted tests to study the effect of net-tension damage in more detail to observe the damage that triggers a catastrophic failure.

An interrupted test in one specimen, from a geometry that finally failed in net-tension, showed that damage in the joint initiated with matrix cracking near the hole edge perpendicular to the loading direction to the specimen sides and at maximum load forms a damage zone of approximately 2-3 mm length. Kontolatis [10] expressed that the damage zone consists of local failures due to discontinuities, including 0° fibre tow fractures and shear matrix cracking. There were also limited amounts of splitting and delamination formed on both sides of the hole and exhibit a fairly constant width. Exceeding the critical load, the damage zone eventually

[a] Corresponding author: hilton@uthm.edu.my

caused catastrophic failures. Kontolatis [10] also conducted optical microscopy in an attempt to highlight fibre tow fracture in the top and bottom reinforcement layers. The damage zone is localised along the path defined by the crack and the surrounding area seems to be unaffected. The limited amount of splitting may be associated with lateral constraint in torqued conditions in which even the tows that have fractured are retained in place.

In a cross-ply woven fabric laminate, it can be deduced that similar damage growth mechanisms occurred as given in quasi-isotropic lay-ups but due to different weave architectures there is some differences. First, the damage length in the cross-ply lay-up might be less than quasi-isotropic laminates as a result of a high proportion of tows in 0° direction compared to quasi-isotropic laminates with identical thickness. This behaviour also found by Manger [12] in the open-hole problem resulting in reduced critical damage zone length. Secondly, the damage zone tends to initiate at 45° to the loading axis and then changes to 90°, resulting in shear-out failures. As found by Collings [13], sufficient *e/d* must be provided in cross-ply lay-ups to eliminates the shear-out failures.

Net-tension, shear-out and bearing failures are three main types of failures that do occur in bolted joints. The specific mode of failure is dependent on the stresses developed along radial, tangential and shear plane, respectively. The present work has eliminated shear-out failure by providing sufficient end-distance in all joint systems, so reducing the current study to net-tension and bearing failure modes. Radial stresses are associated with bearing (compressive) failures of loaded hole behind bolt shank and tangential stresses are associated with net-tension failures. Current work compares radial stress and tangential stress exhibited in woven fabric CFRP double-lap bolted joints between cross-ply and quasi-isotropic lay-ups. The understanding of stress distribution in composite material lay-ups with bolted joint configurations is essential prior to strength prediction works using finite element analysis.

2 Experimental Observations

Net-tension failure occurred within $2 < W/d < 4$, whereas $W/d = 5$ showed bearing mode. Compared with non-woven coupons the transition to bearing failure occurs at a higher *W/d*. It is suggested that this may be due to the lower tensile strength of the woven systems. For coupons tested in the finger-tight condition, the critical *W/d* is reduced to 4 and local bearing failure prior to ultimate failure is apparent in all samples. The hole elongation is obvious for the coupon with *W/d*=5. Although difficult to see on bare eyes, there was also evidence of local bearing damage in the smaller width coupon. A similar observation with regard to critical *W/d* is given by Cooper and Turvey [14]. Cooper and Turvey [14] tested pultruded fibre reinforced plastic composite in which critical *W/d* increased from 4 in pin joint to W/d = 6 with clamping torque 3 N m and *W/d*=10 with clamping torque 30 N m.

For the two cross-ply woven materials (PX and 5X) net-tension failure was observed for all the joints tested, except for the unclamped, larger hole size with *W/d*=3. In non-woven materials bearing failure might be expected with *W/d*=3. Again, it is the lower strength of the woven material that is believed to be associated with the greater occurrence of net-tension failures. Two types of bearing failures are observed prior to ultimate failures. The first type showed local bearing initially but ultimate failures are given as net-tension mode (failures given in net-tension showed ultimate failures from hole edge and failed catastrophically). This type of bearing failure is more pronounced in thin coupons. Second type showed pure bearing failures which exhibited compression behind the bolt up to certain length (no net-tension failures are observed).

3 Finite Element Modelling

3.1 Determination of 3-D elastic properties

Determination of 3-D properties is considered first. This is followed by description of the modelling stages which include the use of "smeared-out" properties, the development of the FE models and the associated boundary conditions and the implementation of clamp-up.

In-plane and out-of-plane elastic properties of the CFRP and GFRP woven fabric composites are required in 3-D modelling. In-plane elastic properties were determined experimentally and are reproduced from Belmonte's work [11] in Table 1. Out-of-plane elastic properties have been estimated following a simple approach. It seems likely that net-tension failure is governed mainly by the in-plane properties. The out-of-plane elastic constants required are E_z, ν_{zx}, ν_{zy}, G_{zx} and G_{zy}. For the cross-ply lay-ups these properties are estimated from idealisations based on simple CLPT principles and are presented in Appendix A1. This leads to averaged values over the thickness of the laminate. Out-of-plane properties for the quasi-isotropic lay-ups are taken from Akkerman [15] and given in Appendix A2. The resulting in-plane and out-of-plane elastic properties for all lay-ups are listed in Table 1.

Table 1. Elastic properties of CFRP lay-ups investigated in current study incorporating out-of-plane properties

Material	E_x/E_y *(GPa)*	E_z *(GPa)*	ν_{xy}	ν_{yz}/ν_{zx}	G_{xy} *(GPa)*	G_{yz}/G_{zx} *(GPa)*
PX4	51.40	12	0.09	0.10	4.42	3.90
PQ4	37.19	12	0.35	0.30	13.75	3.98

In woven composite modelling, the behaviour at the sub-lamina level is far from fully understood due to the effects of complex fibre architecture. Although much work in predicting mechanical properties of woven fabric systems has been implemented using a unit cell approach, this cannot readily be applied to local failure at a stress raiser. A simple approach that takes into account the orthotropic elastic properties of composite materials is by smearing (averaging) to provide "equivalent homogeneous properties". Using smeared properties for a woven fabric system is inherently a better approximation than for non-woven systems because two separate plies in a pre-preg based laminate are replaced by a single layer in a woven fabric composite.

3.2 Modeling Techniques and Approaches

The elastic properties used in the current model are based on smeared-out properties and include the out-of-plane elastic properties given in Table 1. Cross-ply woven lay-up is arguably better represented as smeared-out than a quasi-isotropic woven system because of similar repetitive sequence in adjacent layers. The other components are made from isotropic stainless steel which has a modulus elasticity of 210 N/mm^2 and Poisson's ratio of 0.3. Six components are assembled in each model of the single-lap joint. Sufficient mesh refinement is used to ensure the strength predictions are mesh independent. Mesh is refined in the vicinity of the hole edge in the composite (and steel) plates and the volume under the washers, as ultimate failure occurred within these regions. The boundary conditions are assigned so that left end is held fixed and displacement is applied to the right end.

Two load steps are implemented which apply the clamping load (Step 1) and the far-field tensile load (Step 2). Each model is assigned a clamping load (finger-tight or T=5 N m). Each contact surface pair is assigned with master-slave interaction which includes an appropriate friction coefficient. This is an important step as load transfer will affect the stress distributions and therefore, strength predictions.

4 Results and Discussion

The stress distribution around the circumference of the bolt hole of cross-ply laminates is somewhat different from quasi-isotropic materials due to the lower ratio of shear modulus-to-longitudinal modulus. In open hole problem, quasi-isotropic lay-up showed a lower stress concentration than cross-ply lay-up. However, in a bolted joint problem, the stress concentration is reduced at low bearing stress due to the effect of friction as a result of lateral load in the vicinity of the hole edge.

To investigate lay-up effects in bolted joints further, two different lay-ups of similar thickness (PX4 and PQ4) are compared giving the results shown in Figure 1. 3-D models are able to model the full friction transfer condition correctly giving zero stress in radial and tangential direction around the hole perimeter for low levels of bearing stress. As can be observed from Figure 1 (a) and (b), increased clamping force will reduce both the radial stress and tangential stress. Both lay-up types does not showed much difference in radial stress distribution around hole boundary where maximum radial stress occurs at 45° from the bearing as given in Figure 1 (a).

The different lay-ups give different tangential stress distributions along the hole boundary as shown in Figure 1(b). Quasi-isotropic laminate has ±45° plies and exhibits a single peak in the tangential stress, which increased from bearing plane to peak at about 85° and reduces subsequently, showing the maximum tensile stress at the end of contact regions. Cross-ply lay-ups have no ±45° plies and showed a minimum stress at angle of 45° from bearing plane. Additionally, two peaks are produced at 0° and 90° associated with $(0/90)_{ns}$ lay-up where the second tensile stress peak was larger than the first peak and occurred at a location where the non-contact point started. These trends are a consequence of the local changes in lay-up geometry around the hole boundary in a polar co-ordinate system.

Figure 1 (c) shows the tangential stress along net-section; for a fixed bearing stress the tangential stress reduced with bolt load. With fixed bearing stress, lower bolting forces increase the tangential stress concentration. Larger tensile stresses at net-tension plane in cross-ply were due to having proportionately more 0° plies. Both lay-ups showed insignificant variation in the radial stress distribution. Similar trends were obtained by Crews *et. al.* [1] who studied the stresses using a 2-D model for a bolted joint with six different (non-woven) laminate lay-ups and found that there was strong anisotropy in magnitude and location of peak tangential stress.

5 Conclusions

A 3-D FEA has been carried out to determine the stress distribution in bolted double-lap joints, as friction load in bolted joint is transferred through-the-thickness, the 3-D finite element model gave more reliable stress distribution. Different lay-ups give different tangential stress distributions along the hole boundary, cross-ply lay-up demonstrates higher tangential stress than quasi-isotropic lay-up counterparts. This is consequence of the local changes in lay-up geometry around the hole boundary in a polar co-ordinate system. On the other hand, both cross-ply and quasi-isotropic lay-ups showed insignificant variation in the radial stress distribution.

Acknowledgements

The authors would like to be obliged to Universiti Tun Hussein Onn Malaysia and Ministry of Education Malaysia for providing laboratory facilities and financial assistance under project Vot no. R034.

References

[1] Crews, J.H., Hong, C.S., and Raju, I.S.,. *Stress Concentration Factors for Finite Orthotropic Laminates with a Pin-loaded Hole.* 1-40, 1981: NASATechnology Paper 1862, 1981.

[2] Eriksson, L.I.,. "Contact Stresses in Bolted Joints of Composite Laminates." *Composite Structures*, 1986: 6:57-75.

[3] Hyer, M.W., Klang, E.C. and Cooper, D.E.,. "The Effects of Pin Elasticity, Clearance and Friction on the Stresses in a Pin-loaded Orthotropic Plate." *Journal Composite Material*, 1987: 21:190-206.
[4] Rowlands, R.E., Rahman, M.U., Wilkinson, T.L., Chiang, Y.I.,. "Single and Multiple-bolted Joints in Orthotropic Materials." *Composites*, 1982: 13:273-279.
[5] Quinn, W.J., Matthews, F.L.,. "The Effect of Stacking Sequence on the Pin-Bearing Strength in Glass Fibre Reinforced Plastic." *Journal of Composite Materials*, 1977: 11:139-145.
[6] Park, H.J.,. "Effects of Stacking Sequence and Clamping Force on the Bearing Strengths of Mechanically Fastened Joints in Composite Laminates." *Composite Structures*, 2001: 53:213-221.
[7] Stockdale, J.H., Matthews, F.L.,. "The Effect of Clamping Pressure on Bolt Bearing Loads in Glass Fibre Reinforced Plastics." *Composites*, 1976: 34-39.
[8] Eriksson, I.,. "An Analysis Method for Bolted Joints in Primary Composite." *AGARD Conference Proceedings 427*. Madrid, Spain, 1987. 6.1-6.19.
[9] Smith, P.A.,. *Aspects of the Static and Fatigue Behaviour of Composite Laminates, including Bolted Joints*. PhD thesis, Cambridge, UK: Cambridge University, 1985.
[10] Kontolatis, A.,. *Failure of Composite Bolted Joints Made from Woven fabric GFRP Composite.* MSc dissertation, Guildford: University of Surrey, 2000.
[11] Belmonte, H.M.S., Manger, C.I.C., Ogin, S.L., Smith, P.A., Lewin, R.,. "Characterisation and Modeling of the Notched Tensile Fracture of Woven Quasi-isotropic GFRP Laminates." *Composites Science and Technology*, 2001: 61:585-597.
[12] Manger, C.I.C.,. *Failure of Notched Woven GFRP Composites: Damage Analysis and Strength Modelling*. PhD Thesis, Guildford: University of Surrey, 1999.
[13] Collings, T.A.,. "The Strength of Bolted Joints in Multi-directional CFRP Laminates." *Composites*, 1977: 43-55.
[14] Cooper, C., Turvey, G.J.,. "Effects of Joint Geometry and Bolt Torque on the Structural Performance of Single Bolt Tension Joints in Pultruded GRP Sheet Material." *Composite Structures*, 1995: 32:217-226.
[15] Akkerman, R.,. "On the Properties of Quasi-isotropic Laminates." *Composites: Part B*, 2002: 33:133-140.

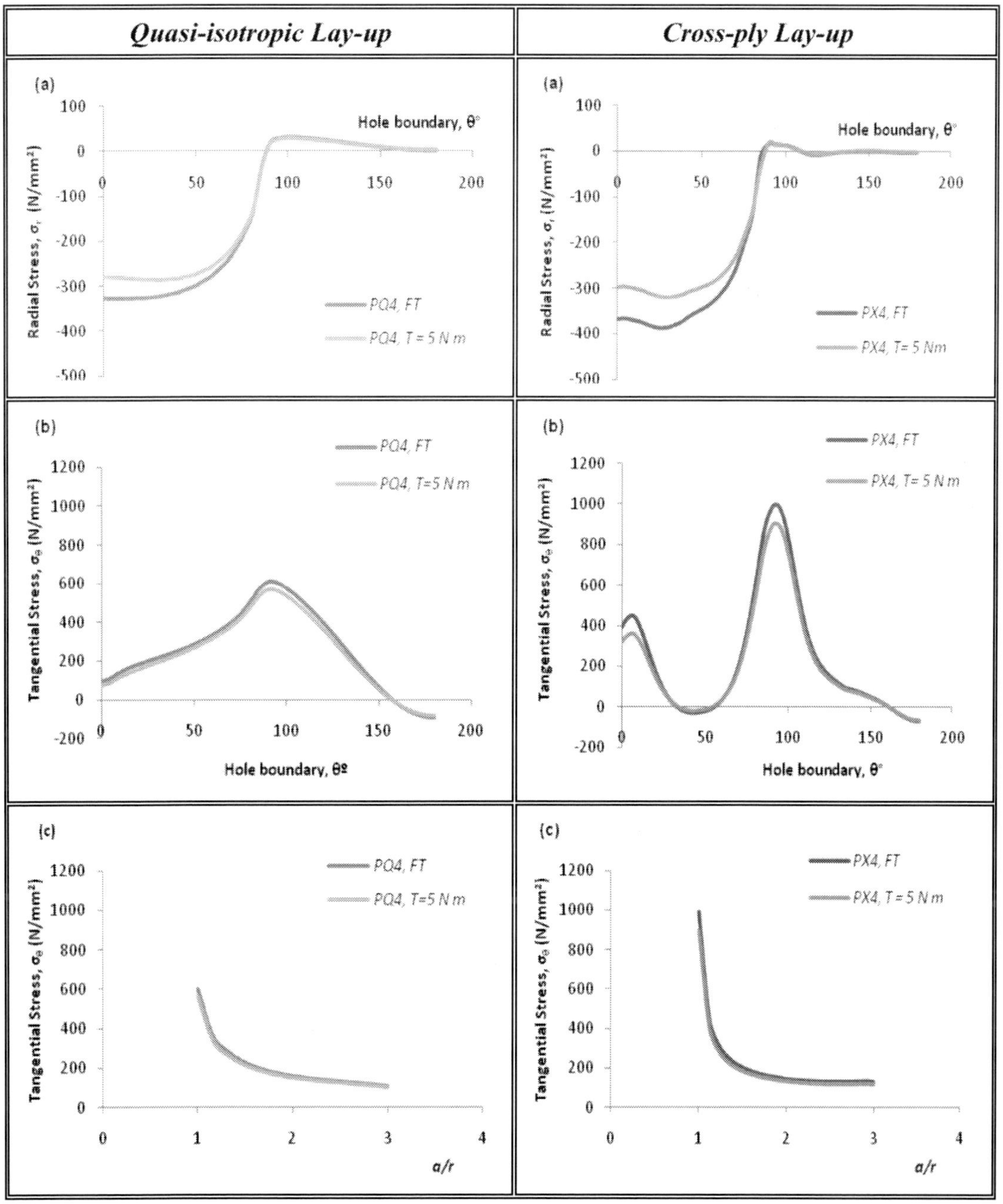

Figure 1. Comparison of FEA stress distributions near loaded hole in quasi-isotropic (PQ4) and cross-ply (PX4) lay-ups with d = 10 mm in the finger-tight and clamped conditions at an applied bearing stress of 400 N/mm^2 (a) Radial stress around hole boundary (b) Tangential stress around hole boundary (c) Tangential stress with normalised distance from the hole edge.

[a] Corresponding author: hilton@uthm.edu.my

DOI: 10.1051/matecconf/20167400003

Appendix A: Determination of out-of-plane elastic constants

A1: Cross-ply laminates

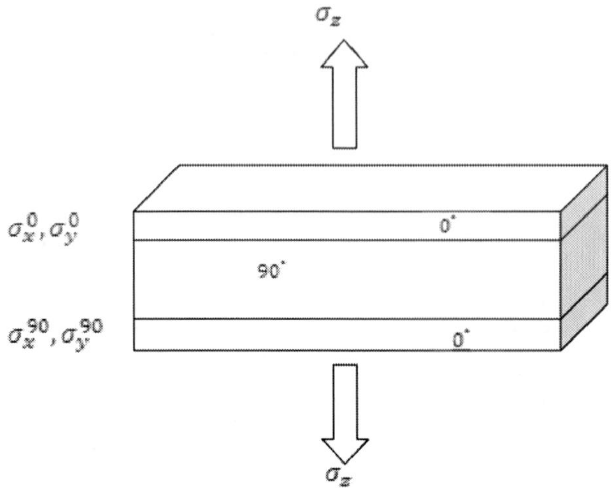

For equal thickness of 0's and 90's

$$\sigma_x^0 + \sigma_x^{90} = 0$$

$$\sigma_y^0 + \sigma_y^{90} = 0 \quad .$$

Now σ_z is constant then the thickness

$$\varepsilon_z^0 = \frac{\sigma_z}{E_3} - \frac{v_{13}\,\sigma_x^0}{E_1} - \frac{v_{23}\,\sigma_y^0}{E_2}$$

$$\varepsilon_z^{90} = \frac{\sigma_z}{E_3} - \frac{v_{23}\,\sigma_x^{90}}{E_2} - \frac{v_{13}\,\sigma_y^{90}}{E_1}$$

Effective stresses in z-direction,

$$\bar{\varepsilon_z} = \frac{\varepsilon_z^0 + \varepsilon_z^{90}}{2}$$

$$\bar{\varepsilon_z} = \frac{\sigma_z}{E_z} = \frac{\sigma_z}{E_3} - \frac{1}{2}\left[\frac{v_{13}}{E_1}\left(\sigma_x^0 + \sigma_y^{90}\right) + \frac{v_{23}}{E_2}\left(\sigma_x^{90} + \sigma_y^0\right)\right]$$

Also have strain compatibility in the x- and y- direction,

$$\varepsilon_x^0 = \frac{\sigma_x^0}{E_1} - \frac{v_{21}\,\sigma_y^0}{E_2} - \frac{v_{31}\,\sigma_z}{E_3}$$

$$\varepsilon_x^{90} = \frac{\sigma_x^{90}}{E_2} - \frac{v_{12}\,\sigma_y^{90}}{E_1} - \frac{v_{23}\,\sigma_z}{E_3}$$

$$\varepsilon_y^0 = \frac{\sigma_y^0}{E_2} - \frac{v_{12}\,\sigma_x^0}{E_1} - \frac{v_{23}\,\sigma_z}{E_3}$$

$$\varepsilon_y^{90} = \frac{\sigma_y^{90}}{E_1} - \frac{v_{21}\,\sigma_x^{90}}{E_2} - \frac{v_{31}\sigma_z}{E_3}$$

As composite shows similar strain in each layer, $\varepsilon_x^0 = \varepsilon_x^{90}$,

$$\frac{\sigma_x^0}{E_1} - \frac{v_{21}\,\sigma_y^0}{E_2} - \frac{v_{31}\,\sigma_z}{E_3} = \frac{\sigma_x^{90}}{E_2} - \frac{v_{12}\,\sigma_y^{90}}{E_1} - \frac{v_{23}\sigma_z}{E_3} \qquad \text{(A1)}$$

$$\frac{\sigma_y^0}{E_2}-\frac{v_{12}\,\sigma_x^0}{E_1}-\frac{v_{23}\,\sigma_z}{E_3}=\frac{\sigma_y^{90}}{E_1}-\frac{v_{21}\,\sigma_x^{90}}{E_2}-\frac{v_{31}\sigma_z}{E_3} \tag{A2}$$

From equilibrium as given by A1 and A2,

$$\sigma_x^0=-\sigma_x^{90} \tag{A3}$$

$$\sigma_y^0=-\sigma_y^{90} \tag{A4}$$

If we substitute (A3) and (A4) into (A1) and (A2), it should be possible to solve for σ_x^{90}, σ_y^{90} in terms of σ_z,

$$\frac{-\sigma_x^{90}}{E_1}+\frac{v_{21}\,\sigma_y^{90}}{E_2}+\frac{\sigma_z}{E_3}(v_{23}-v_{31})=\frac{\sigma_x^{90}}{E_2}-\frac{v_{12}\sigma_y^{90}}{E_1}$$

$$\sigma_x^{90}\left(\frac{1}{E_1}+\frac{1}{E_2}\right)-\sigma_y^{90}\left(\frac{v_{12}}{E_1}+\frac{v_{21}}{E_2}\right)=\frac{\sigma_z}{E_3}(v_{23}-v_{31}) \tag{A5}$$

$$\frac{-\sigma_y^{90}}{E_2}+\frac{v_{12}\sigma_x^{90}}{E_1}-\frac{\sigma_z}{E_3}(v_{23}-v_{31})=\frac{\sigma_y^{90}}{E_1}-\frac{v_{21}\sigma_x^{90}}{E_2}$$

$$\sigma_x^{90}\left(\frac{v_{12}}{E_1}+\frac{v_{21}}{E_2}\right)-\sigma_y^{90}\left(\frac{1}{E_1}+\frac{1}{E_2}\right)=\frac{\sigma_z}{E_3}(v_{23}-v_{31}) \tag{A6}$$

By inspection,

$$\sigma_x^{90}=\sigma_y^0$$

$$\sigma_y^{90}=\sigma_x^0$$

$$\bar{\varepsilon}_z=\frac{\sigma_z}{E_z}=\frac{\sigma_z}{E_3}-\frac{v_{13}}{E_1}\sigma_x^0-\frac{v_{23}}{E_2}\sigma_x^{90}$$

$$=\frac{\sigma_z}{E_3}-\frac{v_{13}}{E_1}\sigma_x^0-\frac{v_{23}}{E_2}(-\sigma_x^0)$$

$$=\frac{\sigma_z}{E_3}+\left(\frac{v_{23}}{E_2}-\frac{v_{13}}{E_1}\right)\sigma_x^0$$

Determination of E_z

From (A5),

$$\sigma_x^{90}\frac{E_1+E_2}{E_1E_2}-\sigma_y^{90}\frac{2v_{12}}{E_1}=\frac{\sigma_z}{E_3}(v_{23}-v_{31})$$

$$-\sigma_x^0\left(\frac{E_1+E_2}{E_1E_2}\right)-\sigma_x^0\frac{2v_{12}}{E_1}=\frac{\sigma_z}{E_3}(v_{23}-v_{31})$$

$$\sigma_x^0\left[\frac{2v_{12}}{E_1}+\left(\frac{E_1+E_2}{E_1E_2}\right)\right]=-\frac{\sigma_z}{E_3}(v_{23}-v_{31})$$

$$\sigma_x^0=-\frac{(v_{23}-v_{31})}{E_3\left[\frac{2v_{12}}{E_1}+\left(\frac{E_1+E_2}{E_1E_2}\right)\right]}\sigma_z \tag{A7}$$

$$\frac{\sigma_z}{E_z}=\frac{\sigma_z}{E_3}-\frac{(v_{23}-v_{31})\left(\frac{v_{23}}{E_2}-\frac{v_{13}}{E_1}\right)}{E_3\left[\frac{2v_{12}}{E_1}+\left(\frac{E_1+E_2}{E_1E_2}\right)\right]}\sigma_z$$

$$\frac{\sigma_z}{E_z}=\sigma_z\frac{1}{E_3}\left[1-\frac{(v_{23}-v_{31})\left(\frac{v_{23}}{E_2}-\frac{v_{13}}{E_1}\right)}{\left[\frac{2v_{12}}{E_1}+\left(\frac{E_1+E_2}{E_1E_2}\right)\right]}\right]$$

$$E_z=\frac{E_3}{\left[1-\frac{(v_{23}-v_{31})\left(\frac{v_{23}}{E_2}-\frac{v_{13}}{E_1}\right)}{\left[\frac{2v_{12}}{E_1}+\left(\frac{E_1+E_2}{E_1E_2}\right)\right]}\right]}$$

Determination of v_{zx}

$$v_{zx}=\frac{\varepsilon_x}{\varepsilon_z}=\frac{Lateral\ strain}{Axis\ strain}$$

$$=-\frac{\frac{\sigma_x^0}{E_1}-\frac{v_{21}\sigma_y^0}{E_2}-\frac{v_{31}\sigma_z}{E_3}}{\frac{\sigma_z}{E_3}-\frac{1}{2}\left[\frac{v_{13}}{E_1}\left(\sigma_x^0+\sigma_y^{90}\right)+\frac{v_{23}}{E_1}\left(\sigma_x^{90}+\sigma_y^0\right)\right]}$$

$$=-\frac{\frac{\sigma_x^0}{E_1}+\frac{v_{21}\sigma_x^0}{E_2}-\frac{v_{31}\sigma_z}{E_3}}{\frac{\sigma_z}{E_3}-\frac{v_{13}\sigma_x^0}{E_1}+\frac{v_{23}\sigma_x^0}{E_2}}$$

From (A7),

$$\sigma_x^0=-\frac{(v_{23}-v_{31})}{E_3\left[\frac{2v_{12}}{E_1}+\left(\frac{E_1+E_2}{E_1E_2}\right)\right]}\sigma_z=\frac{-(v_{23}-v_{31})\sigma_zE_1E_2}{E_3(2v_{12}E_2+E_1+E_2)}$$

$$=-\left[\frac{\frac{-(v_{23}-v_{31})\sigma_zE_2^2E_1}{2v_{12}E_2+E_1+E_2}-\frac{v_{21}(v_{23}-v_{31})\sigma_zE_1^2E_2}{2v_{12}E_2+E_1+E_2}-v_{31}\sigma_zE_1E_2}{\sigma_zE_1E_2+\frac{(v_{23}-v_{31})v_{13}\sigma_zE_2^2E_1}{2v_{12}E_2+E_1+E_2}-\frac{(v_{23}-v_{31})v_{23}\sigma_zE_1^2E_2}{2v_{12}E_2+E_1+E_2}}\right]$$

Diminished the dominator and σ_z

$$v_{zx}=-\left[\frac{-(v_{23}-v_{31})E_2^2E_1-v_{21}(v_{23}-v_{31})E_1^2E_2-v_{31}E_1E_2(2v_{12}E_2+E_1+E_2)}{(2v_{12}E_2+E_1+E_2)E_1E_2+(v_{23}-v_{31})v_{13}E_2^2E_1-(v_{23}-v_{31})v_{23}E_1^2E_2}\right]$$

$$v_{zx}=-\left[\frac{-(v_{23}-v_{31})E_2-v_{21}(v_{23}-v_{31})E_1-v_{31}(2v_{12}E_2+E_1+E_2)}{(2v_{12}E_2+E_1+E_2)+(v_{23}-v_{31})v_{13}E_2-(v_{23}-v_{31})v_{23}E_1}\right]$$

A2: Quasi-Isotropic laminates

*For quasi-isotropic, the exact formulation are given in Akkermann (2002)

$$E_x = E_y \qquad 2\,(1 + v_{xy})G_{xy} \qquad \text{.........(A8)}$$

$$E_z \qquad \frac{E_1 + (1 + 2v_{12})E_2}{(1 - v_{23}^2)\frac{E_1}{E_2} + (1 + 2v_{12} + 2v_{12}v_{23}) - v_{12}^2\frac{E_2}{E_1}} \qquad \text{..........(A9)}$$

$$G_{xz} \qquad 2\left(\frac{G_{12}\,G_{23}}{G_{12} + G_{23}}\right) \qquad \text{..........(A10)}$$

$$v_{xz} = v_{yz} \qquad \frac{E_x}{E_1} \cdot \frac{(v_{12} + v_{23} + v_{12}v_{13}) + v_{12}^2\frac{E_2}{E_1}}{1 + (1 + 2v_{12})\frac{E_2}{E_1}} \qquad \text{...........(A11)}$$

Model reference control of conceptual clutched train substituted for vehicular friction clutch

Wei Huang[1,a], Pakharuddin Mohd Samin[1] and Kamarul Baharin Tawi[1]

[1]*Department of Aeronautical, Automotive and Ocean Engineering, Universiti Teknologi Malaysia, 81310 Johor, Malaysia*

Abstract. In order to essentially reduce heat generation and frictional dissipation carried by friction clutch engagement, conceptual design of clutched train combined with hydrostatic braking system is proposed as a novel substitution for vehicular friction clutch. Potential collateral merits of clutched train may improve service life and control accuracy since less friction heat generated during synchronization process. Parameter of clutched train is obtained by Genetic Algorithm optimization aiming at axial space-saving and light weight. Control-oriented model of proposed concept is derived and used in Model Reference Control development. Based on optimum parameter of clutched train, simulation result has shown the functionality of clutched train on vehicle standing-start, and well-behaved Model Reference Control on smoothing clutched train synchronization process.

1 Introduction

Friction clutch is known as a key mechanical connection and disconnection between an engine and a transmission in a car. The basic function of clutch is supposed to transfer power from engine to vehicle with desired smoothly and efficiently. However, owing to it transmitting torque through friction, some issues particularly in terms of friction dissipation [1] and shift comfort [2-4] are inevitable. For example in case when significant heat generation during slipping phase, high nonlinear fluctuation of friction coefficient [3, 4] would make it difficult to give an accurate signal to actuator for the estimation of transmitted clutch torque.

As a branch of planetary gearset application, concept of clutched train with hydraulic system is proposed in this paper to substitute for friction clutch used in a car and mated to automated mechanical transmission to ensure smooth transfer of torque with a little heat generated by gear-meshed sliding friction [5]. Note that less heating is associated with improving mechanical life, as well as more accurate control due to little variation on friction coefficient. Clutched train utilizes the alternation between one degree-of-freedom (DOF) and two DOFs of rotary motion to determine the output instead of friction. The primary advantages of proposed concept lie in avoiding high non-linearity and significant frictional dissipation with engagement, and in return improving the control veracity in practice.

This paper would present the conceptual design and application of Model Reference Control (MRC) algorithm on clutched train synchronization process control. Simulation results show that the functionality of conceptual design is to connect and disconnect engine and transmission, and the designed MRC could work with conceptual design to reduce vehicle jerk at vehicle standing-start.

[a] Corresponding author: huangweiamoy@gmail.com

2 Conceptual design

As shown in Figure 1, the configuration of conceptual design is comprised of a symmetrical compound planetary gearset and a hydraulic system. The compound planetary gearset can be considered as a combination of single stage Planetary-Gearset-I (PGS-I) and Planetary-Gearset-II (PGS-II), which have the same parameters and share one arm. The PGS-I ring gear is fixed. The PGS-I sun gear is coupled with the driving shaft, which is driven by the output shaft of an engine. The PGS-II sun gear is coupled to a coaxial output shaft which drives an automatic transmission.

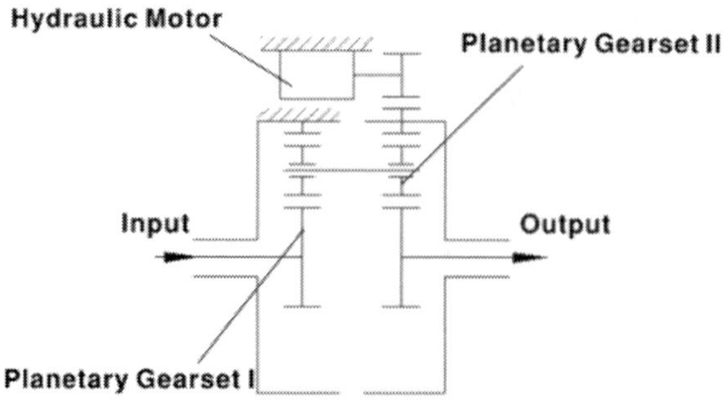

Figure 1. Schematic structure of clutched train.

Figure 2 illustrates the synchronization process of clutched train with hydrostatic braking system from the opening to sticking phase. The mode-switch valve would hand off the DOF of rotational dynamics from two to one or from one to two via controlling the reversible motor-pump free or locked. In the opening and slipping mode, the mode switch is set to connect the reversible motor-pump discharge port to the pressure relief valve for regulating the operating pressure. In the sticking mode, the mode-switch valve is shifted to connect the discharge port to check valve to cut the discharge flow. As an expected consequence, it is supposed to have same capability as a conventional clutch in interrupting and transferring the power from engine to transmission. The torque of output shaft during slipping phase, which depends on the back pressure of reversible motor-pump, can be controlled via the pressure relief valve. Accumulator can be installed to harvest the fluid energy produced by reversible motor-pump for decelerating to zero instead of friction clutch converting into heat. In other words, it makes the reversible motor-pump act as a pump to recover the pent-up momentum-carrying energy. Note that after synchronization the PGS-II ring gear is held by the back pressure of reversible motor-pump without any energy supply in terms of keeping engagement. Once overload is happened on clutched train, pressure relief valve can ensure clutched train working on safe operation by setting maximum allowable pressure.

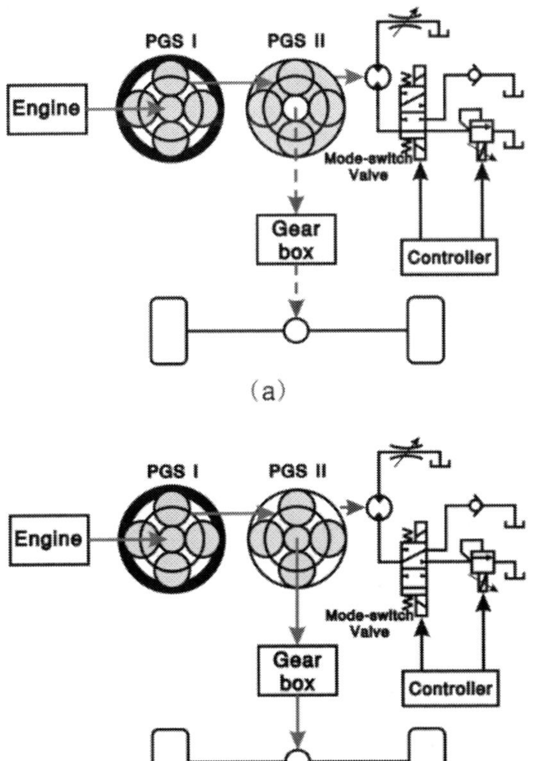

Figure 2. Power flow in driveline equipped with clutched train working in the (a) Opening phase; (b) Sticking phase (PGS-II ring gear locked).

3 Parameter optimization

In view of other merits about friction clutch like compact size and light weight, the parameters correlated to proposed gear train should be taken into account for compact size and light weight as well. Referring to literature[6], Genetic Algorithm (GA) was recognized as an effective and efficient technique to optimize weight design problem of gear train. Several reasonable gear train parameters are selected, and formulated for dimension optimization problem and constrained by the planetary gearset geometry limitations, mating conditions and load capacity, and solve the problem by using GA.

3.1 GA setting

In order to create better new individuals, GA would repeat selection, crossover and mutation until reaching the optimum solution to the problem. The main setting of GA includes:

i) Crossover: 0.85. Determines the probability that an offspring will be created by the crossover of two parents as opposed to a replica of one parent;
ii) Mutation: 0.1. Determines the probability of a spot mutation occurring to each allele in a new population;
iii) Population Size: 400. The population size determines the number of chromosomes that make up a generation.

3.2 Optimum parameter

The optimization result of gear train parameters is listed in Table 1 below. These optimum parameters are used to investigate the functionality and effects on vehicle driving comfort performance of clutched train via simulation.

Table 1. GA optimum parameter.

	Parameter	Value
z_S	Teeth number of sun gear	30
b	Gear face width (mm)	10
m	Gear module (mm)	2
K	Gear ratio between ring gear and sun gear	3

4 Dynamic modeling of clutched train

4.1 Kinematics of clutched train

In this study, the compound planetary gearset is divided into two single stage planetary gearsets which have the same parameters and share one arm. The relationship of angular velocity among sun gear, ring gear and arm in single stage planetary gearset can be determined by the known Willis rule: 'the ratio of the relative (in regard to the arm) angular velocities is equal to their gear ratio' [7].

For the sun gears, ring gears and arm with the PGS-I and PGS-II, these following equations are obtained:

$$\omega_{S1} + K \cdot \omega_{R1} - (1+K) \cdot \omega_A = 0 \tag{1a}$$

$$\omega_{S2} + K \cdot \omega_{R2} - (1+K) \cdot \omega_A = 0 \tag{1b}$$

$$\omega_{R1} = 0 \tag{1c}$$

where, ω_{S1} is the angular velocity of PGS-I sun gear; ω_{S2} is the angular velocity of PGS-II sun gear; ω_{R1} is the angular velocity of PGS-I ring gear; ω_{R2} is the angular velocity of PGS-II ring gear; ω_A is the angular velocity of arm; and K is the gear ratio between ring gear and sun gear, namely z_R/z_S. Since the parameters of PGS-I and PGS-II are totally same, they have the same gear ratio K.

By elimination of ω_A from Eq. (1a) and (1b), the following kinematics equation is obtained:

$$\omega_{R2} = \frac{\omega_{S1} - \omega_{S2}}{K} \tag{2}$$

Note that while the angular velocity of PGS-II ring gear ω_{R2} decelerating to zero via hydrostatic braking system, Eq. (2) demonstrates that $\omega_{S1}=\omega_{S2}$, namely clutched train synchronization.

4.2 Power flow of clutched train

Power losses in gear train (e.g. gear-meshed friction losses) usually is neglected to reasonably simplify the dynamic model on control orient with little consequence. In accordance with the law of energy preservation the sum of powers transmitted by sun gear, ring gear and arm in each single gear train must equal zero [7]:

$$M_{S1} \cdot \omega_{S1} + M_A \cdot \omega_A = 0 \tag{3a}$$

$$M_A \cdot \omega_A + M_{S2} \cdot \omega_{S2} + M_{R2} \cdot \omega_{R2} = 0 \tag{3b}$$

where M_{S1} is the torque of PGS-I sun gear; M_{S2} is the torque of PGS-II sun gear; M_{R1} is the torque of PGS-I ring gear; M_{R2} is the torque of PGS-II ring gear; M_A is the torque of arm.

From the condition of equilibrium of torques, the equations are noted [7]:

$$M_{S1} + M_{R1} + M_A = 0 \tag{4a}$$

$$M_{S2} + M_{R2} + M_A = 0 \tag{4b}$$

Based on Eq. (1a) to (1c), (3a) to (3b) and (4a) to (4b), the relationships of torques among sun gear, ring gear and arm of the PGS-I and PGS-II are derived as follows:

$$M_{S1} = M_{S2} = \frac{1}{K} M_{R2} \tag{5}$$

5 Control-oriented model of slip-stick phase

Since standing-start maneuver has been described as a typical manifestation of phenomenon on longitudinal oscillations excited by friction clutch engagement [8, 9], control algorithm for clutched train synchronization process would be vital important to vehicle driving comfort. Therefore, control-oriented model of clutched train would be built in this section.

5.1 Slipping phase

In light of Eq. (5), torque of sun gear of planetary gearset M_S herein is introduced and denoted as $M_S=M_{S1}=M_{S2}$. In the slipping phase, the driving shaft's angular velocity is not equal to the driven shaft's angular velocity, namely $\omega_e \neq \omega_r$. The vehicle driveline is considered as a lumped mass system. The simplified model of clutched train on controlled orient is set up and shown in Figure 3.

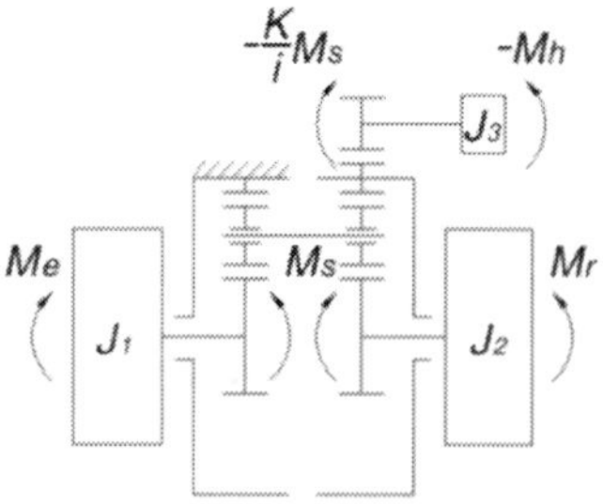

Figure 3. Simplified model of clutched train.

According to the Newton's Second Law, the simplified dynamics of clutched train in the slipping phase is given as follows:

$$J_1 \cdot \dot{\omega}_e = M_e - M_S \tag{6}$$

$$J_2 \cdot \dot{\omega}_r = M_S - M_r \tag{7}$$

$$J_3 \cdot \dot{\omega}_h = \frac{K}{\mu} M_S - \frac{\Delta p \cdot D_v}{2\pi} \eta_m \tag{8}$$

where J_1 is the total moment of inertial of the driving shaft, including the crankshaft, flywheel and PGS-I sun gear; J_2 is the total moment of inertial of the driven shaft, including PGS-II sun gear, driveline, tires and body; J_3 is the moment of inertial of reversible motor-pump shaft; ω_h is the angular velocity of the reversible motor-pump; M_e is engine output torque; M_r is vehicle resistant torque; μ is the gear ratio between the PGS-II ring gear and the gear coupled with reversible motor-pump output shaft, namely z_{R2}/z_h; Δp is the difference of pressure between the hydraulic motor inlet and outlet; D_v is displacement of hydraulic motor; and η_m is mechanical efficiency of hydraulic motor.

Considering the magnitude of J_3 is very small, Eq. (8) can be simplified as:

$$\frac{K}{\mu}M_S \approx \frac{\Delta p \cdot D_v}{2\pi}\eta_m \tag{9}$$

As observed from Eq. (6), (7) and (9), the value of pressure relief valve can be controlled to determine the angular velocity of PGS-II sun gear for less vehicle jerking.

5.2 Sticking phase

Until the angular velocity of PGS-II ring gear declines to zero under hydraulic resistance, according to Eq. (2) the driving shaft's angular velocity is equal to the driven shaft's angular velocity, namely $\omega_e=\omega_r$. Then the mode-switch valve will be shifted to sticking operation position. This will allow the PGS-II ring gear to be locked and keep $\omega_{R2}=0$. When $\omega_e=\omega_r=\omega_m$, the dynamic equation can be rewritten by combining Eq. (6) and (7):

$$(J_1+J_2)\dot{\omega}_m = M_e - M_r \tag{10}$$

6 Model reference control

MRC has been proposed by Chen et al. [8] in friction clutch engagement control to compensate discontinuity of slip-stick transition. However, different feedback control law (regulator) on accommodating clutched train synchronization process is presented. The architecture of MRC including reference model and plant is established in Figure 4.

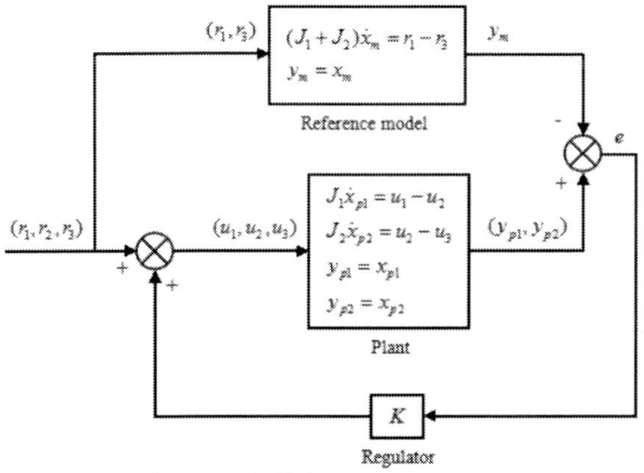

Figure 4. Architecture of MRC.

For the plant, the following variables are introduced and denoted as:

$$x_{p1} = \omega_e; x_{p2} = \omega_r;$$

$$u_1 = M_e; u_2 = M_S; u_3 = M_r$$

Substituting to Eq. (6) and (7), the state equations can be obtained:

$$J_1\dot{x}_{p1} = u_1 - u_2$$

$$J_2\dot{x}_{p2} = u_2 - u_3$$

$$y_{p1} = x_{p1}$$

$$y_{p2} = x_{p2}$$

For the reference model, the following variables are introduced and denoted as:

$$x_m = \omega_m$$

The output of the reference model are denoted as

$$y_m = x_m$$

The state equation is given according to Eq. (7):

$$(J_1+J_2)\dot{x}_m = r_1 - r_3$$

In the proposed MRC scheme, we define that:

i) $r_1(t)$, $r_2(t)$, $r_3(t)$ are reference inputs;
ii) **K** represents the gain for the feedback regulator;
iii) $e_1(t)$ and $e_2(t)$ are two outputs of interest from the closed-loop system, given as:

$$e_1 = y_{p1} - y_m$$

$$e_2 = y_{p2} - y_m$$

The proposed output feedback control algorithm is given as follows:

$$u_1 = r_1 + k_{11}e_1 + k_{12}e_2$$

$$u_2 = r_2 + k_{21}e_1 + k_{22}e_2$$

$$u_3 = r_3 + k_{31}e_1 + k_{32}e_2$$

Then time derivative of the error vector **e** is written as

$$\begin{pmatrix}\dot{e}_1\\ \dot{e}_2\end{pmatrix} = \begin{pmatrix}\dot{y}_{p1}-\dot{y}_m\\ \dot{y}_{p2}-\dot{y}_m\end{pmatrix} = \begin{pmatrix}\frac{u_1-u_2}{J_1} - \frac{r_1-r_3}{J_1+J_2}\\ \frac{u_2-u_3}{J_2} - \frac{r_1-r_3}{J_1+J_2}\end{pmatrix}$$

$$= \begin{pmatrix}\frac{(r_1+k_{11}e_1+k_{12}e_2)-(r_2+k_{21}e_1+k_{22}e_2)}{J_1} - \frac{r_1-r_3}{J_1+J_2}\\ \frac{(r_2+k_{21}e_1+k_{22}e_2)-(r_3+k_{31}e_1+k_{32}e_2)}{J_2} - \frac{r_1-r_3}{J_1+J_2}\end{pmatrix}$$

$$= \begin{pmatrix}\frac{k_{11}-k_{21}}{J_1} & \frac{k_{12}-k_{22}}{J_1}\\ \frac{k_{21}-k_{31}}{J_2} & \frac{k_{22}-k_{32}}{J_2}\end{pmatrix}\begin{pmatrix}e_1\\ e_2\end{pmatrix} + \begin{pmatrix}\frac{1}{J_1}-\frac{1}{J_1+J_2} & -\frac{1}{J_1} & \frac{1}{J_1+J_2}\\ -\frac{1}{J_1+J_2} & \frac{1}{J_2} & \frac{1}{J_1+J_2}-\frac{1}{J_2}\end{pmatrix}\begin{pmatrix}r_1\\ r_2\\ r_3\end{pmatrix}$$

$$= \mathbf{A}\begin{pmatrix}e_1\\ e_2\end{pmatrix} + \mathbf{B}$$

It is marked that:

$$\mathbf{A} = \begin{pmatrix}(\frac{k_{11}-k_{21}}{J_1}) & (\frac{k_{12}-k_{22}}{J_1})\\ (\frac{k_{21}-k_{31}}{J_2}) & (\frac{k_{22}-k_{32}}{J_2})\end{pmatrix}$$

$$\mathbf{B}(r_1,r_2,r_3)=\begin{pmatrix}(\frac{1}{J_1}-\frac{1}{J_1+J_2}) & (-\frac{1}{J_1}) & (\frac{1}{J_1+J_2}) \\ (-\frac{1}{J_1+J_2}) & (\frac{1}{J_2}) & (\frac{1}{J_1+J_2}-\frac{1}{J_2})\end{pmatrix}\begin{pmatrix}r_1\\r_2\\r_3\end{pmatrix}$$

Input u_1 physically stands for the engine output torque M_e, which is determined by the work load and can reflect the driver intention. Hence, u_1 is treated as a predefined input and need not be adjusted by MRC. Therefore,

$$k_{11}=0\,\&\,k_{12}=0$$

Similarly, input u_3 physically stands for the resistant torque M_r imposed on the driven shaft, which cannot be changed subjectively. Hence,

$$k_{31}=0\,\&\,k_{32}=0$$

So,

$$\mathbf{A}=\begin{pmatrix}(\frac{-k_{21}}{J_1}) & (\frac{-k_{22}}{J_1}) \\ (\frac{k_{21}}{J_2}) & (\frac{k_{22}}{J_2})\end{pmatrix}$$

The critical variable is u_2. In order to achieve asymptotic stability, every eigenvalue of **A** should have a strictly negative real part in accordance with Lyapunov stability theory [10]. Then

$$|\lambda\mathbf{I}-\mathbf{A}|=0$$

$$\begin{vmatrix}(\lambda+\frac{k_{21}}{J_1}) & (\frac{k_{22}}{J_1}) \\ (\frac{-k_{21}}{J_2}) & (\lambda-\frac{k_{22}}{J_2})\end{vmatrix}=0$$

$$\lambda^2+(\frac{k_{21}}{J_1}-\frac{k_{22}}{J_2})\lambda-\frac{2k_{21}k_{22}}{J_1J_2}=0$$

According to Vieta's formulas, thus k_{21} and k_{22} should satisfy:

$$\lambda_1+\lambda_2=\frac{k_{22}}{J_2}-\frac{k_{21}}{J_1}<0$$

$$\lambda_1\lambda_2=-\frac{2k_{21}k_{22}}{J_1J_2}>0$$

Regarding to simulation, k_{21} and k_{22} are set to 0.7 and -0.7, respectively.

Note that these three reference inputs r_1, r_2 and r_3 do not affect the system stability but will influence the output response **e**. The main function of the feed-forward regular is to avoid disturbance on the zero-input response, so that the exogenous term $\mathbf{B}(r_1,r_2,r_3)$ can reinforce the error convergence, the feed-forward variables r_1, r_2 and r_3 are therefore subject to the constraints:

$$(\frac{1}{J_1}-\frac{1}{J_1+J_2})r_1+(-\frac{1}{J_1})r_2+(\frac{1}{J_1+J_2})r_3=0$$

and

$$(-\frac{1}{J_1+J_2})r_1+(\frac{1}{J_2})r_2+(\frac{1}{J_1+J_2}-\frac{1}{J_2})r_3=0$$

As a consequence, r_2 can be calculated as:

$$r_2=\frac{J_2r_1+J_1r_3}{J_1+J_2}$$

The overall MRC algorithm can be summarized as:

$$u_2=\frac{J_2r_1+J_1r_3}{J_1+J_2}+k_{21}e_1+k_{22}e_2$$

As mentioned by Chen et al. [8], u_2 should be restricted under certain rate of increase k_f due to physical mechanism limitation, namely

$$u_2^*=k_ft$$

So, the torque of sun gear of planetary gearset M_S is determined by

$$\bar{u}_2=\min(u_2,u_2^*)$$

7 Simulation results

Co-simulation is carried out between ITI SimulationX (as plant) and Simulink/MATLAB (as controller) on the case of vehicle standing-start. The parameters of vehicle model are given in Table 2. The engine characteristics is shown in Figure 5.

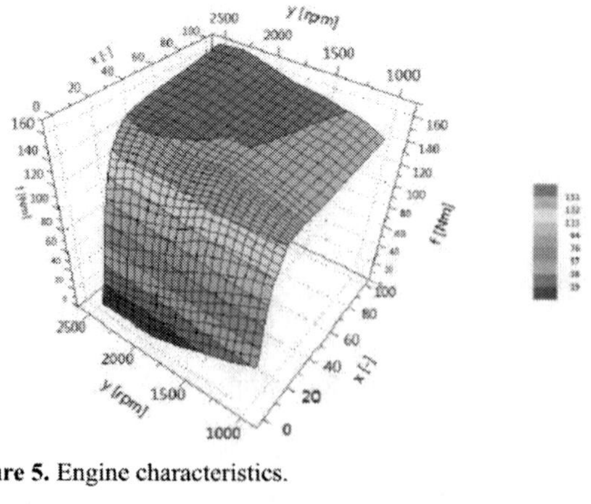

Figure 5. Engine characteristics.

The simulation results of clutched train working with MRC during slip-stick phase are shown in Figure 6, Figure 7 and Figure 8. As illustrated in Figure 6, clutched train can reach synchronization under MRC. Meanwhile, vehicle jerk observed in Figure 7 shows that it can be reduced to below 8.01 m/s^3 and 16.18 m/s^3 with less than 3 Hz and thus satisfy the requirement of driving comfort

[11]. It also indicates that the simplified control-oriented model omitting the moment of inertial of reversible motor-pump shaft has little effect on comfortable synchronization performance. However, as shown in Figure 8, the maximum of measured back pressure from reversible motor-pump are 38.89 MPa. In other words, rated pressure of hydraulic system will become the limitation of maximum transmitted torque in power train system.

Table 2. Parameters of vehicle model.

Parameters		Value
J_1	Total moment of inertial of the driving shaft (kg·m^2)	1.01
J_2	Total moment of inertial of the driven shaft (kg·m^2)	0.97
J_3	Total moment of inertial of the reversible motor-pump shaft (kg·m^2)	0.01
m	Vehicle mass (kg)	1400
i_0	Gear ratio of reducer	4
i_1	Gear ratio of gearbox	3.32
R	Tire radius (m)	0.35
f	Rolling friction coefficient	0.01

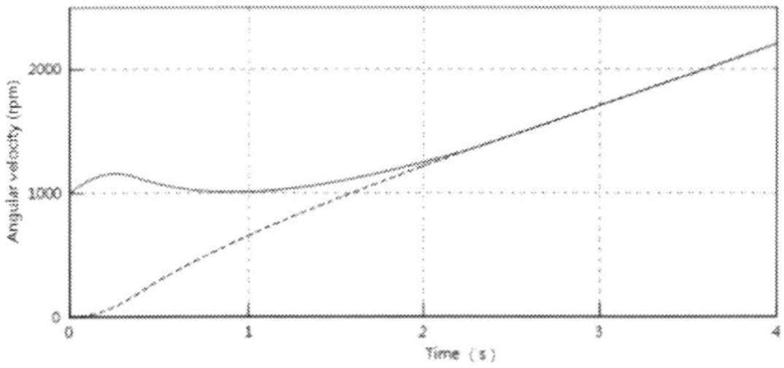

Figure 6. Angular velocity of driving and driven shaft at the slip-stick transition. (solid line: angular velocity of driving shaft, dashed line: angular velocity of driven shaft)

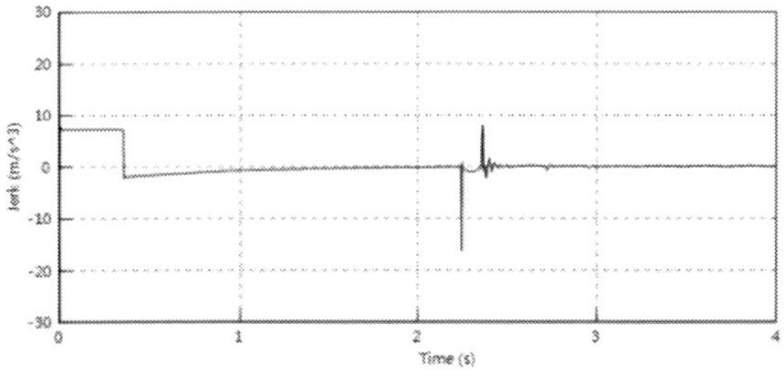

Figure 7. Derivation of vehicle acceleration, namely jerk.

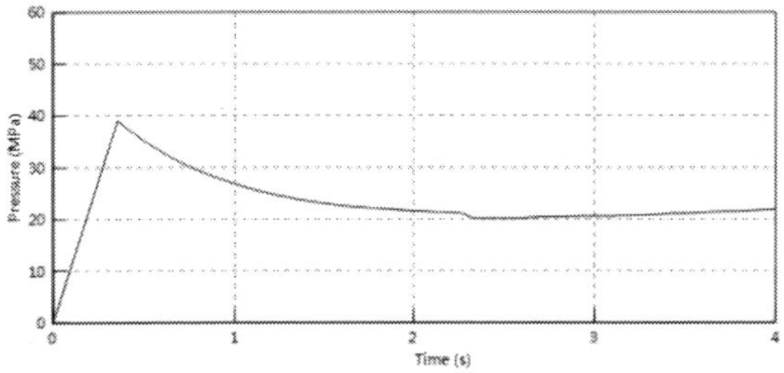

Figure 8. Pressure of hydraulic system.

8 Conclusion

Conceptual design of clutched train with hydrostatic braking system is provided in this paper. Control-oriented model and MRC on clutched train synchronization process are developed. Simulation results show that conceptual proposal has the desired function to connect and disconnect engine with transmission. In the period of clutched train slip-stick, vehicle jerk can be accommodated by MRC for comfortable driving experience of vehicle standing-start. However, the maximum transmitted torque of proposed concept will be primarily limited by the rated pressure of hydraulic system and result in limitation of relevant application in heavy vehicle.

References

1. V. Aravind, Engagement characteristics of a friction pad for commercial vehicle clutch system. Sadhana, 2010. **35**(5): p. 585-595.
2. N. Cappetti, M. Pisaturo, A. Senatore. Modelling the cushion spring characteristic to enhance the automated dry-clutch performance: The temperature effect. P I Mech Eng D-J Aut, 2012. **0**(0): p. 1-11.
3. P. Marklund, R. Larsson. Wet clutch under limited slip conditions - simplified testing and simulation. P I Mech Eng J-J Eng, 2007. **221**(5): p. 545-551.
4. T. Petrun, M. Kegl, J. Flašker. Development and Validation of a Friction Model for Simulation of Friction Clutch Dynamics in a Multi-Body System, in *Tribology and Design II*, M. Hadfield and C.A. Brebbia, Editors. 2012, WIT Press: Southampton. p. 77-88.
5. W. Huang, P.M. Samin, K.B. Tawi, B. Supriyo. Frictional dissipation and mechanical efficiency analysis of clutched train engagement. J Appl Sci & Agric, 2014. **9**(18): p. 238-244.
6. Y. Takao, T. Takeaki, G. Mitsuo. A solution method for optimal weight design problem of the gear using genetic algorithms. Comput Ind Eng, 1998. **35**(3–4): p. 523-526.
7. T. Jelaska. *Gears and gear drives*, 2012, John Wiley & Sons: Chichester.

8. L. Chen, G. Xi, C.L. Yin. Model referenced adaptive control to compensate slip-stick transition during clutch engagement. Int J Auto Tech-Kor, 2011. **12**(6): p. 913-920.
9. D. Hrovat, W. E. Tobler. *Bond graph modeling of automotive power trains.* J Franklin I, 1991. **328**(5–6): p. 623-662.
10. A. Karl W. Björn. *Adaptive* control, 2013, Courier Corporation: New York.
11. J. G. Zhang, Y.L. Lei, X. Hua, J. Wang, A. Ge. Proposed shift quality metrics and experimentation on AMT shift quality evaluation. in *3rd International Conference on Natural Computation*. 2007.

MATEC Web of Conferences 74, 00005 (2016) DOI: 10.1051/matecconf/20167400005
ICMER 2015

Gas turbine gas path diagnostics: A review

Amare Desalegn Fentaye[1,a], Syed Ihtsham Ul-Haq Gilani [1] and Aklilu Tesfamichael Baheta[1]

[1]*Department of Mechanical Engineering, Universiti Teknologi PETRONAS, 31750 Tronoh, Perak, Malaysia*

Abstract. In this competitive business world one way to increase profitability of a power production unit is to reduce the operation and maintenance expenses. This is possible if the gas turbine availability and reliability is improved using the appropriate maintenance action at the right time. In that case, fault diagnostics is very critical and effective and advanced methods are essential. Gas turbine diagnostics has been studied for the past six decades and several methods are introduced. This paper aims to review and summarise the published literature on gas path diagnostics, giving more emphasis to the recent developments, and identify advantages and limitations of the methods so that beginners in diagnostics can easily be introduced. Towards this end, this paper, identifies various diagnostic methods and point out their pros and cons. Finally, the paper concludes the review along with some recommended future works.

Introduction

Gas turbine consists of complex, highly expensive, and precise elements operating at high pressure and temperature gas conditions. Failure or performance degradation of the engine strongly affects its operation. Performance of a gas turbine highly relies on the performance of gas path components, namely, compressor, combustion chamber, and turbine [1]. Component performance can be degraded attributable to the establishment of different fault cases, such as, fouling, erosion, blade tip clearance, corrosion, and object damages. The detailed description of these faults and their effects on the gas turbine performance including suitable maintenance advises can be found in the study led by Mejer-Homji et al. [2]. The occurrence of these faults result deviations of measurable engine parameters. To avoid these faults and maintain its best performance the gas turbine should have a mechanism that can detect, classify, and quantify developing engine faults as early as possible. This will help the operators to take action so that the engine is restored to the normal performance.

The change of the performance could be due to off-design operation. That is to say because of load variation, ambient temperature or fuel delivery changes. Equivalently, the change of performance could be due to performance deterioration or malfunction. Hence, the operators should know the cause(s) of the change. In such cases the performance monitoring tool should be able to identify the cause(s). For instance, degradations are manifested by change of gas path measurable parameters or health indices. To predict the gas path parameters performance simulation model is required. Once the gas path parameters are predicted, the non-measurable health indices would be computed for diagnostic purpose.

Diagnostics is the process of fault detection, isolation, and identification of a component or a system for an appropriate maintenance advice [3]. Specifically, gas turbine diagnostics can be for the sensor system, control system, fuel and oil system, vibration system, and gas path system [4]. Among them gas path diagnostics is the most critical tool and it works on the basis of engine performance deterioration [5].

Previously, large number of gas turbine diagnostics researches have been published focusing on component performance. As an example, Ogaji and Singh [6] developed advanced gas path fault diagnostic scheme for a single spool stationary gas turbines that can approximate both component faults and sensor faults, simultaneously. In the same way, Marinai [7] developed a diagnostic algorithm for gas path components of a gas turbine engine that can isolate single, dual and multiple component faults in the presence of noisy data. For more detail, an extensive review is available in [8]. However, recent researches [9-11] indicate that the existing algorithms still need an improvement and the area has some rooms for further study.

Thus, the aim of this paper is to present the literature survey on gas path diagnostics with a special focus on the current works and future trends. Section 2 point out the available advanced engine fault diagnostic approaches. In this part, the improvement of the methods as well as the recent research works utilizing them have also been presented. Then, section 3 provides the comparison among the approaches on the advantages and limitations. Lastly, section 4 belongs to the conclusion of the review, along with their implication for future works.

[a] Corresponding author: dadorthot@gmail.com

Gas turbine diagnostics methods

In the field of gas turbine diagnostics several methods have been devised by the research community over the years [12-14]. Those methods can be classified into two main groups, namely model-based and data driven approaches [15]. Under these classifications a critical survey of the most commonly used methods and their applications are summarized here after.

Model based diagnostics methods

Model based diagnostics approaches rely on the thermodynamic model of a gas turbine. According to this approach the relationship among the gas turbine dependent parameters (e.g. pressure, temperature, fuel mass flow rate, and engine horse power) and the performance parameters (efficiency, pressure ratio and flow capacity) is determined through an explicit mathematical and thermodynamic equations [9].

In the field of model based diagnostics different approaches are available. Among these methods the gas path analysis (GPA) scheme is investigated very intensively by several authors, particularly, by those who worked on gas turbine gas path diagnostics [8]. Even though, gas turbine models for diagnosis are highly non-linear, Urban [16] in 1967 introduced a simplified linear description, referred to as "linear gas path analysis" (LGPA), neglecting the measurement uncertainty and the non-linearity nature of the engine behavior. After that, the same author [17] and other authors [18, 19] proposed linear GPA algorithms with some modifications. Figure 1 illustrates the schematic representation of GPA methods.

According to the principle of linear GPA approach, the thermodynamic relationship among measurement parameters and performance parameters can be expressed as:

$$\vec{Z} = H \bullet \vec{X} \tag{1}$$

where $\vec{Z} \in R^M$ is engine measurable parameter vector and M is the number of measurable parameters, $\vec{X} \in R^N$ is engine components performance parameter vector and N is the number of component performance parameters, and **H** is Influence Coefficient Matrix (ICM).

In the same fashion, the engine measurable parameter changes and component performance changes can be related using Eq. (2).

$$\Delta\vec{Z} = H \bullet \Delta\vec{X} \tag{2}$$

The calculation of the performance parameter changes can be performed using the inverse of the linear ICM which is referred to as "Fault coefficient Matrix (FCM)" given by Eq. (3). when M=N.

$$\Delta\vec{X} = H^{-1} \bullet \Delta\vec{Z} \tag{3}$$

when M > N, Eq. (3) is over-determined and $\mathbf{H^{-1}}$ should be substituted by the pseudo-invers as defined in Eq. (4).

$$\Delta\vec{X} = \left((H \bullet H)^{-1} \bullet H^T\right) \bullet \Delta\vec{Z} \tag{4}$$

In fact, gas turbine performance deviation is not linear. To consider the non-linear behavior of the engine, an iterative Neuton Rapson method is applied to the linear GPA until the solution converges. For the first iteration, a small delta on the component performance is introduced and the ICM is generated followed by FCM. Then, using the FCM the performance parameter deviation vector is calculated. From the calculated results, a new ICM and FCM is generated and the procedure is repeated until the process converges. The convergence of the root mean square (RMS) value of the difference between the predicted performance value and the actual performance value can be checked using Eq. (5).

$$RMS = \sqrt{\frac{\sum_{i=1}^{M} \left|Z_{i,p} - Z_{i,a}\right|^2}{M}} \leq \sigma \tag{5}$$

where $Z_{i,p}$ is the predicted measurable parameter value, $Z_{i,a}$ is the actual measurement and σ is a very small target RMS value.

The nonlinear gas path analysis (NLGPA) approach is introduced by Escher [20]. Since then several algorithms with some improvements have been contributed by others [21, 22]. Marinai and Singh [5], have made a detailed review of NLGPA.

Recently, Li [23] developed a novel gas turbine performance and health state estimation method using adaptive GPA approach. According to this paper the proposed method can identify performance deteriorations accurately within a few seconds. The diagnosis effectiveness of three different GPA methods have been investigated using different test fault cases for double shaft gas turbine engine by Stamatis [24]. Similarly, the effectiveness of GPA methods in comparison with advanced artificial intelligence approaches have been examined and their pros and cons identified based on case studies for twin shaft gas turbine by Kong [25]. Also, Emil Larsson [26] developed a systematic design procedure to construct non-linear model based fault diagnosis method for industrial gas turbines. Jasmani [27], mentioned that effective measurement parameter selection is a very critical issue for GPA based engine diagnostics. To tackle this problem he introduced a new system by combining analytical approach and measurement subset concept. More recently, Chen et al. [28] proposed an approach that can select the optimal number of engine measurements for diagnostics. However, GPA techniques can diagnose gas turbine faults if and only if noise and bias doesn't exist and they cannot classify clearly the gas path component faults [25].

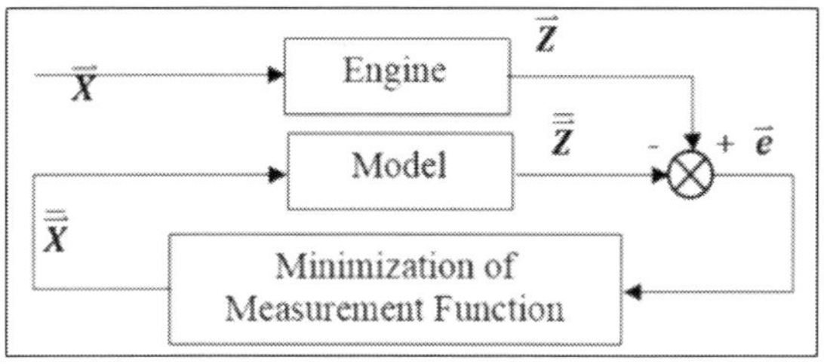

Figure 1. Schematic framwork of GPA method [8]

Data driven diagnostics methods

In this method, in order to develop the diagnostic tool, operational data with an appropriate quality, quantity, and type or performance data from model simulation, when operational data is not available, is required. In the field of gas turbine diagnostics data-driven methods are severely utilized by many researchers in the past several years. They can be categorized into two groups, namely, Artificial Intelligence (AI) and statistical methods [14]. According to Konar [29] AI can be defined as "*the simulation of human intelligence on a machine, so as to make the machine efficient to identify and use the right piece of "Knowledge" at a given step of solving a problem*". Alternatively, it can also be defined as "*a subject dealing with computational models that can think and act rationally*" [30]. There are different AI methods used to develop diagnostics techniques. They are, Expert Systems, Genetic Algorithm (GA), Artificial Neural Networks (ANNs), Fuzzy Logic (FL) and their combinations. As far as statistical methods are concerned, engine faults are diagnosed by applying fundamental principles of statistics based on the available condition monitoring information (historical data). Even though, so far, both techniques have been extensively studied, but the most powerful and popular types of gas turbine diagnostic algorithms are from AI methods [9]. Thus, a comprehensive survey of this approach is presented hereafter.

Artificial Neural Network (ANN) methods

According to Simon [31], neural network can be defined as "*a massively parallel distributed processor made up of simple processing units, which has a natural propensity for storing experimental knowledge and making it available for use*". Usually the structure of neural network consists of an input layer with single or multiple neuron (s), one or more hidden layers with single or multiple neurons, and one output layer with single or multiple neurons. The numbers of input and output neurons are determined by the input and output parameters used. Furthermore, the optimal number of hidden neurons is determined based on a convergence criterion and the input-output mapping relationship characteristics. The neurons in the input, hidden and output layers are joined together using weighted connections. Then the network is trained to learn the relationship among the dependent and independent parameters by applying suitable error minimization technique [32]. Figure 2 illustrates the schematics of the topology of artificial neural networks. And the detailed description of different neural network techniques is available in [31].

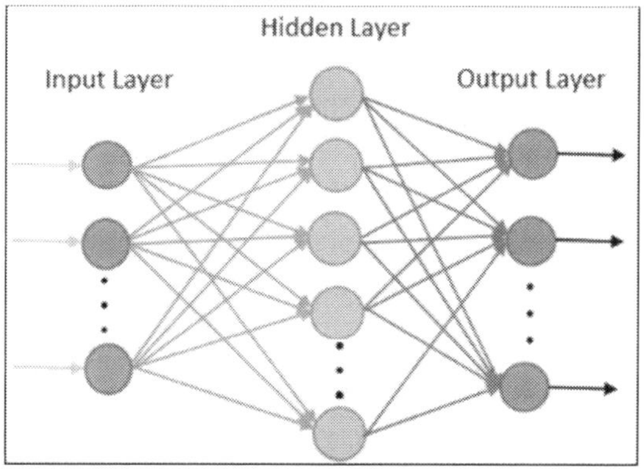

Figure 2. Architecture of a Feed Forward Back Propagation Neural Network

ANNs for gas turbine diagnosis is introduced by Denney et al. [33] in 1965. Since then a lot of ANN based diagnostic algorithms have been produced. For instance, Multilayer Perceptron Neural Network (MLPNN), Feed Forward Back Propagation (FFBP) Neural Network, Dynamic Neural Network [DNN], Auto-Associative Neural Network (AANN), Probabilistic Neural Network (PNN), and Nested Neural Network (NNN) are the most commonly used types [9, 31].

In the past decades, several studies have been done on engine diagnostics based on ANN techniques. For example, Ogaji [34] conducted a PhD research work on single, dual, and triple spool stationary gas turbines to develop a diagnostic algorithm that can diagnose both sensor and component faults simultaneously. The proposed mechanism can detect, isolate and identify single dual and triple sensor and component faults. Although, practically, more than three components can be faulty at the same time, in this research only three simultaneous component faults were considered. Ogaji and Singh [35] as well developed a sensor fault diagnostic system for a 2-shaft stationary gas-turbine engine using NNN. The network is capable to detect and isolate single and dual sensor faults even with the presence of noise and bias. However, the maximum combinations of sensors considered to be faulty at the same time were limited to two. Fast et al. [12], proposed a gas turbine fault diagnostic scheme using multi-layer FFBP-ANN. As per the report, artificial neural network is a suitable approach to develop a performance prediction model and fault diagnostic algorithm if an operational data is available in the required quality and quantity. The proposed tool can detect compressor fouling and is useful for an effective washing schedule. Similarly, Matuck [36] introduced MLPNN based engine diagnostics algorithm that can recognize a single, double and triple faults. Likewise, to evaluate component performance and sensor faults, hybrid diagnostic method was developed by Sampath [37] using GA and ANN techniques. It is shown that the accuracy, reliability and consistency of the results obtained from the hybrid technique is better than ANN and GA based algorithms.

To answer the question why neural networks are more popular than the other AI methods, Patan et al. [38] conducted a research work on two different feed forward multilayer perceptron programs using DNN models taking into account the nonlinear behavior of the gas turbine and the robustness of a fault diagnosis scheme with respect to modelling uncertainty. It was pointed out that the fault diagnostic algorithms belonging to ANN were having smaller false alarms during detection, much more sensitive to early fault detection, high fault classification rate, and are efficient in fault identification than the other AI techniques.

Recently, Mohammadi et al. [39] modelled a fault detecting mechanism for dual-spool gas turbine engine using DNN technique based on model simulation data. The authors stated that, although, DNN is more known by its capability in nonlinear fault identification, in this research its identification capability was used for fault detection and it was effective. Sina et al. [40] also developed a new fault detection and isolation scheme using multiple layer perceptron dynamic neural network (MLPDNN) for a dual spool turbofan engine. It is presented that the proposed method can detect all ranges of fault cases (that is to say, low severity, medium severity, and high severity fault levels) with in an average time of 15.2 seconds. Similarly, to learn the nonlinear behavior of an engine, diagnostic mechanism for a dual spool aircraft jet engine has been modelled by Vanini et al. [9] using MLPDNN and multiple model approach. For each fault cases a corresponding network has been constructed in order to determine the occurrence time and location of the fault. It is shown that fault levels as low as 3% can be detected within 16 seconds. But the method has limitations on multiple fault isolation. Moreover, Vanini et al. [11] proposed an algorithm using AANNs to alleviate the limitations of both sensor and component fault detection and isolation tasks even with the existence of component and sensor faults concurrently. The detection and isolation performance of the new method has been checked for sensor faults, component faults, concurrent sensor and components faults, and concurrent component faults. And finally its performance has been validated using a known DNN based diagnostic algorithm. As per the results, the developed method is capable of sensor and component fault detection both one at a time and concurrently but it has a problem on concurrent fault isolation.

Generally, ANNs have many advantages, such as, they are effective in approximation for non linier models, can solve optimization problems, are cost effective for sensor diagnosis, have high prediction accuracy, have high computational speed after learning, and can perform well even with measurement biases, noisy data and missing data [34]. However, they have also some difficulties, for instance, their complexity increases for complex structures and it is difficult to understand how the training process is taking place, they have high computational time during training, and usually large amount of training data sets are required [12, 26, 40]. For more detail have a look at Table 1.

Fuzzy Logic (FL) methods

Fuzzy logic is one of the most commonly used AI methods to approximate the relationship between dependent and independent parameter vector values based on a set of if/then statements [34]. Fuzzy logic approach for diagnostics application consists of four basic components: fuzzy rules (sets of if-then statements), fuzzifier (the mechanism which maps numbers of input signals into the fuzzy set), inference engine (the technique used to determine the ways in which the fuzzy sets are combined with each other) and defuzzifier (the mechanism used to calculate the output values) [25]. The schematic representation of rule based fuzzy logic system shown in Figure 3. Thus, for the gas turbine diagnostics sets of measurement parameter deltas are given as input to the fuzzy logic algorithm in order to calculate the performance parameter deltas [7].

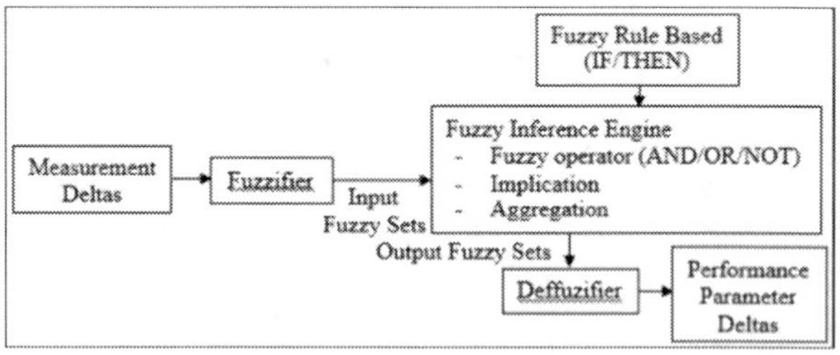

Fig. 3. Framework of a rule based fuzzy system.

After the earliest use of fuzzy logic by Fuster et al. [41] in 1997, several fuzzy logic based gas turbine gas path diagnostic techniques have been modelled by other researchers. Among these, Marinai [7] contributed a diagnostic model for a Rolls-Royce Trent 800 engine that can isolate both single and multiple component faults in the presence of measurement noise and sensor bias. The investigation of the method for multiple faults diagnosis was limited to dual component faults only. Similarly, Ganguli [42], developed engine measurements trend shift detection mechanism using median filters and fuzzy logic. Simulated data sets for clean and faulty engine cases were used to test the fault detection capability of the model and results showed that the detection based on filtered data was very accurate with negligible missed alarms and no false alarms. Guanguli [43], used fuzzy logic to develop a gas turbine engine gas path fault isolation algorithm with an accuracy of 95% and can work with poor quality data. Ogaji [44] also proposed a diagnostic algorithm for a modern military turbofan engine that can identify single component faults. Recently, Kyriazis et al. [45] proposed a fuzzy logic based gas turbine compressor fault diagnostic system using performance data. Its diagnostics effectiveness was compared with pattern recognition and PNN methods. It is mentioned that the developed method has good generality and effective in fault diagnostics like the other two methods.

However, the available fuzzy logic based diagnostic algorithms has limitations regarding multiple fault diagnostics and their effectiveness and performance needs more improvements [7].

Genetic Algorithm (GA)

GA is a method which mimics nature on the basis of Darwin's evolutionary theory of 'survival of the fittest' [46]. The GA is applied as an effective optimization tool to obtain a set of component parameters that produce a set of predicted dependant parameters, through a nonlinear gas turbine model that leads to predictions that best match the measurements [8]. The solution is obtained when an objective function, which is the measure of the difference between predicted values and measured values, achieves the minimum value.

The application of GA for gas turbine fault diagnosis was started in 1999 by Zedda [47] and have been investigated by several scholars for a couple of decades. It is mentioned that GA and ANN methods used as modelling technique to develop an integrated engine fault diagnostic tool for identifying shifts in component performance and sensor faults [36]. Recently, a comparative study was carried out by Kong et al. [48] on GPA and GA based diagnostic systems of a dual-spool turbofan engine. He stated that the diagnostic model based on GA is better than the model based on GPA particularly when sensor noise and bias are considered. In a similar manner, Kong [25] investigated the diagnostic effectiveness of GA in comparison with GPA and fuzzy-neuro techniques based on case studies.

Expert Systems (ES)

An expert system is a computer program that used to capture human expert knowledge in the form of facts and rules to solve a problem or giving an advice [5, 8]. The architecture of expert systems consist of four basic elements, user interface (used to acquire information and display results), inference engine (deals with all the reasoning operations of the system based on known facts and rules), knowledge base (contains facts and rules about the problem to offer the appropriate decision) and developer (stores information about current education) [25]. The configuration of an expert system is shown in Figure 4. Expert systems for engine diagnostics application can be classified as knowledge base and rule base. Several expert system based gas turbine diagnostic techniques are available in the open literature [8, 49]. However, these systems require precise inputs and rely entirely on knowledge of experts and extensive database of rules.

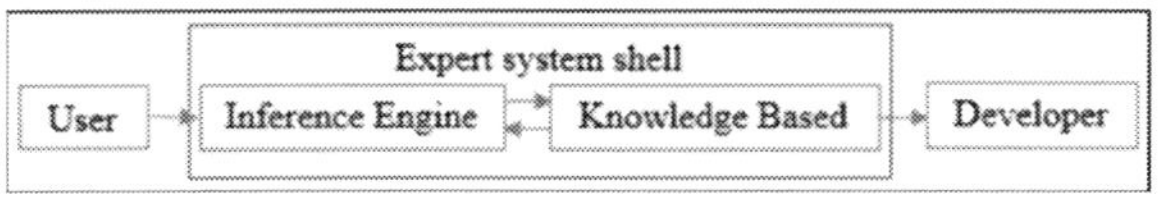

Fig. 4. Configuration of an expert system [25]

Combined/hybrid AI methods

In spite of the fact that, there is no single technique which addresses all the issues, it would be interesting to try and combine more than one technique to offset the limitations of one with the advantages of another. With regard to this, many combined AI techniques are proposed by different authors including fuzzy–neural networks (FNNs), neural–fuzzy systems, Genetic fuzzy logic, Genetics neural network [3, 33, 37]. Combined CI techniques are applied for diagnostics purpose and showed a better performance [37, 50, 51, 52, 53].

Comparison of the methods: advantages and limitations

In general, AI methods have many advantages than GPA methods. For example, they are capable to correlate input and output parameters efficiently even in the existence of limited data, they are efficient for complex systems with large amounts of equations and unknown variables, once they are developed, they have faster diagnosis time, and they can handle stochastic problems very efficiently. In addition, all AI based methods are capable to diagnose engine faults in the presence of noise and bias and are computationally more efficient. On the other hand, GPA based techniques have more advantages in terms of on-board implementation considerations than AI based methods. Also, they apply the real physics of the gas path and have low model complexity than AI based methods. Furthermore, during development, GPA based algorithms take less time than AI based systems.

As far as limitation is concerned, most of AI based techniques cannot give confidence limits on the output. In addition, pre-processing is required to limit the number of data inputs and unable to diagnose faults outside the range of data to which they have been exposed during the training time. Likewise, GPA based engine diagnostic methods have different problems including model uncertainties, model noise, model mismatch, need for more sensors, the existence of a very limited information in the public domain, and modelling difficulties at various operational conditions [12, 15, 40].

To have clear view of the pros and cons of the aforementioned diagnostic methods, a comparison summary of major attributes of these methods is delineated in Table 1.

Table 1. Summary of advantages and limitations of the diagnostic methods

Methods	Advantages	Disadvantages
ANN	☞ complex, multi-dimensional, non-linear systems can be modelled efficiently ☞ if there is operational data, system knowledge is no more required ☞ computer software is available for ANN	✎ no standard (fixed optimal) data set for training and testing ✎ long training time is required ✎ selecting the most suitable model is to a great extent trial and error and takes time

	modelling ☞ diagnostics is possible with only a few measurements (but with large data) [5] ☞ can isolate multiple sensor and component faults partially [5, 34, 35] ☞ can isolate concurrent faults [11] ☞ capable to solve complex gas path diagnostics problems ☞ able to deal with noise and bias ☞ more preferable than others when a large amount of noise is occurred ☞ more preferable when an accurate diagnosis result is required with a limited system knowledge or lacking totally.	✂ since unable to perform outside the range of data to which the network has been trained for, a large amount of training and test data sets are required. ✂ Retraining is required to recognize, new data sets, sensor noise level changes, and even operating condition changes after engine overhaul. ✂ the process inside the network is a "black box" ✂ A unique model is required to take care of only a particular issue.
FL	☞ few numbers of rules are required compared to other rule based ESs ☞ some fuzzy models can provide confidence limits for their outputs ☞ can manage diagnostics with only limited numbers of measurements ☞ relative to ANNs, they are much simpler to translate and realize ☞ can isolate single dual and triple sensor and component faults concurrently [7] ☞ can tackle high levels of measurement uncertainties and sensor biases ☞ able to deal with the non-linear nature of gas turbine performance ☞ can provide an approximate solution to a diagnostics process ☞ easy to set-up and fast computational capability [7]	✂ Rules depend on the knowledge of subject expert and diagnosis accuracy depends on the available rules ✂ large quantity of rules and training data sets are required ✂ they cannot recognize new data sets that the model doesn't see in the training process ✂ are inefficient in diagnostics with a very limited number of data ✂ does not admit model robustness ✂ difficult to define exact queries that identify specific faults
GA	☞ able to deal with noise and bias ☞ more preferable with other model based methods	✂ long computation time model complexity
ES	☞ simple to develop and easy to understand ☞ preferable when the diagnostic problem is well understood, stable, and human experts are available to develop the knowledge base (rules) ☞ more suitable technique for stable and predictable engine operating conditions, if potential faults can be defined easily	✂ model accuracy highly depends on the up to date knowledge of the human expert ✂ significant number of comprehensive set of rules are required ✂ precise inputs required ✂ give approximate results are not effective for highly variable operating condition cases
GPA	☞ more suitable for on-board application ☞ the algorithm have direct relationship with the gas turbine gas path physics ☞ fairly high computational speed ☞ capable to handle abrupt changes in measurements due to operational damage ☞ capable to multiple components fault isolation [7]	✂ has a limitation on clear component fault isolation [5] ✂ for better prediction large number of sensors are required ✂ face a problem when N>M ✂ can diagnose engine faults if and only if noise and bias doesn't exist ✂ modeling difficulties due to the existence of limited information in the open domain ✂ For large variations in component faults the programme has convergence problem [34]

Conclusions

In the evolution of maintenance, fault diagnostics is a very critical advancement and works on the bases of condition. The existing gas turbine gas path diagnosis approaches have two major categories, namely, model based and data driven. In this paper, the research works using advanced diagnostic methods, GPA and AI, are reviewed. According to the review, the most powerful gas path diagnostic algorithms belongs to AI methods. Summary of pros and cons of these methods is given in Table 1. It is stated that the current AI based algorithms can successfully diagnose single component/sensor faults

(SCF/SSF) both one at a time and simultaneously and multiple component/sensor faults (MCF/MSF) partially, even in the presence of noise and bias. However, the more recent works suggested that the following topics need future investigation.

1. Diagnostic tool that can address the entire gas turbine problems in an integrated manner
2. Method that can efficiently validate the performance and effectiveness of diagnostic algorithms.
3. Further improvement of gas path diagnostic methods concerning with complexity, computational efficiency, measurement uncertainties.
4. More efficient algorithm that can isolate concurrent faults and multiple faults.
5. Effective combined diagnostics and prognostics methods
6. Combining two or more AI methods and using for diagnostics (hybrid techniques)

Acknowledgment

This work was supported by a Yayasan University Technology PETRONAS (YUTP) project grant (Project Cost Centre No: 0153AA-A84).

References

[1] G. Vachtsevanos et al., *Intelligent Fault Diagnosis and Prognosis for Engineering Systems*, 1-434, (2007).

[2] C.B. Meher-Homji, A. Bromley, Gas Turbine Axial Compressor Fouling and Washing 33rd Turbomach. Symp., Texas A & M University, (2004).

[3] Marco Mucino and Y. G. Li, A diagnostic system for gas turbines using GPA-Index, COMADEM-C007-(2005).

[4] I. Loboda, Gas Turbine Condition Monitoring and Diagnostics, National Politechnique Institute, Mexico, (2010).

[5] L. Marinai, D. Probert, and R. Singh, Prospects for aero gas-turbine diagnostics: A review. Appl. Energy, **79**(1): p. 109-126, (2004).

[6] S.O.T. Ogaji, R. Singh, Advanced engine diagnostics using artificial neural networks. Appl. Soft Comput., **3**(3): p. 259-271, (2003).

[7] L. Marinai, Gas-path diagnostics and prognostics for aero-engines using fuzzy logic and Time series analysis" Cranfield University, PhD thesis, UK, (2004).

[8] Y.G. Li, Performance-analysis-based gas turbine diagnostics: A review. Proceedings of the Institution of Mechanical Engineers, Part A: Int. J. Power Energy Syst, **216**(5): p. 363-377, (2002).

[9] Z.N.S. Vanini, K. Khorasani, N. Meskin, Fault detection and isolation of a dual spool gas turbine engine using dynamic neural networks and multiple model approach. Information Sciences, **259**(0): p. 234-251, (2014).

[10] C. Romesis, Y.G. Li, Condition based maintenance (CBM) for gas turbines plants, Laboratory of Thermal Turbomachines National Technical University of Athens, Greec & Department of Power and Propulsion, Cranfield University, UK, (2013).

[11] Z.N.S. Vanini, N. Meskin, K. Khorasani, Multiple-model sensor and components fault diagnosis in gas turbine engines using autoassociative neural networks. J. Eng. Gas Turbines Power, **136**(9), (2014).

[12] m. Fast, M. Assadi, S. De, Development and multi-utility of an ANN model for an industrial gas turbine. Appl. Energy, **86**(1): p. 9-17, (2009).

[13] D. An, N.H. Kim, J.H. Choi, Practical options for selecting data-driven or physics-based prognostics algorithms with reviews. Reliability Engineering and System Safety, **133**: p. 223-236, (2015).

[14] A.K.S Jardine, D. Lin, D. Banjevic, A review on machinery diagnostics and prognostics implementing condition-based maintenance. Mech. Syst. Sig. Process., **20**(7): p. 1483-1510, (2006).

[15] R. Verma, N. Roy, R. Ganguli, Gas turbine diagnostics using a soft computing approach. Appl. Math. Comput., **172**(2 SPEC. ISS.): p. 1342-1363, (2006).

[16] L.A. Urban, *Gas Turbine Engine Parameter Interrelationships*, HSD UTC, Windsor Locks, Ct., 1st edition, 1967, 2nd edition, (1969).

[17] L.A. Urban, Parameter Selection for Multiple Fault Diagnostics of Gas Turbine Engines. J Eng Power Trans ASME, 97 Ser A(2): p. 225-230, (1975).

[18] D.L. Doel, TEMPER – A Gas Path Analysis Tool for Commercial Jet Engines, J. Eng. Gas Turbines Power, Vol. **116**, No. 1, pp.82-89, (1994).

[19] Volponi, The use of optimal estimation techniques in the analysis of gas turbines PhD thesis, Cranfield Univerrsity, UK, (1994).

[20] P. C. Escher, Pythia: An object-oriented Gas Path Analysis Computer Program for General Applications, PhD thesis, Cranfield University, UK, (1995).

[21] D.L. Doel, TEMPER - a gas-path analysis tool for commercial jet engines. J. Eng. Gas Turbines Power, **116**(1): p. 82-89, (1994).

[22] A. Stamatis, K. Mathioudakis, K. Papailiou, Optimal measurement and health index selection for gas turbine performance status and fault diagnosis. J. Eng. Gas Turbines Power, **114**(2): p. 209-216, (1992).

[23] Y.G. Li, Gas Turbine Performance and Health Status Estimation Using Adaptive Gas Path Analysis. J. Eng. Gas Turbines Power, **132**(4): p. 041701-041701, (2010).

[24] A.G. Stamatis, Evaluation of gas path analysis methods for gas turbine diagnosis. J. Mech. Sci. Technol., **25**(2): p. 469-477, (2011).

[25] C. Kong, Review on advanced health monitoring methods for aero gas turbines using model based methods and artificial intelligent methods. Int. J. Aeronaut. Space Sci., **15**(2): p. 123-137, (2014).

[26] Emil Larsson, Model Based Diagnosis and Supervision of Industrial Gas Turbines, Linköping University, PhD thesis, Sweden, (2014).

[27] M.S. Jasmani, Y.G. Li, Z. Ariffin, Measurement selections for multicomponent gas path diagnostics using analytical approach and measurement subset concept. J. Eng. Gas Turbines Power, **133**(11), (2011).

[28] M. Chen, H. L. Quan, H. Tang, An Approach for Optimal Measurements Selection on Gas Turbine Engine Fault Diagnosis. ASME. J. Eng. Gas Turbines Power.;**137**(7): 071203-071203-9. doi:10.1115/1.4029171, (2015).

[29] B. Paul, A. Konar, A. K. Mandal, Fuzzy ADALINEs for gray image recognition, Neurocomputing, vol. **24**, pp. 207-223, (1999).

[30] A. Konar, *Artificial intelligence and soft computing : behavioral and cognitive modeling of the human brain*, ISBN: 978-0-9778582-3-1, India, (2008).

[31] M. Kubat, *Neural networks: a comprehensive foundation by Simon Haykin*, Macmillan, 1994, ISBN 0-02-352781-7. The Knowledge Engineering Review, (2003).

[32] R. Ganguli, *Gas Turbine Diagnostics Signal Processing and Fault Isolation*, CRC Press (2013) Print ISBN: 978-1-4665-0272-7.

[33] G. Denney, F16 jet engine trending and diagnostics with neural networks. Proc. SPIE 1965, Applications of Artificial Neural Networks IV, **419** (September 2, 1993); doi:10.1117/12.152545.

[34] S.O.T. Ogaji, Advanced Gas-path Fault Diagnostics for Stationary Gas Turbines, Cranfield University, PhD thesis, UK, (2003).

[35] S.O.T. Ogaji, R. Singh, S.D. Probert, Multiple-sensor fault-diagnoses for a 2-shaft stationary gas-turbine. Applied Energy, **71**(4): p. 321-339, (2002).

[36] G.R. Matuck, et al. Multiple faults detection of gas turbine by MLP neural network. in Proceedings of the ASME Turbo Expo, (2009).

[37] S. Sampath, R. Singh, An integrated fault diagnostics model using genetic algorithm and neural networks. J. Eng. Gas Turbines Power, **128**(1): p. 49-56, (2006).

[38] K. Patan, M. Witczak, J. Korbicz, Towards robustness in neural network based fault diagnosis. Int. J. Appl. Math. Comput. Sci., **18**(4): p. 443-454, (2008).

[39] R. Mohammadi, et al. Fault diagnosis of gas turbine engines by using dynamic neural networks. in Midwest Symposium on Circuits and Systems. (2011).

[40] T.B.S. Sina, Z.N.S. Vanini, K. Khorasani, Dynamic neural network-based fault diagnosis of gas turbine engines. Neurocomputing, **125**(0): p. 153-165, (2014).

[41] P. Fuster, A. Ligeza, J. Aguilar Martin, adductive diagnostic procedure on an AND/OR/NOT graph for expert behaviour : Application to a gas turbine , 10th international congress and exhibition on condition monitoring and diagnostic

[42] R. Ganguli, Data rectification and detection of trend shifts in jet engine path measurements using median filters and fuzzy logic. J. Eng. Gas Turbines Power, **124**(4): p. 809-816, (2002).

[43] R. Ganguli, Application of fuzzy logic for fault isolation of jet engines. J. Eng. Gas Turbines Power, 2003. **125**(3): p. 617-623.

[44] S.O.T. Ogaji, et al., Gas-turbine fault diagnostics: a fuzzy-logic approach. Appl. Energy, **82**(1): p. 81-89, (2005).

[45] A. Kyriazis, et al. Gas turbines compressor fault identification by utilizing fuzzy logic-based diagnostic systems. in 9th European Conference on Turbomachinery: Fluid Dynamics and Thermodynamics, ETC 2011 - Conference Proceedings. (2011).

[46] Y.G. Li, An Genetic Algorithm Approach to Estimate Performance Status of Gas Turbine, ASME Paper No. ASME GT2008-50175, (2008).

[47] M. Zedda, Gas-turbine engine and sensor fault diagnosis, PhD Thesis, School of Mechanical Engineering, Cranfield University, UK, (1999).

[48] C.D. Kong, M.C. Kang, G.L. Park, Study on Condition Monitoring of 2-Spool Turbofan Engine using Non-Linear Gas Path Analysis Method and Genetic Algorithms, IJMMM, Vol. **1**, No. 2, pp. 214-220, (2013).

[49] H. Abbasi Nozari, et al., Model-based robust fault detection and isolation of an industrial gas turbine prototype using soft computing techniques. Neurocomputing, **91**(0): p. 29-47, (2012).

[50] Aiad Gannan, Cascade Testing and CFD Applied to Gas Turbine Performance Improvement with Compressor Cleaning, Cranfield University, PhD thesis, UK, (2012).

[51] A.L. Tamiru, Intelligent Fault Detection and Diagnosis System for A Cogeneration And Cooling Plant., Universiti Teknologi PETRONAS, PhD Thesis, Malaysia, (2012).

[52] B. Shahriari, A. N. Shahrbabaki, A. Shahriari, "Gas Turbine Fault Detection and Isolation Using Adaptive Neurofuzzy Inference System (ANFIS)", Advanced Materials Research, Vol. **1016**, pp. 721-725, Aug. (2014).

[53] R. Bettocchi, M. Pinelli, P. R. Spina, and M. Venturini, "Artificial Intelligence for the Diagnostics of Gas Turbines—Part II Neural-Fuzzy Approach ," J. Eng. Gas Turbines Power. Vol . **129**, No. 3;,pp.720-729, (2007).

Early tube leak detection system for steam boiler at KEV power plant

Firas B. Ismail[1a], Deshvin Singh[1], N. Maisurah[1] and Abu Bakar B. Musa[1]

[1]*Power Generation Research Centre, College of Engineering, Universiti Tenaga Nasional (UNITEN), Putrajaya Campus, Jalan IKRAM-UNITEN, 43000.*

Abstract. Tube leakage in boilers has been a major contribution to trips which eventually leads to power plant shut downs. Training of network and developing artificial neural network (ANN) models are essential in fault detection in critically large systems. This research focusses on the ANN modelling through training and validation of real data acquired from a sub-critical boiler unit. The artificial neural network (ANN) was used to develop a compatible model and to evaluate the working properties and behaviour of boiler. The training and validation of real data has been applied using the feed-forward with back-propagation (BP). The right combination of number of neurons, number of hidden layers, training algorithms and training functions was run to achieve the best ANN model with lowest error. The ANN was trained and validated using real site data acquired from a coal fired power plant in Malaysia. The results showed that the Neural Network (NN) with one hidden layers performed better than two hidden layer using feed-forward back-propagation network. The outcome from this study give us the best ANN model which eventually allows for early detection of boiler tube leakages, and forecast of a trip before the real shutdown. This will eventually reduce shutdowns in power plants.

1 Introduction

A thermal power plant works using steam to drive the turbines. These power plants consist of various essential components working together including the boiler, condenser, burner, turbines and also coal bunkers. All these systems has to work together in order to ensure smooth operations in the plant. One of the main components that is prone to damages and failures is the boiler. The boiler consists of various sections including the economisers, steam drum, superheaters, fluidising grid and boiler tubes. These tubes contain large amount of water travelling through inside at high velocity. Water is heated at extremely high temperatures and pressures to ensure efficient and reliable rotation of the turbines and generation of stable electricity. Eventually the boiler tubes wear off and can no longer withstand the high pressures and temperatures. Any debris or foreign object flowing in the tubes at high velocity will easily cause tube damage or failure. Besides that, water-side corrosion, fire-side corrosion and lack of quality control may also cause tube failures. Undetected tube failures, may cause even more damages to other tubes.

Artificial Neural Network (ANN) is a computational model which is considered a part of statistical learning algorithms. Generally, the ANN is inspired from the biological neural network available in human brains. ANN is considered one of the main branches of artificial intelligence, apart from genetic algorithms, expert systems and fuzzy logic. ANN as a processor has natural ability to keep knowledge and large amount of data. These data are required to ensure that training is done effectively. Synaptic weights also known as inter-neuron connection strengths stores knowledge during the learning process. Artificial neural networking does not require detailed information on the system. Instead it is a continuous learning process which learns about the relationship between the input parameters and variables. The learning process acquires information from the large spreadsheet of data available. The three sections involved in the multilayer feed-forward neural network are the input layer, hidden layers and output layer. The neurons in an ANN are all connected with suitable synaptic weights. The training process basically modifies the set of connection weights through storing of knowledge. An input is introduced, the desired outputs are determined and the weights are adjusted during the learning process of the network. he incoming connections allow the sending of weighted activation of other nodes to a particular node. The activation function eventually calculates the weightage of each node introduced. Combinations of linear transfer function, log-sigmoid transfer function and tan-sigmoid transfer function will produce varied output values.

A node is activated as the input nodes enter the hidden layers. As the input nodes increase, they pass through activation functions. The output is produced when the network uses the input pattern and reads it. Difference in the connection weights causes error. The connection weights have to be modified in order to decrease such errors. The input patterns are run through all over again until the error value reaches the goal value

[a] Firas B. Ismail : firas@uniten.edu.my

set. RMSE function is used to calculate the error difference between actual output and predicted output.

In a study on artificial intelligent system for steam boiler diagnosis based on superheater by Firas [1], training was done using prepared data. As a result, the FDDNN is able to detect existing fault even quickly. The output is considered faulty if the value is close to 1. In another study on multidimensional minimisation training algorithms for steam boiler drum level trip using artificial intelligent monitoring system, Firas [2] used Two different interpretation algorithms. They are the Broyden-Fletcher-Goldfarb-Shanno quasi-Newton and Levenberg-Marquart. Real data is obtained from a coal-fired thermal power plant in Perak. Firas and Vinesh [3] implemented the feed-forward back-propagation network (BP) as supervised system. From this algorithm, weights that maximises performance is identified. In this case, the input and output is provided. Jankowska [4] in an approach to early boiler tube leakage detection with artificial neural network, used two main structure in building the ANN model of flue gas which are the linear nets (LIN, without hidden layers), and feedforward Multilayer Perceptron (MLP). Shahul [5] in a study on automatic detection and analysis of boiler tube leakage system, used BPNN algorithm. From the results of this case study, tube leakages from holes of different diameters and distances were successfully detected. The methods involved include; data recording and preprocessing, and feature extraction.

Based on a study on detection and location of sound emission caused by tube by Jenshan et al. [6] , a 660MW boiler is taken as a model. The localisation of different positions of the leakage source in the furnace of the boiler's body is simulated. Yang Zhao [7] in a study on pattern recognition-based chillers fault detection method using support vector data description (SVDD) proposed a pattern recognition-based chiller fault detection method using a novel one-class classification algorithm. Basically, the objective is to find a minimum-volume hypersphere in a high dimensional feature space. Both SVDD-based method and PCA-based methods are used and compared. Using pattern recognition, it is easier to apply the same pattern for different cases. The algorithms and theories that backs up pattern recognition is quite reliable with a large amount of documented information and tools available. Jayanta[8] in a research on use of artificial neural network in pattern recognition, summarises and compares some of the well-known methods used in various stages of a pattern recognition system using ANN and identify research topics and applications included in the field. For this study, automatic recognition, description, classification and grouping of patterns were important factors taken into account. From this study it is known that there are pros and cons in the application of pattern recognition. The application of ANN in each pattern recognition case has always performed better result than that of without implementing ANN. Zafei [9] proposes time domain features as a proper alternative to frequency features. The ANNs are trained with subset of experimental data for known machine conditions. Through sufficient training, the efficiency of the proposed method is evaluated using remaining set of data.

The main objective of this research is to design an ANN model for detection and diagnosis of fault in a boiler unit. This would eventually minimise forced outage and trips in power plants caused by tube failures and damages. Apart from that this study allows understanding on the behaviour of boiler operation variables and common causes of tube leakages. This research focuses on the early detection of boiler trip due to boiler tube leakage by implementing ANN modelling through training and validation. The ANN model is designed to forecast a trip before the real shutdown. This will give time for plant operators to take measureable actions to avoid the real shut down.

2 Data preparation

Acquiring real data is essential in order to proceed with the training and simulation using the Neural Network. The real data is acquired from a working unit of a coal fired power plant. This power plant consists of 2 oil and gas fired units and 4 coal, gas and oil fired units with a maximum power generation capacity of 2400 MW. The unit selected was Unit 4 boiler which is part of a sub-critical unit. Based on the trip reports, the most suitable and boiler-related trip is chosen. There had been a shutdown due to boiler tube leakage from 30th July 2013 to 6th August 2013. Data is collected from one day before the trip and one day after the trip with time interval of 1 minute. The large amount of data is processed and is shortlisted to 26 important variables. They are shortlisted based on plant operator experience on identifying the essential variables that contributed to boiler trips of the particular unit. The sequence of data processing is shown in Figure 1.

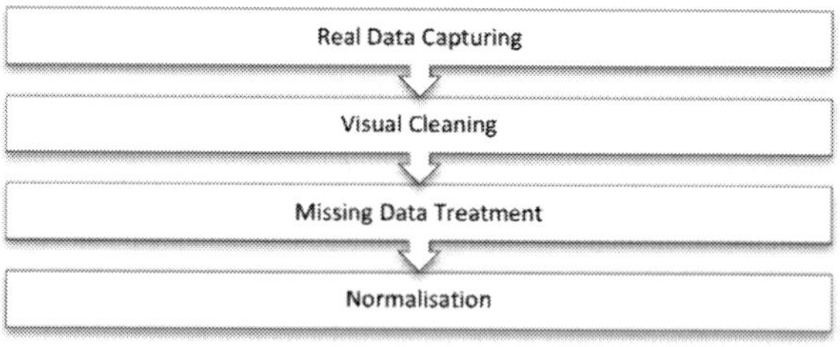

Figure 1. The sequence of data processing.

The real data collected from the power plant unit is enormously large and contains thousands of values. From the big spreadsheet of data, there were missing data and invalid values available. These values are irrelevant to be input into ANN directly. Therefore these invalid and missing data are visually screened and identified. The missing data were treated with missing data treatment. In this process, the invalid and missing values identified were replaced with values using interpolation of available values. The interpolation was done using the following formula.

$$V_1 at\ t_1 = t_0 + (t_3 - t_0)/time\ intervals \quad (1)$$

In order to ensure that the data can be fed to the neural network, the values in hundreds are normalised into values between 0 and 1 using the normalisation equation.

$$\text{Normalization equation} = X' = \frac{(X - Xmin)}{Xmax - Xmin} \quad (2)$$

In this particular study, training and validation is classified into 70% training and 30% validation. Only 70% of data from the real data is used to train the ANN model. The ANN model is trained and validated using different combinations of training algorithms, activation functions, number of hidden layers and number of neurons. Through training, the training algorithm with the minimal root mean square error (RMSE) will be selected for the next step, validation.

3 ANN Topologies

The feed-forward NN structure selection will be done based on the numbers of hidden layers of the network, mainly one hidden layer and two hidden layer. 1 to 10 neurons for each hidden layers will allow variable outputs. These combinations of hidden layers and number of neurons result in different outputs.

Only up to 2 hidden layers are used because using greater number of hidden layers may cause constant values of RMSE. For each hidden layer, different combinations of number of neurons are applied. Each hidden layer can fill up to 10 neurons as more than 10 neurons caused RMSE value of more than 0.5, which is the set goal RMSE.

3.1. Training algorithms

The feed-forward back-propagation training algorithm is the most effective due to its non-linear solutions. These training algorithms are few of the faster and reliable algorithms available in the NN MATLAB. The training algorithms are Resilient Backpropagation, Levenberg-Marquardt, Scaled Conjugate Gradient, and BFGS quasi-Newton. The Resilient Backpropagation has moderate memory requirement, is faster than standard steepest descent algorithm and is able to eliminate effects of having small slope at extreme ends of sigmoid transfer function. The Levenberg-Marquardt algorithm provides solution to minimising a function in least square curve fitting and nonlinear programming. Scaled conjugate gradient has faster convergence than steepest descent directions. BFGS quasi-Newton on the other hand requires more computation in each iteration, and more storage than the conjugate gradient methods.

3.2 Activation functions

In this study, three activation functions were used to define the outputs from the given inputs. The three activation functions include linear transfer function, log-sigmoid transfer function and tan-sig transfer function. Linear transfer function returns the value passed to it to calculate the neuron's output. The training of the neuron allows the neuron to learn a function and find a linear approximation to a nonlinear function. However a linear network is not able to perform a nonlinear process. Being the most common transfer function in back-propagation networks, log-sigmoid function is differentiable. The network outputs can take on any value in case of linear output neurons. The activation functions carries out various calculation using the inputs, and generating the output and error. Different inputs shall produce different outputs and thus the network is modified.

Simulations tests were done on the network to calculate the outputs from the original inputs. Through the RMSE equation, RMSE values are achieved which equates to the weights that minimises the error. Thus MATLAB is used to retrieve the output. The combination with the minimal root mean square error (RMSE) was selected for the validation process.

$$RMSE = \frac{1}{Q}\sum\big(t(k) - a(k)\big)^2 \quad (3)$$

Where k is the number of iterations, Q is the total number of iterations (epochs), t is the target output, and a is the actual output.

4 Results and discussion

The data which has been shortlisted to 26 variables (Table 1) has undergone the process of data preparation. This includes visual cleaning, missing data treatment and normalisation. This is important to allow proper input of these large set of data into MATLAB for training and validation. The fault introduced for each variable was identified by comparing the fault introduced graphs of each variable. All the variables are evaluated and grouped as 'influenced' (I) and 'most influenced' (MI). This grouping is done based on the tripping time closest to the real shutdown.

Table 1. Fault Introduced For Each Variable

Variables	Minute	Group
Total Combined Steam Flow	1435	
Feed Water Flow	1547	I
Boiler Drum Pressure	1499	
Superheater Steam Pressure	1544	I
Superheater Steam Temperature	1547	I
High Reheater Outlet Temp	1536	I
Final SH Outlet Left Temp	1426	
Final SH Outlet Right Temp	1574	I
Drum Level Corrected Left	-	
Drum Level Corrected Right	-	
Boiler Feedwater Pump Pressure	1536	I
Low Temp Superheater Outlet A	1609	I
Low Temp Superheater Outlet B	1560	I
Reheater Outlet Steam Pressure	1537	I

Economiser Inlet Pressure	1499	
Economiser Inlet Temp.	1536	I
Economiser Outlet Temp.	1536	I
Economiser Outlet Pressure	-	
Reheater Inlet Steam Temp.	1602	I
Reheater Inlet Steam Pressure	1537	I
Final SH Inlet Steam Temp.	1590	I
Platen SH Outlet A Temp.	1577	I
Platen SH Outlet B Temp.	1580	I
Primary SH Metal Temp. 1	1617	MI
Final SH Metal Temp. 1	1536	I
Platen SH Metal Temp. 1	1580	I

4.1 Analysis of the ANN training results

It can be seen in Figure 2, Variable 24 (Primary Superheater Metal Temperature 1) was identified as the most important variable. The closest variable to the real shutdown is at 1617 minutes where the fault is introduced. The real shutdown occurs 10 minutes after that which is at 1627 minutes. Therefore this variable is the most important variable among the 26 variables as it caused the boilers to trip immediately after the fault was detected.

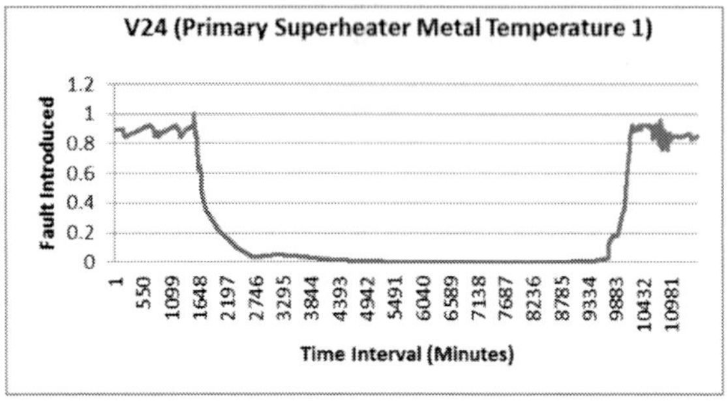

Figure 2. Fault Introduced for Primary Superheater Metal Temperature 1

Construction of the ANN model is divided into 2 different number of hidden layers. The first model of one hidden layer consists of two activation functions. The activation functions are purelin(P), logsig(L) and tansig(T). Total combinations of 9 activation functions are possible for one hidden layer. The combinations are P+L, T+L, P+T and so on. These activation functions are constructed with up to 10 neurons each. Any number of neurons more than 10 will cause the root mean square (RMSE) to be more than 0.5 which is the RMSE goal set in the training. The activation functions and neurons are trained using four different training algorithms which are 'trainlm', 'trainbfg', 'trainrp', and 'trainscg'.

Two hidden layer model consists of three training activations when simulated. Thus a total of 27 combinations are possible for a two hidden layer model. Only up to two hidden layers are trained because three hidden layers produce similar RMSE as the model with 2 hidden layers.

Each combination produced different RMSE under different hidden layers, activation functions, number of neurons and training algorithm. After several training, the combination that produces the smallest RMSE is considered as the best combination for the ANN model. The results are graphed as in Figure 3 to ease comparison of RMSE values.

For 2 hidden layers there are 27 combinations of activation functions which include purelin, logsig and tansig, the lowest RMSE of each training algorithm is chosen and compared in the graphs below. Through visual comparison and analysis of all the 27 combinations, these best combinations with lowest RMSE are tabulated to allow data analysis. Comparison of training functions of 1 hidden layer and 2 hidden layers is shown in Figure 3.

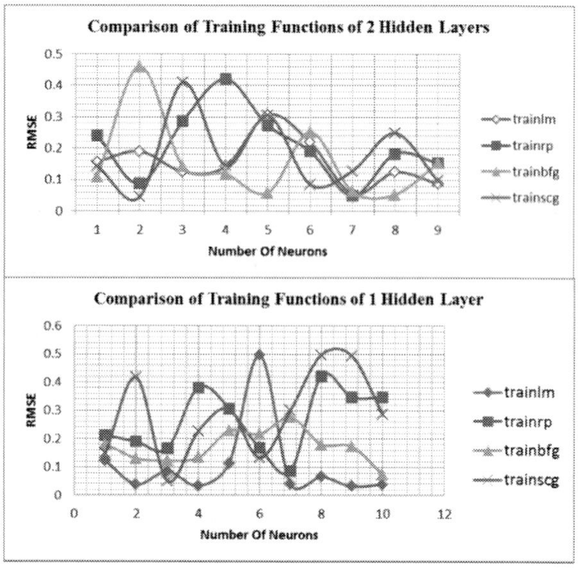

Figure 3. Comparison of training functions of 1 hidden layer and 2 hidden layers

Based on the RMSE for each training algorithm, the lowest RMSE produced for 2 hidden layers is the Trainscg algorithm, under the activation function of 'logsig', 'purelin', 'logsig' (LPL). A RMSE of 0.0451 is produced with 7 neurons in the first hidden layer, and 3 neurons in the second hidden layer. Comparing the RMSE results of 1 hidden layer, and 2 hidden layers, the Trainlm in 1 hidden layer model produced the smallest RMSE value. The model chosen is 1 hidden layer , Trainlm algorithm with 'purelin' and 'purelin' activation functions with RMSE 0.0328.

4.2 Analysis of the ANN validation results

The model with least RMSE value is considered the best model combination, therefore that particular model is used for the validation process. A different set of coding is used to achieve the final forecasted graph. This is essential to prove that the model allows for trip forecasting, and is able to detect trip before the real shutdown. The produced results are presented in Figure 4.

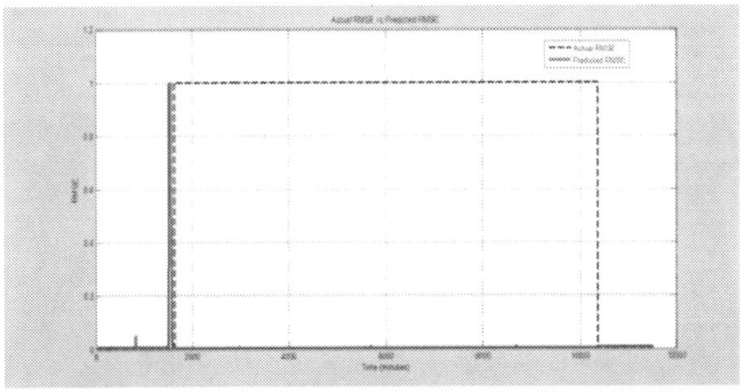

Figure 4. Graph of Actual RMSE vs Predicted RMSE

The chosen ANN model consisting of 1 hidden layer is used for the validation process and to come up with the forecasting graph. The first graph consists of the actual RMSE values, whereas the second one displays the predicted RMSE values. In Figure 4 , both the actual RMSE and predicted RMSE are plotted in the same graph. This is essential in order to identify the timing of the actual and predicted RMSE values.

Actual RMSE graph shows the real trip that has occurred. This is done based on the trip recorded by the plant itself. The predicted RMSE graph reflects the forecasted model. As seen in Figure 4 the trip will continuously occur without the ANN system implementation, eventually affecting the plant normal operations. As shown in Figure 4, the sudden increase in value is observed at the 1617th minute. Comparing the actual RMSE (blue lines) and predicted RMSE (green line) values in the forecasted graph, the predicted trip is detected at the point of 1617th minute. This allows for a gap of 10 minutes before the actual trip at 1627th minute.

The classification of data classifies a trip when it reaches the trip value which is „1". The difference between actual and predicted RMSE which is 10 minutes, has proven that the real trip could be eliminated successfully. The time gap available allows the plant operator to prevent real trip by taking prevention actions within the available time. The ability to forecast a trip, ensures the continuous operations of the plant. Being a continuous-learning system, the ANN system will be able to detect and trips that will occur in the future.

5 Conclusion

This study has focused on constructing the best ANN combination model. This study has come up with the best combination of ANN model through ANN simulation training and validation. The combination allows for effective forecasting. The training algorithm, activation functions, number of hidden layers and number of neurons are few of ANN topologies that are discussed in this study. Every time the number of neurons are set for each hidden layer and the coding is run, the results vary. The number of neurons is set to a certain limit of range to avoid repeatedly occurring RMSE values. Higher number of neurons causes lower RMSE.

Through this study the behaviour of boiler operation variables and common causes of tube leakages were understood. The variables are identified and the Variable 24 (Primary Superheater Metal Temperature 1) is considered as the main contributor of the shutdown. Apart from that the fact that the current study uses offline data received from the plant. Future studies may focus on developing the online data input system which would allow data to be transferred directly from the plant's control room to the ANN system. The methodology established in this study can further be developed using other artificial intelligence systems to identify and detect boiler tube leakage in power plants.

References

1. Firas B. Ismail, Artificial intelligent system for steam boiler diagnosis based on superheater monitoring, *Journal of Applied Science* (2011)
2. Firas B. Ismail, Multidimensional minimisation training algorithms for steam boiler drum level trip using artificial intelligent monitoring system, *Intl. Conf. Control, Automation, Robotics and Vision* (2010)
3. Firas B. Ismail, Vinesh Thiruchelvam, Back propagation artificial neural network and its application in fault detection of condenser, *Iop Science* (2013)
4. A. Jankowska, Approach to early boiler tube leakage detection with artificial neural network Institute Of Autom. And Robotics, *The Warsaw University Of Technology Warsaw, Poland* (2007)
5. S.Shahul, D. N. Jamal, M. Shajahan, Automatic detection and analysis if boiler tube leakage system, *Int. Journal of Comp. App.* **84** (2013).
6. Jenshan J. & Lin W., research on the detection and location of sound emission caused by tube leakage in boilers, *Power and Energy Engineering Conference (APPEEC)* (2012)
7. S. Yang Zhao, Pattern recognition-based chillers fault detection method using Support Vector Data Description (SVDD), *Applied Energy*, pp. 1041-1048 (2013)
8. K. Jayanta. 2010. Use of artificial network in pattern recognition, *Journal of Software Engineering and Its Application*, **4** (2010)
9. J. Zarei, Induction mototrs bearing fault detection using pattern recognition, *Expert Systems with Application*, pp. 68-73 (2012)

FUZZY ACTIVITY BASED LIFE CYCLE COSTING FOR REPAIRABLE EQUIPMENT

Freselam Mulubrhan[a], Ainul Akmar Mokhtar, Masdi Muhammad,
Mechanical Engineering Department, Universiti Teknologi PETRONAS
Bandar Seri Iskandar, 31750 Tronoh, Perak, Malaysia
frity4u@gmail.com

Abstract: Life-cycle cost (LCC) is the much known method used for decision making that considers all costs in the life of a system or equipment. Predicting LCCs is fraught with potential errors, owing to the uncertainty in future events, future costs, interest rates, and even hidden costs. These uncertainties have a direct impact on the decision making. Activity based LCC is used to identify the activities and cost drivers in acquisition, operation and maintenance phase. This activity based LCC is integrated with fuzzy set theory and interval mathematics to model these uncertainties. Day–Stout–Warren (DSW) algorithm and the vertex method are then used to evaluate competing alternatives. A case of two pumps (Pump A and Pump B) are taken and their LCC is analysed using the developed model. The equivalent annual cost of Pump B is greater than Pump A, which leads the decision maker to choose Pump A over Pump B.

Key words: - fuzzy set theory, activity based life cycle costing, interval mathematics

1 Introduction

Life cycle cost (LCC) is used as a decision support tool to aid decision makers to propose, compare, and select the cost effective alternatives for maintenance, renewal, and capital investment [1]. LCC consists of acquisition cost which incurred when the asset is purchased and ownership cost which incurred throughout the asset's life [2]. Studies show that the engineering system ownership cost can vary from 10 to 100 times higher than the original acquisition cost [3]. The cost estimation method used for determining LCC has a higher impact on producing an accurate and efficient result and needs to have the ability to grip the uncertainties raised. Predicting the future LCCs is fraught with potential errors, owing to uncertainties in; future events, failure rate, life span of equipment, future costs, interest rates, and so on [1]. Activity Based Costing (ABC) model deals with all the activities that will incur cost and it has the best capability to deal with the uncertainties [4]. It was first introduce by Kaplan and Cooper in 1988[5]. In ABC cost estimating techniques it is necessary to identify each activity in all the life cycle stages; acquisition, installation, operation, and disposal.

LCC can be estimated through deterministic, probabilistic or soft computing approaches. Eric and Timo [6] found that 83% of manufacturing industries used deterministic nature of LCC analysis, while only 17% employed probabilistic model. In deterministic approach, uncertainties in the input data were ignored. The probabilistic approach on the other hand accounts for the uncertainty and variation associated with input values [7]. Probabilistic method requires the defining of a probability distribution for every uncertain variable. Defining probability for every uncertain variable requires adequate historical data. If historical data is available the method can give realistic results. However if there is no adequate historical data, which happens in most real cases, soft computing will be a more suitable method. Even if there are adequate historical data, it is necessary to polish these input data by the judgment of the experts. Since these estimates are often based upon uncertain, ambiguous subjective and sometimes incomplete information, soft computing techniques emerge as a very appealing alternative for developing LCC tools [8].

Lately the application of fuzzy set theory in modeling uncertainties has an upward trend due to its appropriateness handling situations where human reasoning, human perception, or human decision making is inextricably involved [10]. Fuzzy theory is based on fuzzy sets, which is the expansion of crisp sets. Fuzzy theory overthrows the two/dual value (yes or no) so that its multi-value could be pressed close to reality. Defining fuzzy variable is a less complicated that defining random number if there is limited information.

The main objective of this research is to develop fuzzy activity based LCC is by incorporating fuzzy set theory, interval mathematics and activity based costing. This model is adapted from Mohammad Ammar, et al. 2013 [11]. In this research there are four types of uncertainties defined; cost, interest rate, life span of the equipment and number of failure. Since there is availability of historical data from existing plant number of failure is estimated using a probabilistic method. Engineering economic concept of equivalent annual cost is used to transform all the present and future costs to annual cost. Day–Stout–Warren (DSW) algorithm and the vertex method are then used to evaluate competing alternatives.

2 Methodology

In this section the develop decision support model is discussed in detail. The general framework is shown below in Figure 1.

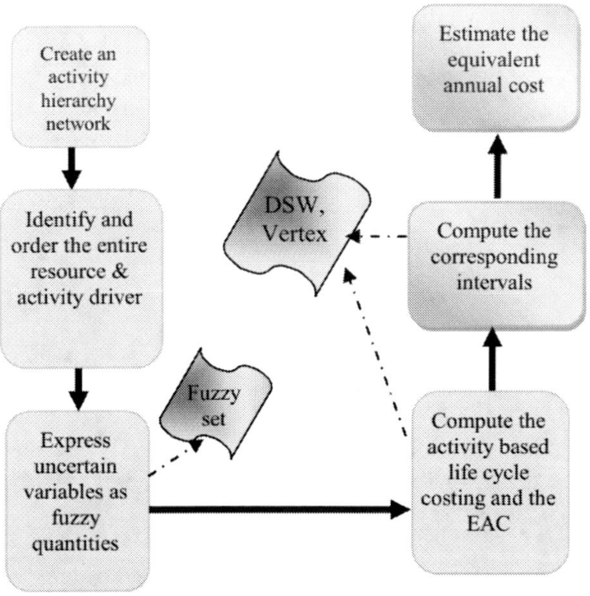

Figure 1. Schematic representation for decision support model

2.1 Create an activity hierarchy network

The model is applied to two different pumps manufactured by the concerned pump manufacturer (pump A and pump B). The design service life for pump A is 45 years and for pump B is 60 yrs. The data required for the analysis is extracted from [12] [13]. The principle of ABC is that products or services consume activities; activities consume resources and resources generate costs. This makes the identified activities to be cost drivers. All the activities which are the driver of the cost and their relation are identified as shown in the Table 1 below.

Table 1. Pump life cycle activities and drivers

Activities	Cost driver
Operation	
Day to day supervision	Number of hours of pump operation, personnel, and labour
Day to day operation	Cost of energy and number of hours of pump operation
Corrective maintenance and repair	
Access to the failed component	Time to gain access to failed component, personnel, and tools used
Diagnosis	Fault isolation time, personnel, manuals, technical data, test equipments, and tools used
Repair/replacement	Actual hands on time to repair/replacement, personnel, equipments and tools used
Verification and alignment	Time spent, personnel and tools used

2.2 Express the uncertainty variables as fuzzy quantities

There are different types of fuzzy number, triangular, trapezoidal, gaussian, sigmoid and bell shape membership functions are some of it. Triangular fuzzy number is widely used in many applications.

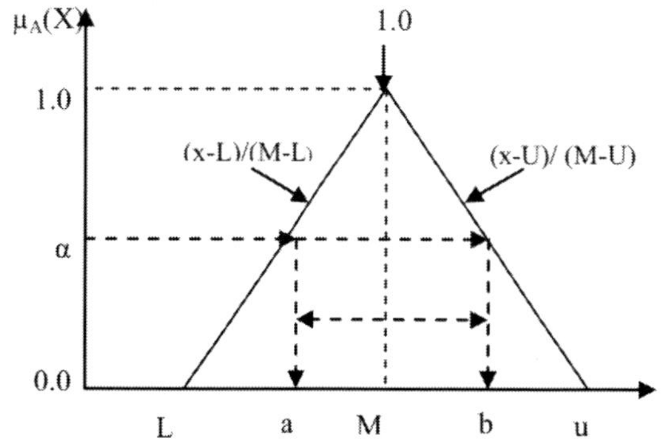

Figure 2. Triangular membership functions with α-cut

The membership function ($\mu_A(x)$) of a triangular fuzzy number which associated with a real number in the interval [0, 1] can be defined as:

$$\mu_A(x) = \begin{cases} \dfrac{x}{M-L} - \dfrac{L}{M-L}, \ldots\ldots\ldots\ldots\ldots x \in [L, M], \\ \dfrac{x}{M-U} - \dfrac{U}{M-U}, \ldots\ldots\ldots\ldots x \in [M, U], \end{cases} \quad (1)$$

The acquisition cost, the operation and the corrective maintenance cost is given in a fuzzy triangular number with the lower, modal and upper value as shown in Table 2. All given costs are in USD.

Table 2. Costs in triangular fuzzy number

	Pump A			Pump B		
	L	M	U	L	M	U
Acquisition cost						
cost of pump	2080	2600	3120	4844	5190	5363
cost of motor	1375	1500	1688	21375	23750	28500
cost of base frame	167	200	233	482	550	619
Cost of coupling	25	30	35	213	250	313
Operation cost						
C_{pw} is cost per input power	0.08	0.1	0.12	0.08	0.1	0.12

($/kw),						
C_l is cost of labour per hour	0.8	1	1.2	0.8	1	1.2
Maintenance cost						
$C_{s.p}$ is cost of spare part	160	200	240	160	200	240
C_t is cost of tools						
✓ *for access to the failed, component, diagnosis, verification & alignment*	2.7	3	3.4	2.7	3	3.4
✓ *for activity repair/replacement*	4	6	7.5	4	6	7.5
l is cost per labor	7.5	8	9.5	7.5	8	9.5

2.2 Compute the activity based LCC

In this section the fuzzy activity based LCC is developed.

$$LCC = \widetilde{C}_{aq} + \widetilde{C}_{op} + \widetilde{M}_c + \widetilde{D}_c \quad (2)$$

where $\widetilde{C}_{aq}$ is acquisition cost, $\widetilde{C}_{op}$ is operating cost, $\widetilde{M}_c$ is maintenance cost and $\widetilde{D}_c$ is decomposition cost. Due to unavailability of data decomposition cost is not covered.

The acquisition cost contains $\widetilde{C}_1$ cost of pump, $\widetilde{C}_2$ cost of motor, $\widetilde{C}_3$ cost of base frame, and $\widetilde{C}_4$ cost of coupling. The general expression for the acquisition cost is;

$$\widetilde{C}_{aq} = \sum_{j=1}^{j=4} \widetilde{C}_j \quad (3)$$

The high impact cost drivers in the operation are number of operation hours, personnel, and cost of energy. By integrating these factors the following equation is developed to estimate the operation cost. Mathematically, it is expressed as shown in Eq. (4):

$$\widetilde{C}_{op} = t * (\widetilde{C}_e(KW) + \widetilde{C}_l) \quad (4)$$

where t is the design life of pump (h), $\widetilde{C}_e$ is cost of energy ($/KWH) and $\widetilde{C}_l$ is cost of labour per hour. Energy consumption is calculated by gathering data on the pattern of the system output. Cost of energy for pump can be estimated as shown below [12] [13].

$$C_e = C_{pw}\left[\left(\frac{Q.H}{366.\eta_p.\eta_m}\right)\right] \quad (5)$$

where C_{pw} is cost per input power ($/kw), Q is the pump flow rate (m^3 /h), H is the pump head (m), η_p is the pump efficiency, η_m is the motor efficiency. These parameters for pump A & B are shown in Table 3.

Table 3. Parameters of Pump A and B

Parameter	Pump A	Pump B
pump flow rate Q (m^3 /h)	300	205
The pump head H (m)	90	300
the pump efficiency η_p	76	75
the motor efficiency η_m	92	90

One of the main factors which affect the reliability of an equipment of a system is proper maintenance. Uncertainties arise from maintenance cost determination, because the failure of the system can happen stochastically. The general equation of maintenance cost is as follows;

$$\widetilde{M}_C = \widetilde{C}_c N \quad (6)$$

where $\widetilde{M}_C$ is the maintenance cost, N is the number of failure, and $\widetilde{C}_c$ is the cost of repair. The corrective maintenance is conducted whenever there is a failure and the cost of repair is estimated by the activities it perform

$$\widetilde{C}_c = \widetilde{C}_{s.p} + \widetilde{C}_t + MTTR(l * n) \quad (7)$$

where $\widetilde{C}_{s.p}$ is cost of spare part for repairing a failure, if the pump is repaired without replacing any parts $\widetilde{C}_{s.p}$ is going to be zero. $\widetilde{C}_t$ is cost of tools; *MTTR* is mean time to repair, l is cost per labor and n is number of labor for each activity. Maintenance cost is therefore,

$$\widetilde{M}_c = N(\widetilde{C}_{s.p} + \widetilde{C}_t + MTTR(l * n)) \quad (8)$$

The number of failures can be determined from failure probability. The failure probability in this paper is determined using parametric recurrent data analysis. The number of failure for repairable system depends on the repair assumption taken. For repairable systems, generally there are two main repair assumptions, either "as good as new" or "as bad as old", but in actual practice the equipment lies somewhere in between these two conditions which is "better than old, but worse than new" [14]. Kijima and Sumita suggested a new approach called general renewal process (GRP) which is capable to cover all the three possible repair assumptions of repairable system [15]. The parametric RDA approach is based on GRP model, which provides a way to define the recurrence rate of repairable system failure overtime by considering the repair effect on succeeding failure.

There are two types of GRP models, type I and type II. In GRP type I model the system age of only the previous failure epoch i.e., the time between the previous two failures is improved. Let V_i denote the virtual age of the repairable system after the ith corrective maintenance action, and let $V_0 = 0$. GRP type I virtual age model indicates that,

$$V_i = V_{i-1} + qT_i \tag{9}$$

where q is maintenance effectiveness $0 < q < 1$. Under this model, each corrective maintenance action removes a portion, $1 - q$, of the age accumulated during the most recent period of repairable system function. GRP type II model is extension of GRP type I model. The difference between them is on assumption about impact of repairing on the damaged incurred [16]. GRP typeII reflects the reality where the maintenance action reduces the cumulative damage of all the previous failures. It is governed by the following equation:

$$V_i = q(V_{i-1} + T_i) \tag{10}$$

Note that if $q = 0$, then corrective maintenance is perfect, if $q=1$, then it is minimal and if $0 < q < 1$, then the corresponding repair assumption is somewhere in between perfect and minimal [17]. Under this model, the time to repairable system failure is a Weibull random variable having scale parameters $\beta > 1$ and shape parameters λ.

As stated in Wahyu [15] maximum likelihood estimation (MLE) is more considered to estimate GRP model parameter (q, β, λ). Greater the MLE value of the model, the best will be the statistical fit for the given data.

2.3 Compute the corresponding intervals

In some complex systems, it would be more reliable to give an interval estimate than a point estimate for many quantities. Interval number I is defined as an ordered pair of real numbers $[a, b]$, with lower bound a and upper bound b. When $b = a$, the interval number $[a, a]$ degenerates to a real number a [18]. The α-cut of a fuzzy set may also be represented as an interval, such that $A_\alpha = [a, b]$, as shown in Figure 2.

The goal of interval computation is to find the minimum and the maximum of the function when the different possible values of the variables range in their intervals. The arithmetic operations on any two intervals [a, b] and [c, d] are [19],

$$\lambda.[a,b] = [\lambda a, \lambda b], \quad \lambda \text{ is a constant} \tag{11}$$
$$[a,b]+[c,d] = [a+c, b+d], \tag{12}$$
$$[a,b]-[c,d] = [a-c, b-d], \tag{13}$$
$$[a,b].[c,d] = [ac, bd], \tag{14}$$
$$[a,b]/[c,d] = [a/d, b/c], \tag{15}$$

Dong et al.1985 [11] propose the Day Stout Warren (DSW) algorithm, which is based on the α-cut representation of a fuzzy number. The idea behind the DSW algorithm is to choose values for the $\mu_{(x)}$ and computes the corresponding intervals in $X_1, X_2, \ldots, X_n$. For further details on DSW algorithm and the vertex method refer to [12] and [13]

2.4 Fuzzy Equivalent annual cost

Present worth method discount future amounts to the present by using the interest rate over the appropriate study period therefore before conducting an AB-LCC, the analyst must define the general parameters, such as analysis period and discount rate.

$$\widetilde{NPW}_i = \widetilde{IC}_i + \sum_{j=1}^{NF_i} [\widetilde{FC}_i * 1/(1+\tilde{i})^{\tilde{t}_{ij}}] \tag{16}$$

where $\widetilde{NPW}$ is the net present worth, $\widetilde{IC}$ is the initial cost $\widetilde{FC}$ is the future cost, $\tilde{i}$ is the interest rate and $\tilde{t}_{ij}$ the time. The equivalent annual cost (EAC) is obtained by converting the equivalent value (at a specified time, usually the present) of a given set of cash flows into a series of uniform annual payments. The Equivalent annual cost (EAC) is,

$$\widetilde{EAC}_i = \widetilde{AC} + \widetilde{NPW}_i * \tilde{i}(1+\tilde{i})^{\tilde{T}_i} / [(1+\tilde{i})^{\tilde{T}_i} - 1] \tag{17}$$

where $\widetilde{AC}_i$ is the annual cost and $\tilde{T}$ is time

The fuzzy equivalent annual concept is adapted from Mohammad A et.al (2013) [11]. In this paper the initial cost is the accusation cost ($\tilde{C}_{aq}$); the future cost is the maintenance cost ($\tilde{M}_c$) and the operation cost($\tilde{C}_{op}$) is taken as annual cost, the equivalent annual cost is therefore

$$\widetilde{EAC}_i = \tilde{C}_{op} + \{\tilde{C}_{aq} + \sum_{j=1}^{NF_i} [\tilde{M}_c * 1/(1+\tilde{i})^{\tilde{t}_{ij}}]\} * \tilde{i}(1+\tilde{i})^{\tilde{T}_i} / [(1+\tilde{i})^{\tilde{T}_i} - 1] \tag{18}$$

The DSW algorithm is used to determine the interval of the fuzzy values. Representing fuzzy values by interval is computationally more effective than other methods [9]. Using a series of α-cuts, the corresponding intervals of problem fuzzy parameter $(\tilde{r}, \tilde{C}_{op}, \tilde{M}_c, \tilde{C}_{aq}, t, \tilde{D}_c, \tilde{T})$ can be easily determined. The vertex method is then used to determine the equivalent annual cost.

3. RESULT AND DISCUSSION

3.1 Acquisition operation and Maintenance cost

Using Eq. (3); the acquisition cost for pump A is found to be (3647, 4330, and 5076) and for pump B (26914, 29740, 34795).

The energy cost per KWh and the labour cost of operation is given as a fuzzy value in Table 2. The total cost of operation for the life cycle is then (831650.4,

1039563, 1247476) for pump A and (1864350, 2330437, 2796525) for pump B.

Since there are no historical data the failure data is taken from another plant that uses a similar type of pump. The failure data is collected for four years and eight failures are recorded within this time for Pump A, which are given in hour; 545, 1945, 3119, 3799, 4631, 5081, 6024 and 6900. Four failures are recorded for pump B (hour); 3026, 4759, 5874, and 7015. The scale and shape parameters of Weibull and the meantime between failure and number of failure of pump are calculated. Since the solution cannot be obtain analytically, the numerical methods is applied by Weibul ++8 software. Assuming 8 working hour per day and 243 working days per year it is found that the total time used for pump A is 87480 hour and 116640 hour for pump B. It is found that the number of failure for Pump A is 112 and for Pump B are 57.

Given the mean time to repair for activities; access to the failed component, diagnosis, and verification and alignment is 3hr and for activity repair/replacement it is 6hr respectively, number of personnel for all the activities is 4. The cost of spare part, tooling cost and labour cost is given in a triangular fuzzy number as shown in Table 2.

In practice it is unlikely to change all the component of a pump for one failure which is the assumptions of perfect repair which results in a higher expected cost. Similarly the imperfect or the minimal repair will cause a minimum expected cost but will lead to an increasing number of failure. However under the combination strategy parts which face major failure will get perfect repair while other parts like pump casing, shaft, housing, motor, and coupling which face minor failure will face an imperfect repair. This strategy seems to be a more practical and financially attractive choice so the maintenance cost is estimated for this assumption using Eq. (8) is (66830.4, 78064, and 93284.8) for pump A and (34012, 39729, 47475.3) for pump B. Using the above given data and equations, all costs are summarized in Table 4.

Table 4. Summary of cost for all activities

Parameter	Pump A			Pump B		
	L	M	U	L	M	U
Acquisition cost	3647	4330	5076	26914	29740	34795
Operation cost	831650.4	1039563	1247476	1864350	2330437	2796525
Maintenance cost	363766.4	375000	390220.8	66830.4	78064	93284.8
Interest rate	4	6	8	4	6	8
Service life	45 yr	50 yr	55 yr	55 yr	60 yr	70 yr

3.2 Interval analysis and equivalent annual cost

In this paper the alpha cuts used are (0, 0.2, 0.4, 0.6, 0.8, and 1). The number of α-cuts depends on the function to be calculated and the degree of accuracy needed. The interval for the 6 alpha cut value is calculated using DSW algorithm which is shown in Table 5.

Table 5. Interval Values of for all alpha cut for Pump A (PA) and Pump B (PB)

		Acquisition cost		Operation cost		Maintenance cost		Interest rate		Service life (hr)	
		P A	P B	P A	P B	P A	P B	P A	P B	P A	P B
0	a	3647	26914	831650	1864350	66830	34012	4	4	45	55
	b	5076	34795	1247476	2796525	93284	47475	8	8	55	70
0.2	a	3783	27479	873232	1957565	69077	35155	4.4	4.4	46	56
	b	4926	33784	1205893	2703307	90240	45926	7.6	7.6	54	68
0.4	a	3920	28044	914815	2050783	71323	36298	4.8	4.8	47	57
	b	4777	32773	1164311	2610090	87196	44376	7.2	7.2	53	66
0.6	a	4056	28609	956398	2144000	73570	37442	5.2	5.2	48	58
	b	4628	31762	1122728	2516872	84152	42827	6.8	6.8	52	64
0.8	a	4193	29174	997980	2237218	75817	38585	5.6	5.6	49	59
	b	4479	30751	1081146	2423655	81108	41278	6.4	6.4	51	62
1	a	4330	29740	1039563	2330435	78064	39729	6	6	50	60
	b	4330	29740	1039563	2330437	78064	39729	6	6	50	60

In order to explain the model α-cut 0.6 for pump A is taken and the EAC for this value is calculated as shown below. The present worth factor is calculated using vertex method for interest rate of [5.2, 6.8] and service life [48, 52]. The PWF is determined for the four combinations of interest rate and service life [5.2, 48], [5.2, 52], [6.8, 48], and [6.8, 52] using Eq. (17) the PWF is found to be [0.08775, 0.07164, 0.04252, and 0.03268].

The minimum and maximum value is [0.03268, 0.08775]. The fuzzy equivalent annual cost of pump A& B for the entire alpha cut is shown in Table 6 below.

Table 6. EAC result Pump A and Pump B

α-cut value	Pump A		Pump B	
	a	b	a	b
0	832172.3	1248204	1972006	3074885
0.2	874579.4	1206154	2078474	2960066
0.4	915879.3	1164619	2185396	2846055
0.6	957242.4	1123096	2292770	2732854
0.8	998655.2	1081591	2400596	2620461
1	1040107	1040107	2508875	2508877

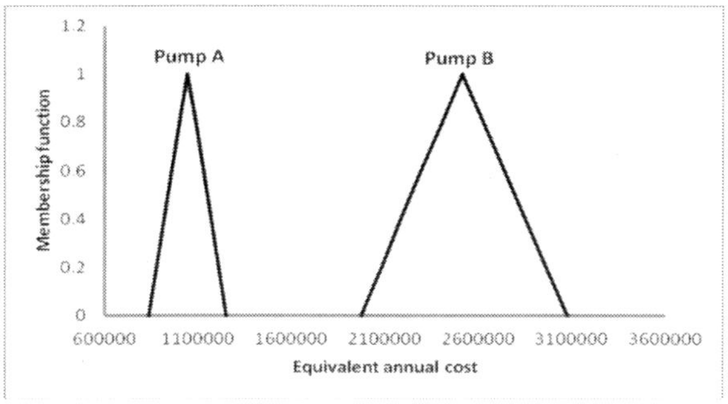

Figure 3. EAC result for all α-cut

The value of EAC for Pump B is greater than the EAC of Pump A in all the alpha-cut values. Therefore Pump A is more preferable than Pump B.

4. CONCLUSION

A decision support model is developed by integrating the concept of ABC-LCC, fuzzy logic, interval analysis, DSW and vertex method. Activity based method is used to identify the activities and cost drivers in acquisition, operation and maintenance phase. Fuzzy logic is used to incorporate uncertainties in the cost value. The DSW and the vertex method are helpful in to extending ordinary algebraic operations to fuzzy algebraic. The decision made using fuzzy activity based costing is accurate since subjectivity of expert can be easily considered and thus the uncertainty can be handled. Two Pump sets, Pump A and Pump B is taken as a case for this paper. The acquisition cost of both pumps as shown in Table 4, is very small amount of the total LCC, which shows that it is necessary to have a long-term outlook to the investment decision-making process rather than trying to save money in the short-term by simply purchasing assets with lower initial acquisition costs. From the output of the fuzzy ABC-LCC, it is found that Pump B has higher equivalent annual cost in which pump A is more preferable than pump B.

REFERENCES

[1] Rahman S, Vanier.D.J. Life cycle cost analysis as a decision support tool for managing municipal infrastructure. Toronto, Ontario 2004

[2] Dhillon B. S. Design reliability. Canada 1999: TA174.D4929

[3] Dhillon B. S. Life Cycle Costing for Engineers. Taylor and Francis Group, LLC 2010

[4] Emblemsvåg J, Bras B. Activity-Based Cost and Environmental Management – A Different Approach to ISO14000 Compliance. 2001 Boston: Kluwer

[5] Robin C, Robert S. Activity-Based Systems: Measuring the Costs of Resource Usage. Accounting Horizons/September 1992

[6] Korpi E, Ala-Risku T. Life cycle costing: a review of published case studies. Helsinki University of Technology Department of Industrial Engineering and Management Finland

[7] Vanier S, Rahman D.J. Life cycle cost analysis as a decision support tool for managing municipal infrastructure" National Research Council Canada-46774

[8] Chen, C., Flintsch, G.W., & Al-Qadi, I.L. Fuzzy Logic-Based Life-Cycle Costs Analysis Model for Pavement and Asset Managemen. 6th International Conference on Managing Pavements 2004

[9] Ross T. Fuzzy logic with engineering applications. McGraw-Hill 1995

[10] Kishk M, and Al-Hajj A. A fuzzy model and algorithm to handle subjectivity in life-cycle costing based decision-making. J. Financial Manage. Property Constr, 2002; 5(1–2), 93–104

[11] Mohammad A, Tarek Z, and Osama M. Fuzzy-Based Life-cycle Cost Model for Decision Making under Subjectivity. American Society of Civil Engineers 2013

[12] Waghmode L, Sahasrabudhe A, Kulkarni P. Life Cycle Cost Modelling of Pumps Using an Activity Based Costing Methodology. Journal of Mechanical Design by ASME, Vol. 132 / 121006-1, 2010

[13] Hennecke FW. Life cycle costs of pumps in chemical industry. ELSEVIER Chemical Engineering and Processing 38, 511–516, 1999

[14] Doyen L. Repair efficiency estimation in the ARI1 imperfect repair model. Modern Statistical And Mathematical Methods in Reliability, 2005;10:153.

[15] Wahyu W, Haryono. On Approaches For Repairable System Analysis. Laboratory of Statistical Industry, Department of Statistics FMIPA ITS Surabaya

[16] Annaruemon P, Wanida R, Supapan C, Pisal Y. A Fuzzy Time-Driven Activity-Based Costing Model in an Uncertain Manufacturing Environment. Proceedings of the Asia Pacific Industrial Engineering & Management Systems Conference 2012

[17] Bülent B. A Critique on the Consistency Ratios of Some Selected Articles Regarding Fuzzy AHP and Sustainability. 3rd International Symposium on Sustainable Development, 2012, Sarajevo

[18] Dong W, Shah H, Wong F. Fuzzy computations in risk and decision analysis. Civ. Eng. Syst. 1985; 2(4), 201–208.

[19] Dong W, Shah H. Vertex method for computing functions of fuzzy variables. Fuzzy Sets Syst;1987: 24(1), 65–78

Prominence of new Luxury car sales in Malaysia

Dr Muneer Sultana[1*], Mohd Hizam Hashim[2] and Samsudin Shafii[3]

*Department of Business and Management[*1, 2, 3]*
DRB-HICOM University of Automotive,
DRB-HICOM, Automotive Complex, Peramujaya, Industrial Area,
*Pekan, Pahang. Malaysia. ; *[1]Email:muneer_sultana@icam.edu.my ; roshnisaziz@gmail.com*

Abstract. Contemporary customers purchase belongings to compensate themselves, to please mental requirements or to make themselves touch well. Up-to-date customers purchase belongings because they are luxurious. They purchase belongings to make a declaration, to show off their character or to increase their self-confidence. Extraordinary superior merchandises, decorative packing, high-class stock localities, high marketing limitations, luxurious upgrades, promotion operations, and brand terms are all the involvement to the developed charges of luxury things. Corporations make great reserves of these machineries in direction to mark their goods promptly identifiable and acquainted. Vehicle transportation is perhaps the utmost regularly recycled method of carriage. The luxury goods will mislay their scarcity and superiority features if they are not valued high. Luxury products are costly in comparative and complete relations. Luxury car is a marketing word for an automobile that delivers luxury — enjoyable or wanted features outside severe requirement — at augmented outlay. The period proposes a car with advanced superiority equipment, superior presentation, additional exact structure, luxury, advanced strategy, technically inventive contemporary, or structures that take an appearance, brand, status, or prestige, or any other 'discretionary' feature or amalgamation of them. In such a case a study of luxury car sales has been conducted in Malaysia. The primary facts have been collected through oral interviews with stalwarts of automotive industry. The secondary facts are composed from the 30th November 2014 of Sunday Star, newspaper. Outcome of the research study divulges that despite disparity in income and wealth of the ethnic people, there are more luxury cars sales in Malaysia.

Key Words: *Automotive, Industry, Luxury, Car, sales, Malaysia, Auto parts*

1. INTRODUCTION

A fixed of notions that made intellect a period ago formed the contemporary perception on shop. Nearly of these out-of-date notions are public shop for money worth, which spending choices make real intelligence or that spending is regularly about obtaining desirable things and facilities. The original method of Under standing in the spending atmosphere is not extended appropriately to recognize modest consumer claim and try to please them. Contemporary customers purchase belongings to compensate themselves, to please mental requirements or to make themselves touch well. Up-to-date customers purchase belongings because they are luxurious. They purchase belongings to make a declaration, to show off their character or to increase their self-confidence. Extraordinary superiority merchandises, decorative packing, high-class stock localities, high marketing limitations, luxurious upgrades, promotion operations, and brand terms are all the involvement to the developed charges of luxury things. Corporations make great reserves into these machineries in direction to mark their goods promptly identifiable and acquainted. . Vehicle transportation is perhaps the utmost regularly recycled method of carriage. Each time, public usage of car transportation to stay their residences of effort, their home environment and as well as for the other purposes. Automobile transportation can be used initially for marketable determinations. For instance, car transportation practices the asset of attractiveness for numerous corporations such as carrier, wood and similarly for transportation companies. Vehicle transportation is similarly used for entertaining drives.

Currently it is precise significant facility that benefits locomotive traders and similarly for customers to handover autos on or after single apartment to alternative. Maximum Vehicle transport corporations are consistent, completely protected and deliver nationwide door to door distribution. The facilities are generally available accessible. Public might interpret completely the facilities provided by numerous vehicle transport corporations and they could even interpret and associate charges too at the identical time. Vehicle transport corporations take charge of moving not only cars but ships, motorbikes and even household moving facilities are obtainable. Producers or consumers could be promoted by immediate quotations, which are obtainable accessible. Producers could associate charges and decent facilities from the highest automobile shipment corporations. According to the Oxford American Dictionary, luxury is the state of countless well-being and excessive breathing. This is really what each separate serious for. The idea of luxury is restricted to separately distinct and what luxury means for them. In general luxury products are extremely valued goods frequently in the position marketplace of products. People consume these products for diverse determinations. These could comprise diamond commitment sphere, self-rewarding with Louis Vuitton bag or luxurious car to equal the neighbour's BMW. The luxury goods will mislay their scarcity and superiority features if they are not valued high. Luxury products are costly in comparative and complete relations. Luxury car is a marketing word for an automobile that delivers luxury — enjoyable or wanted features outside severe requirement — at augmented outlay. The period proposes a car with advanced superiority equipment, superior presentation, additional exact structure, luxury, advanced strategy, technically inventive contemporary, or structures that take an appearance, brand, status, or prestige, or any other 'discretionary' feature or amalgamation of them.

2. Statement of the problem

The middle class is characterised as having reasonable amount of discretionary income, claiming that roughly one third of their income after paying for basic necessaries and remaining is left for discretionary spending. Discretionary income allows people to buy consumer goods, improve their health care and provide their children's education. A car is one of the important goods of necessity in life. Despite having middle income group, the importance of luxury cars sales has increased in Malaysia. This is because of dreaming of buying a luxury car, brand perception and also changing city life. In such a circumstance, the study has conducted to know the reason behind the importance of new luxury car sales in Malaysia.

3. Need and Significance of the Study

The world has been fetching an additional prosperous residence; there is an additional and extra prosperous society. They are attractive progressively brand perception. The development of sale of luxury cars would be a combination advent of original representations and gradual development as consumers aimed to owe a luxury vehicle when wealth is augmented. Sales of luxury cars have been progressively rising from an additional than a period, with the greatest producers selling in market divisions they haven't usually contested in. The bulk of that development has been from strong competitiveness from German brands Mercedes-Benz, BMW and Audi. Rich consumers favour to have luxury cars owing to having additional influential machine primary to advanced supreme rapidity and earlier hastening rate and productivity in subordinate fuel consumption, extra consistency, superiority, stability, security, safety and luxury. Thus there is a need and significance to study new luxury cars sales due to increase in popularity of luxury brand.

4. Objectives of the Study

4.1: To Examine individual median monthly salaries and wages by ethnicity and strata in 2013 in Malaysia

4.2 :To scrutinize the total number of new luxury cars sold in Malaysia in the year 2013.

5. Scope of the Study

Study covers the new Luxury car sold in Malaysia to the extent of GDP of all states and individual median monthly wages and salaries in Malaysia.

6. Limitations of the Study

The study of new luxury car sales in Malaysia is equipped grounded on secondary data and the secondary sources were inadequate to the amount of the monthly salaries of individuals and luxury cars sales in Malaysia

7. Previous Studies

1. According to Judin Abdul Karim 2008)[1] stated that, it is obvious that road transportation controls Malaysia's transport segment, whereby the road system shields 96% of the financial events for moving things and customers.

2. On an added note, a public review directed by Abdalla *et al.* (2007)[2] ,has established out that native road operators have great fondness for the use of their individual cars slightly than community transport due to the suitability reason.

3. According to Judin Abdul Karim (2008)[3], opined that, the portion of the traveller segment is extra separated into private cars and community transportations. The previous has the largest portion with 65%, while the final takes up the remaining30%.

4. According to Li Wei et al.(2008)[4], A great mandate of isolated cars may also clarify the inspiring situations in national automotive production. The national motor car marketplace signifies roughly 30% of the total demand in the five largest ASEAN economies— Indonesia, Malaysia, the Philippines, Singapore and Thailand.

5. Additionally according to Mohd Rosli (2006)[5] , car sales in Malaysia are measured by native car producers by quality of administration strategies that defend the local automotive industry i.e. via tariff and non-tariff barriers.

6. Lily Amelia et al. (2009)[6] pointed out that, the main native car producers in Malaysia—Proton, Perodua and Naza—had clutched about 60% of the market share in 2007. Also, the growth of the national automotive industry has facilitated the development of the connected trades, comprising Small and Medium Enterprises (SMEs) as part producers or sub-assemblies.

7. MIDA (2010)[7] reported that, Since Malaysia does not only create cars but also make automotive apparatuses, native car holders should not face any problems in obtaining aftermarket goods. These goods that comprise car components, substances and fixtures should be effortlessly available from the market place and also valued comparatively subordinate than introduced goods. Though, the topics of car preservation do not merely rotate everywhere the foundations of the aftermarket goods, since the status quo of Malaysia's automotive environment proposes that many openings static necessity to be talked.

8. According to Gall, Jared (August 2010)[8] explained that, Bentley and Rolls-Royce have newly stimulated into more reasonable amount groups with new models priced significantly subordinate than their out-dated contributions (Bentley Mulsanne and Rolls-Royce Phantom and Maserati Quattroporte). These new cars, the Bentley Flying Spurand Continental GT sister cars and Rolls-Royce Ghost, shaped what Car and Driver described as the "entry-opulent segment".

9. According to Berger, Michael L. (2001) [9] opined that, Automobile producers market precise creates and representations that are directed at precise socio-economic programmes, and thus "social position came to be related further with a specific car than possession of a car per se. "Consequently, automakers distinguish among their creation lines in "collusion" with the car-buying community. While an extraordinary price is the greatest normal aspect, it is "designing, manufacturing, and smooth public estimation which cars had the uppermost and lowermost position related with them."

10. According to Wendler, Andrew (January 2012)[10] claimed that, Luxury vehicle manufacturers may moreover be stand-alone corporations in their individual true, such as BMW and Mercedes-Benz, or a division/subsidiary of a mass market automaker (e.g., Lexus is part of Toyota). Badge engineering is frequently used for price reserves, for example, the Lincoln cars that are created on Ford platforms or Acura representations resulting from Honda.

8. RESEARCH METHODOLOGY

8.1 : Sources of Data

The strength of any research is reliant on the organized method of information assortment sources and its analysis. The study is equipped grounded on primary information's and secondary information's. The primary facts have been collected through oral interview with stalwarts of automotive industry. The secondary facts are composed from the 30th November 2014 Sunday Star, newspaper and all existing literature have been obtained from internet automotive websites, auto business magazines, and e-auto journals.

9 : RESULTS AND DISCUSSION

Table.1. portrays the Individual median monthly salaries and wages by ethnicity and strata in Malaysia for the year 2013. The above table shows huge disparities in Income and spending patterns in the country. Majority of households earn less than RM 6000 and it highlights wealth inequality. Further more, the most of ethnic Chinese seem to be somewhat earning better than the other ethnic people because of their interest in business and trades in domestic and abroad. Therefore, it can be concluded that there is urban-rural disparities of income and spending pattern in Malaysia.

Table.1. Individual median monthly salaries and wages by ethnicity and strata in Malaysia for the year 2013

S.no.	Ethnicity/strata	Median monthly salaries and wages (RM)
1	Malaysia (total)	1,700
2	Bumiputera	1,600
3	Chinese	2,000
4	Indian	1,500
5	Others	900
6	Urban	1,680
7	Rural	1,040

Source: Sunday Star 30th November 2014.

Table .2. Total number of new luxury cars sold in Malaysia for the year 2013.

S.no.	Brand Name	No. of new Luxury cars sold
	Volkswagen	9,538
2	BMW	7,057
3	Mercedes	5,550
4	Audi	3,102
5	Lexus	1,336
6	Land Rover	1,003
7	Mini Coper	437
8	Porsche	275

Source: Sunday Star 30th November 2014

Table.2. displays the total number of new luxury cars sold in Malaysia for the year 2013. Out of total eight brands i.e. Volkswagen, BMW, Mercedes, Audi, Lexus, land Rover, Mini Coper and Porsche of luxury car sales , Volkswagen sales are more compared to other brands of luxury cars sales in Malaysia due to increase in dealer network and healthy car market segment of wealthy people. The main source of income is solely from their salaries as the most do not have assets. The main bulk of their expenses go to car instalment and the main liabilities for the most of the Malaysians are car loans to purchase luxury cars due to adopt city life style.

The following are the suggestions for the study of prominence of new luxury cars sales in Malaysia.

1. 1.Outcome of the research study divulges that despite disparity in income and wealth of the ethnic people, there are more luxury cars sales in Malaysia, specially the sale of Volkswagen is more compared to other brands, in such a circumstances the manufacturers of luxury cars sales should still introduce and increase several new models keeping in view of the wealthy people preferences, create more customer confidence and consideration for the brand and increase more dealer network to avoid any delay in supply.

2. It is also suggested that the business of automotive segment of luxury cars will remain to be its foremost revenue creator,

which determines a majority of the auto company's proceeds is still anticipated to be seen as growing and moving forward notwithstanding the occurrence of extra business parts such as oil and gas that is rising in consequence and weightage on its complete financials. The sales of luxury car segment will contribute profits for every financial year and also expand its market share.

10. CONCLUSION

Therefore, Producers of Prominent luxury brands like BMW and Mercedes are now endeavouring to equal their brand with post-modern method specifically 'brand as company' and brand as strategy. Vendors should be conscious of customer pessimism and careful with their movements which could affect purchaser and community standards completely or undesirably. Vendors should emphasise on consumer value-building and this will lead the consumer to procurement frequently or furthermore to commend that trader to families and survive the promoting success of other producers.

REFERENCES

Judin Abdul Karim (2008), *Road infrastructure development in Malaysia,* Public Works Department, Malaysia, Accessed in November 2009, from http://rakan1.jkr.gov.my/csfj/editor/files/File/Dokumentasi/International%20 Paper/ROAD%20MANAGEMENT%20AND%20MAINTENANCE2pdf.

Abdalla, NK, Riza Atiq, OKR & Amiruddin, I (2007), Why do people use their cars: a case study in Malaysia, *Journal of Social Sciences*, 3(3):117–122.

Judin Abdul Karim (2008), *Road infrastructure development in Malaysia,* Public Works Department, Malaysia, Accessed in November 2009, from http://rakan1.jkr.gov.my/csfj/editor/files/File/Dokumentasi/International%20 paper/ROAD%20MANAGEMENT%20AND%20MAINTENANCE2. pdf.

Li Wei, K, Mat Saman, MZ, & Meng Chiao, L (2008). *Development of green design framework for Malaysian automotive industries,* Proceedings of the 2nd Regional Conference on Vehicle Engineering & Technology, KualaLumpur.

Mohd Rosli (2006), The automobile industry and performance of Malaysian auto production, *Journal of Economic Cooperation,* 27: 89–114.

Lily Amelia, Wahab, DA, Che Haron, CH, Muhamad, N & Azhari, CH (2009), Initiating automotive component reuse in Malaysia, *Journal of Cleaner Production,* 17:1572–1579.

MIDA (Malaysian Industrial Development Authority) (2010), *Business opportunities: Malaysia's automotive industry,* August 2010, Accessed in December 2010, fromhttp://www.mida.gov.my/en_v2/uploads/Publications_pdf/BO_MalaysiaAutomotive/Automotive_FA.pdf.

Gall, Jared (August 2010). "2010 Bentley Continental Flying Spur Speed vs. 2011 Rolls-Royce Ghost". *Car and Driver*. Retrieved 2014-03-11

Berger, Michael L. (2001). The automobile in American history and culture: a reference guide. Greenwood. p. 160. ISBN 978-0-313-24558-9. Retrieved 2014-03-11

Wendler, Andrew (January 2012). "2013 Acura ILX Concept - This concept previews a production Civic-based compact for Honda's luxury brand". Car and Driver. Retrieved2014-03-11.

Microstructure and Mechanical Properties of Austenitic Compacted Cast Iron with Additive Manganese

K. M. Ahmad[1,a], M. R. Maarof[1], M. Ishak[1], M. S. Huzairi[1]

[1]Faculty of Mechanical Engineering, Universiti Malaysia Pahang, 26600 Pekan, Pahang, Malaysia,

Abstract. High composition of austenitic matrix in microstructure is contributed to its special characteristic in reducing cracking and distortion. This special alloy will be in austenitic matrix when high temperature and also at room temperature which make it successful avoid matrix volume changes. In this study, austenitic compacted iron with minimum nickel contain of 12wt% is produced by using in-mould magnesium treatment. Magnesium Ferro Silicon (MgFeSi) used in determining the graphite form. Some of nickel replace by manganese in range of 9wt% to 12wt% Mn while Nickel remain at 12wt% Ni at all parameter. This research focuses on to investigate the effect of alloying elements on microstructure and mechanical properties. The outcome compared to standard of unmodified austenitic cast iron as in ASTM standard. The matrix obtained was all in austenitic matrix with the presence of compacted graphite distribution throughout the matrix. It is appeared that result shows increment of manganese at the centre region between graphite in the microstructure. It is occurred when percentage of manganese addition was increased. There are increases in hardness compared to unmodified austenitic iron. Tensile strength decreases when the manganese content increases in alloy.

1 Introduction

Compacted Graphite cast iron (CGI) is also referred as vermicular graphite. It has inadvertently been produced in the past as a result of insufficient magnesium or cerium levels in molten intended to produce spheroidal graphite iron [1]. CGI with rounded ends graphite irons possesses mechanical properties intermediate between those of grey and ductile irons. Its unique properties have led it to be used in many applications which are unsuitable for grey or ductile iron especially in automotive industry. CGI has high fluidity and low solidification shrinkage, its tensile strength are around 1.5 – 2 times of gray cast iron [2]. In comparison with ductile cast iron (SGI), CGI has better damping properties, better thermal conductivity and good castability. Based on these characteristics CGI fulfils the requirements for many light weight constructions especially automotive engine blocks.

However, modification levels of graphite must be controlled strictly in the CGI production. According to ASTM A842-85 standard requirement specifications for CGI castings state that the percentage of nodular graphite is 20% maximum and the formation of flake graphite must be avoided in the main parts of a casting [3,4]. In microstructure, 80% minimum of all graphite is needed to be in the form of compacted (vermicular) according to ASTM A247 (type IV) [5].

[a] Corresponding author: khairul_muzafar@yahoo.com

In recent years, there has been an increasing interest in Austenitic type of cast iron which can be an economic alternative to stainless steel as it is easier to cast and therefore suitable for precision casting of complicated shaped parts [6]. Therefore, austenitic cast iron also offers an outstanding combination of properties that meet a variety of industrial demands in withstanding the effects of corrosion, heat and wear [7]. However, so far to date there has been little attention given to Austenitic compacted iron (ACI) compared to austenitic gray iron (AGI) and Ductile Ni-resist (DNR). Most studies in ACI only been carried out in a small number of areas limited to compacted iron only (without austenitic matrix). So much so, there is no ASTM standard for this kind of special material.

ACI is produced by taking a normal melt grey iron addition with controlling the carbon and silicon at lower levels by adding various alloys to produce a stable austenitic structure at ambient temperature. Controls of magnesium treatment and addition of anti-nodular element to the melt such as copper and Manganese very important to promote graphite in the form of compacted. Anti nodular agent will interfere in nodularisation process during solidification. Nickel cannot be neglects as the main alloying element who acts as austenitic promoter either in cast iron or ductile iron. Nickel becomes the principle reason why the casting alloy has an austenitic structure during solidification. With sufficient amount of alloy addition, an austenitic microstructure may occur and remained stable at room temperature.

Eventually, there are also suggestions using manganese and copper as austenitic matrix promoter since they have same function as nickel and at the same time they are cheaper than nickel. However the uses of manganese will promote carbide while cooper is strong anti nodular agent. Hence, this investigation will study manganese as one of the addition alloying element use together with nickel as austenitic stabilizer.

A consider amount of literature has been published describing role of manganese in cast iron austenitic stabilizer. Previous studies reported that manganese will promote carbide in cast iron during the solidification process. In ductile cast iron manganese will segregate at cell boundary [8-11]. In previous studies by different authors had found that, Instead of becomes a carbide formation manganese also effect the cast iron in nodularity [12].

Therefore, the aim of this study was to evaluate and validate the effect of manganese in austenitic compacted iron on mechanical properties and microstructure.

2 Experimental Set Up

2.1 Material preparation

The material used in this study is ACI. Sample preparation of 200 mm solid cylinder shape castings prepared by using a pattern as in figure 1. Materials will melt in the frequency induction furnace. Magnesium treatment used in the casting was applied using In The Mould Treatment Technique as it has high magnesium recovery percentage [13,14], by employing 1.0 Fe-Si-Mg in the reaction chamber. Experimental alloy of ACI divided into four groups as 9Mn-12Ni, 10Mn-12Ni, 11Mn-12Ni, 12Mn-12Ni wt%. Pure nickel and steel scrap were added into the furnace to increase Ni percentage and to control carbon content respectively. All charging material was melted in the induction furnace, during base material is melting Ferro manganese was added to the charging material in the furnace. The molten then stirred to make sure the alloyed elements smoothly diffused. The melt then was tapped using ladle before pouring into green sand mould.

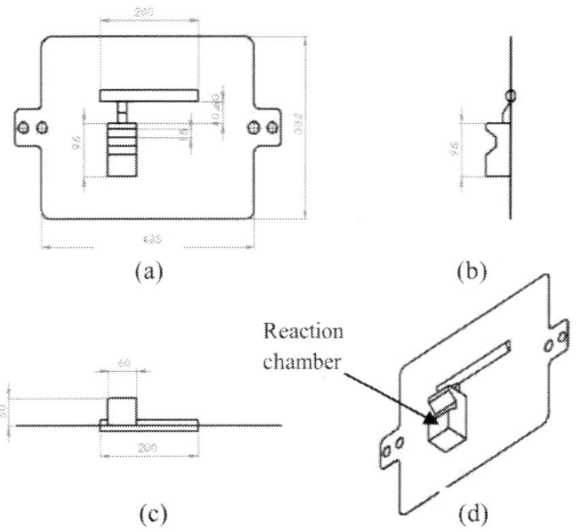

Figure 1: Solid cylinder shape castings pattern in this experiment (a) plan view (b) side view (c) front view (d) isometric view.

2.2 Mechanical test

Tensile test were carried out using universal testing machine 50 kN was used according to ASTM E8M at room temperature. CNC lathe machine was used to shape the cast alloy specimen to produce dog bone shape as shown in figure 2. The samples were further examined using scanning electron magnetic (SEM) to assess the microstructure after fracture. Samples from the specimens of 9wt%, 10wt%, 11wt% and 12wt% were taken for the hardness test. The Rockwell hardness test machine was used to determine the macro hardness. Five reading was taken from each sample and averaged. Every batch of samples was named according to its manganese weight percentage (wt %) temperature.

2.3 Microstructure analysis

Optical Microscopic (OM) and SEM were used for the microstructural characterization of the casting. Metallographic samples were cut from the broken halves of the tensile specimen. Microstructural analysis was come out using optical microscope model OLYMPUS BX60F5, while fractographic and fracture conducted using scanning electron microscopic model PHILIPS XL40.

Table 1: Chemical composition of raw material.

	Element										
	C	Si	Mn	P	S	Mg	Ni	Ca	Cr	R.E	Fe
Pig iron	4.5	1.73	0.309	0.180	0.150	0.310	0.236	-	-	-	balance
Steel	0.19	0.15	0.54	0.90	0.02	-	-	-	-	-	balance
Nickel	-	-	-	-	-	-	99.0	-	-	-	balance
FeMn	-	-	86.00	-	-	-	-	-	-	-	-
Nodulant	-	44.00	-	-	-	5.00	-	2.0	-	1.90	10.00
Inoculant	-	70.00	-	-	-	-	-	2.0	-	-	balance

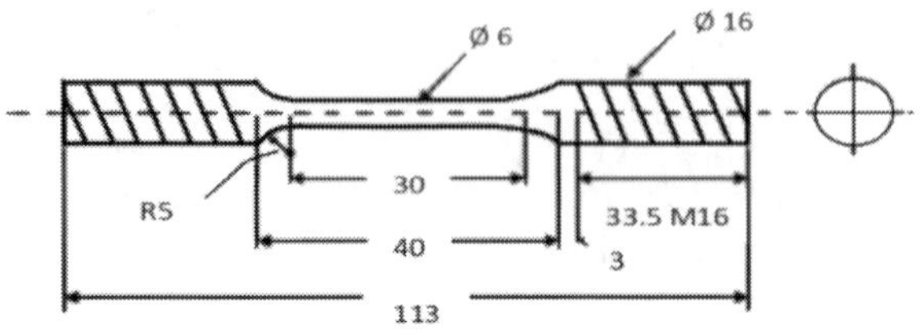

Figure 2: Dimension of tensile test specimen.

3 Result and Discussion

3.1 Microstructure Analysis

In figure 4 presents a typical as-cast matrix images for each sample at 100x magnifications. All samples are in austenitic microstructure with single bright matrix and the graphite precipitated in the form of vermicular (compacted). Generally, the figure shows dark compacted graphite distributed and embedded throughout in the austenite matrix. Austenitic compacted irons must contain sufficient amount of nickel and manganese in order to promote austenitic characteristic. Literature indicates that 18 wt% Ni for standard types of austenitic ductile iron [15] and 13.5 wt% Ni together with presence of 5.5 wt% Cu for standard types of austenitic gray cast iron [3]. Through this study, the minimum used is 12 wt% with addition of manganese to assist in promoting austenite matrix which is lower than the previous study [16].

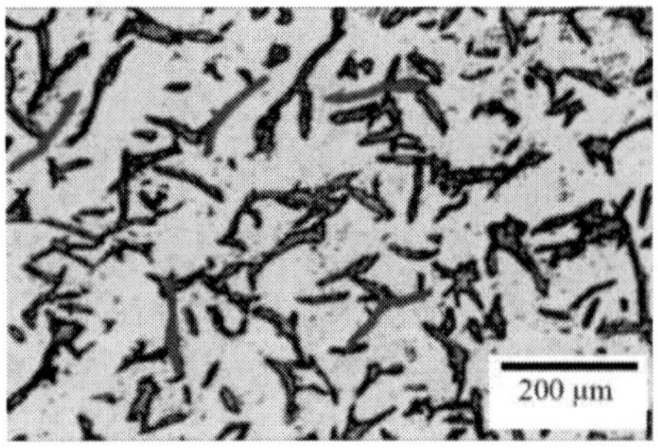

(a)

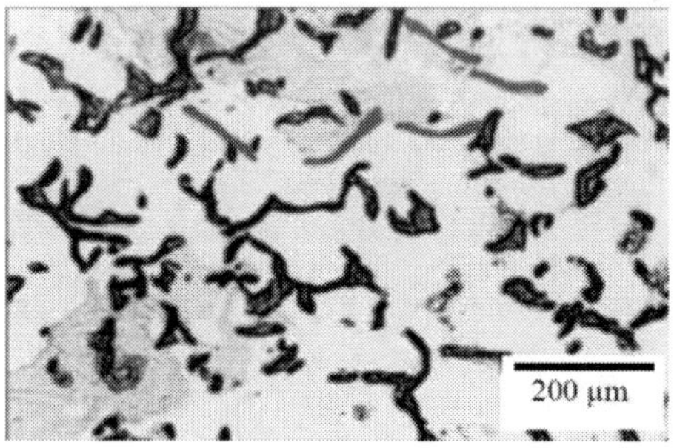

(b)

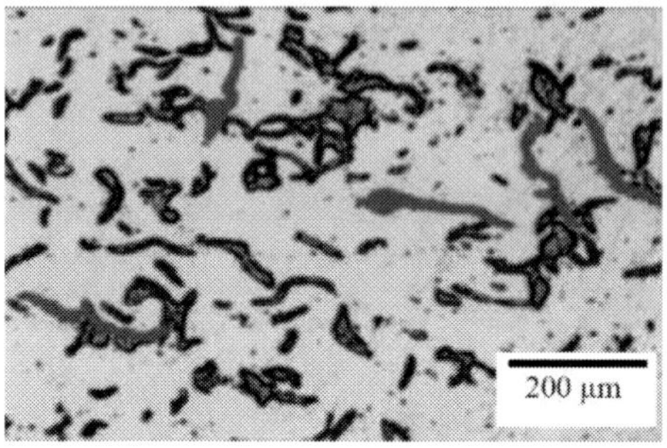

(c)

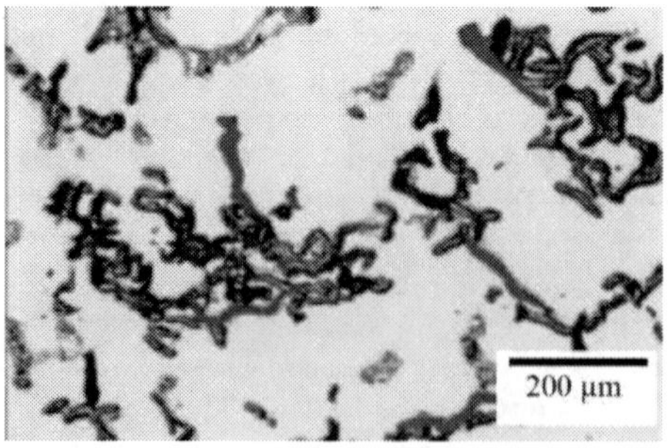

(d)

Figure 3: Microstructure of the modified CNR: (a) 9Mn-12Ni, (b) 10Mn-12Ni, (c) 11Mn-12Ni, (d) 12Mn-12Niwt%.

It is suggested that when high manganese addition alloyed, iron shows decreasing pattern of the graphite appearance in the microstructure. The graphite looks becomes distantly each other such as shown in figure 4(a) compared to figure 4(d). The formations of the carbide at this region push the graphite aside. It is making the length between graphite increases. Carbide formation will decrease the machinability of the iron and increase the production cost. Therefore it is preferred to be low in casting process. During the solidification of iron, manganese move away from the solid phase (graphite) which is first nucleate and solidifies. Manganese is the last element solidifies after the nucleation of graphite. It has been grouping and segregated away from the graphite. So, if the manganese increases the carbide will increase as suggested by Rasidi and Hasbullah [11]. This occurrences indicates segregated carbide suppress the existences of free graphite. Even though the combination of higher manganese wt% and less nickel wt% stabilized the austenite matrix, it is also appears that those combination promote carbide formation.

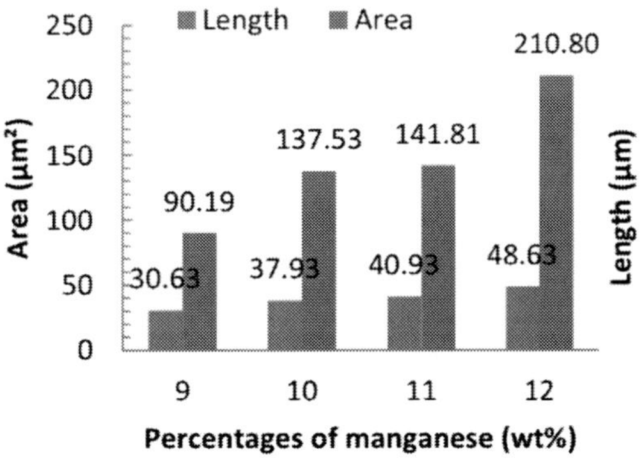

Figure 4: Graph Comparison between Percentages of Manganese (%), Length of Graphite (µm) and Area of graphite (µm²).

An analysis is conducted using A ProgRes Capture Pro v28.8 Imaging Analysis Software and Video Test-Structure by randomly selected the flake or worm shape graphite and calculate the length and area for each percentages of Mn. The red highlight color in figure 3 is the selected graphite for the measuring process. The comparison between Manganese percentages and the Length of graphite is being made to determine the effect of the Manganese percentages on the characteristic of graphite. The average of length data was collected from 9% Mn, 10% Mn, 11% Mn and 12% Mn is compared and graph that produce is shown in the Figure 4. Figure shows that the highest length of graphite is at 12% Mn which is 48.63 µm. The lowest value of average length is at 9% Mn which is 30.63 µm. Based on graph above, there are increasing trend on the length of graphite proportional to the percentages of manganese. When more manganese is added to the alloyed iron, there are increasing in the length of graphite. The highest area of graphite is at 12% of Manganese which is 210.80 µm² and 9% Mn contain in alloyed iron has the lowest area value which is 90.19 µm². The graph shows same pattern as length graph, the increasing trend of area proportional to the increasing of manganese.

3.2 Tensile properties for austenitic compacted cast iron

Table 2 shows the influence of manganese contains range from 9 wt% Mn to 12 wt% Mn in tensile properties of austenitic compacted cast iron for all heats. The result shows that when a lot of manganese is added to the alloyed iron, ultimate tensile strength, proof stress and elongation shows the decreasing trend as shown in figure 4(a)(b)(c). This is because of increasing of manganese led to an increases carbide phase amount at inter space between compacted graphite and resulted to low properties of mechanical. There are several effects for this phenomenon such as a lot of carbide formation clique together rather than small free form carbide. Hence, this will encourage cracking characteristic that responsible for the weakening overall material strength.

Table 2: Mechanical properties of alloyed irons in the experiments.

materials	Room temperature test			
	Ultimate tensile strength (MPa)	Proof Stress at 0.2 % offset (MPa)	Elongation (%)	Macro hardness (HV)
9 Mn-12 Ni	265.42	129.16	13.56	80.2
10 Mn-12 Ni	260.21	120.82	11.94	83.1
11 Mn-12 Ni	248.02	118.92	12.61	86.6
12 Mn-12 Ni	190.56	118.49	11.12	87.8

From the figure 5 test result, it is evident that material was relatively strong than the standard of DNR and AGI. The highest tensile strength is 265.42 MPa higher than austenitic gray cast iron which is 207 MPa and lower than austenitic ductile iron (DNR) 379 MPa. It is cannot be argued the high value tensile strength of DNR related to its nodule graphite formation compare to compacted graphite which well known as crack arrester. As the result, nodule graphite successful increase the mechanical properties compare to compacted graphite.

Table 3: The EDX checking Data.

Element	Point (%)							
	a	b	c	d	e	f	g	h
Si K	1.67	1.86	1.68	1.73	1.74	1.71	2.13	2.19
Mn L	4.46	5.98	4.87	3.98	5.46	5.96	4.18	3.75
Ni L	10.06	10.84	12.28	12.64	12.37	12.1	11.65	8.14

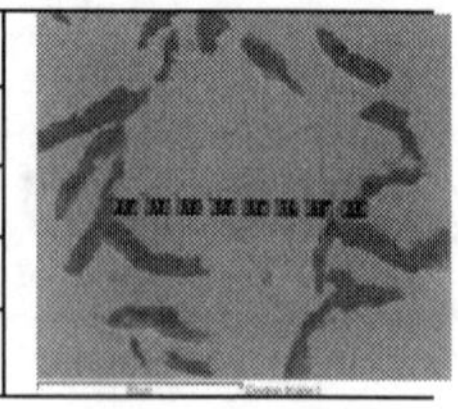

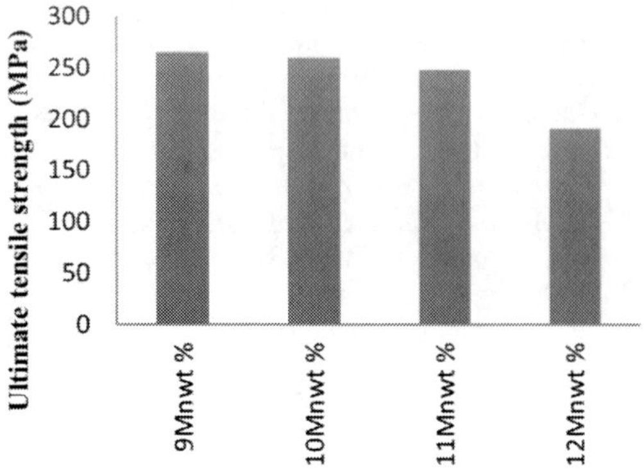

(a)

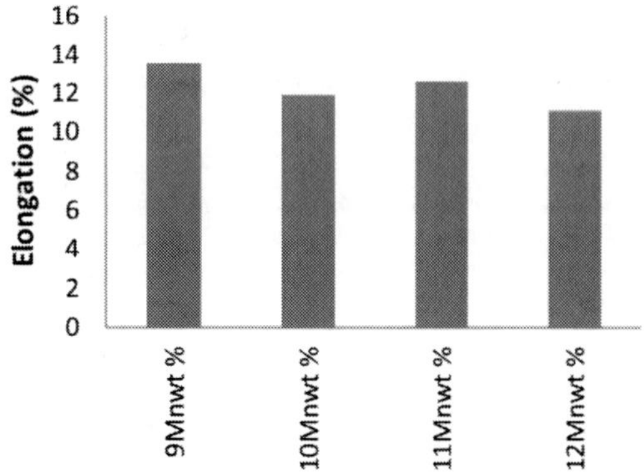

(b)

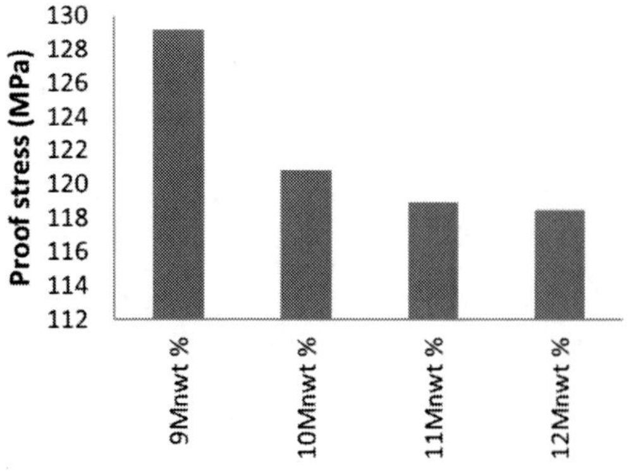

(c)

Figure 5: Comparison of tensile properties and macro-hardness of the alloyed iron: (a) ultimate tensile strength, (b) proof stress, (c) elongation.

Table 3 shows element of silicon, manganese and nickel at the region between graphite. It is shows increasing reading pattern for manganese at a certain point. Overall, the highest reading is at point b and point f second. Most of the high reading located at the centre of the region. Point c and d shows some of the inconsistence reading. Nickel reading shows increasing pattern until it reaches a peak reading, then decreasing graph pattern start. Blue spot in the picture in table 3 shows where the position of Energy Dispersion X-Ray Spectrometer (EDX) reading obtained, from right start with point a to point h.

3.3 Hardness test

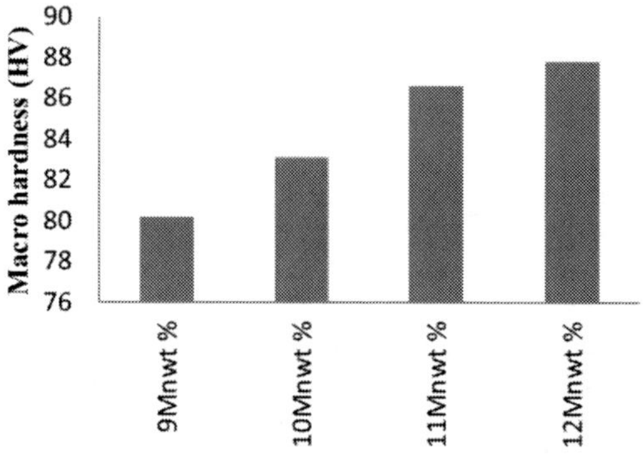

Figure 6: macro-hardness.

Manganese one of the carbide promoter and it is already proved. The existences of carbide in the iron will increase the hardness reading as shown in figure 6. The higher hardness reading is 87.8 HV but lower than any types of DNR which is the lower is in range 121 HV to 171 HV. But it was surprising that the percentage of elongation of the austenitic compacted iron lower than DNR type D-2C and D-5.

 DOI: 10.1051/matecconf/20167400009

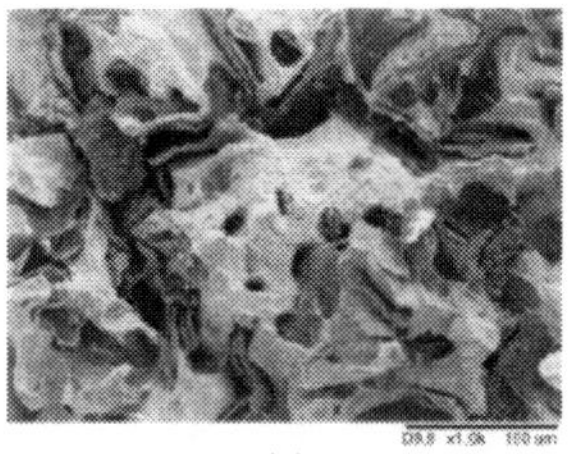

(a)

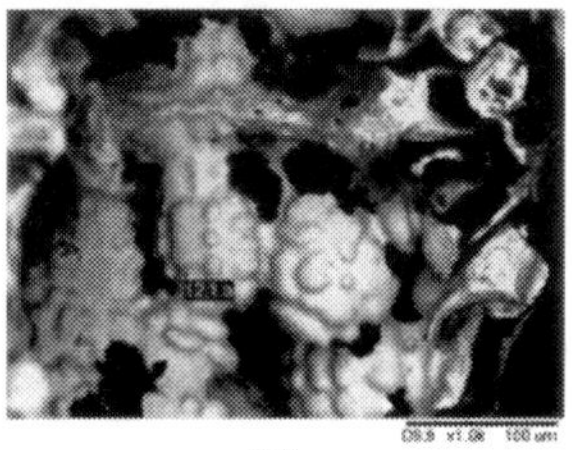

(b)

Figure 7: SEM micrograph of fracture surface after tensile test (a) 9 wt% and (b) 12 wt% Mn.

Figure 7 (a) and (b) is the SEM micrographs taken from the broken halves of the room temperature tensile test specimen of alloyed iron. Examination of deep-etched SEM micrographs shows that the individual 'worms shaped' graphite is connected to their nearest neighbours within the eutectic cell for figure 7(a) instead of separate for figure 5 (b). This complex coral-like graphite morphology, together with the rounded edges and irregular bumpy surfaces, results in strong adhesion between the graphite and the iron austenite matrix. This compacted graphite morphology inhibits crack initiation and growth and is the source of the improved mechanical properties relative to grey iron.

4 Conclusion

1. Base on the study the conclusion of following, Compacted graphite length and area tend to increase when the manganese content increase.

2. 12 wt% manganese has the lowest tensile test properties but highest macro hardness while 9 wt% manganese evidently has the highest tensile strength with lowest macro hardness.

3. It is suggested that the formation of carbide due to element segregation governs for the most part they averaged macro hardness obtained.

4. Besides that, the influence of solid solution by manganese in alloyed iron could reduce strength and yield strength of late to freeze (LTF) region formation.

5. These results prove that the alloyed iron with 9 wt% Mn has the best mechanical properties and alloyed iron with 12 wt% Mn has the weakest mechanical properties.

Acknowledgements

The authors would like to be obliged to Universiti Malaysia Pahang and Malaysia Government for providing laboratory facilities and financial assistance under project no. RDU 140135.

References

1. Stefanescu, D.M., *Compacted Graphite Iron*, in *ASM Handbook Vol 01 Properties and Selection Irons, Steels, and High-Performance Alloys*. 1996, ASM International. p. 102 – 129.
2. Brown, J.R., *Foseco Ferrous Foundryman's Handbook*. 2000: Butterworth-Heinemann. 84 - 89.
3. A436, A., *Standard Specification for Austenitic Gray Iron Castings*, in *Annual Book of ASTM Standard, Vol 01*, A. International, Editor. 2009: Annual Book of ASTM Standard, Vol 01. p. 1-5.
4. Sun, X.J., Y.X. Li, and X. Chen, *Identification and evaluation of modification level for compacted graphite cast iron*. Journal of Materials Processing Technology, 2008. **200**(1-3): p. 471-480.
5. A247, A., *Standard Test Method for Evaluating the Microstructure of Graphite in Iron Castings*, in *Annual Book of ASTM Standard, Vol 01*. 2009, ASTM International: united state. p. 1-5.
6. Elisabeth Weidmann, A.G., *Application Notes: Metallographic preparation of Cast iron*, in *Struers*, S. Inc, Editor. 2006, Struers A/S: http://struers.com/. p. 1-6.
7. Gallagher, J.J., *Ductile Iron For Engineers*. 1998, Metals Internaltional Limited: www.ductile.org. p. 61-89.
8. M. Nili Ahmadabadi, E.N., J. Echigoya, *Transmission electron microscopy study of high temperature bainitic transformation in 1 wt.% Mn ductile iron*. Materials Science and Engineering A 1995. **194**: p. 87-98.
9. Mohd Rashidi, M. and M.H. Idris, *Effect of inoculation on microstructure, mechanical and corrosion properties of high manganese ductile Ni-resist alloy*. Materials & Design, 2013. **51**: p. 861-869.
10. Mohd. Rashidi, M. and M. Hasbullah Idris, *Microstructure and mechanical properties of modified ductile Ni-resist with higher manganese content*. Materials Science and Engineering: A, 2013. **574**: p. 226-234.
11. Rashidi, M.M. and M.H. Idris, *The effects of solidification on the microstructure and mechanical properties of modified ductile Ni-resist iron with a high manganese content*. Materials Science and Engineering: A, 2014. **597**: p. 395-407.

12. Roula, A. and G.A. Kosnikov, *Manganese distribution and effect on graphite shape in advanced cast irons.* Materials Letters, 2008. **62**(23): p. 3796-3799.
13. ASA, E., *Magnesium Treatment Processes*, in *Technical Information*, F.P. Elkem ASA, Editor. 2004, Elkem ASA. p. 1-2.
14. Skaland, D.T., *Ductile Iron Production - A Comparison of Alternative Treatment Methods.* Elkem ASA, 1999: p. 2-19.
15. A439, A., *Standard Specification for Austenitic Ductile Iron Castings*, in *Annual Book of ASTM Standard, Vol 01*, A. International, Editor. 2009: united state. p. 1-6.
16. Fatahalla, N., A. AbuElEzz, and M. Semeida, *C, Si and Ni as alloying elements to vary carbon equivalent of austenitic ductile cast iron: Microstructure and mechanical properties.* Materials Science and Engineering: A, 2009. **504**: p. 81-89.

Validation Effectiveness Of Develop Maintainability Allocation On Aircraft Mechanical Components

W.M.S. Wan Husain[1], S. Aman Shah[2]

[1]Aerospace Section, Universiti Kuala Lumpur, Malaysian Institute of Aviation Technology, Selangor.
[2]Avionics Section, Universiti Kuala Lumpur, Malaysian Institute of Aviation Technology. Selangor.

Abstact. Maintainability Allocation is a process to identify the allowable maximum task time for each individual component. Consequently, this provides clear pictures to the designers to design and identify potential design improvement within allowable maintenance allocation time limits. During the design process elements such as missteps or misapplications most commonly occur. Here, the authors propose having the maximum target for each individual maintainability component. The main objective of this paper is to present the validation process of developed Maintainability Allocation to potentially eliminate previous problems. The process of validation begins with analysed all the data collected from Service Difficulty Reports (SDR) for selected aircraft. This is to understand the problems from existing aircraft before a new design is proposed through the process of Maintainability Allocation prediction. The validation processes have discovered the importance of utilising historical information such as feedback information. The second area is looking at the element of quantifying the data collected from aircraft feedback information which contains various types of information that could be used for future improvement. Validation process shows that feedback information has helped to identify the critical and sensitive components that need more attention for further improvement. The study shows that the aircraft maintenance related feedback information systems analyses were very useful for deciding maintainability effectiveness; these include planning, organising maintenance and design improvement. There is no doubt that feedback information has the ability to contribute an important role in design activities. The results also show that maintainability is an important measure that can be used as a guideline for managing efforts made for the improvement of aircraft components.

1 Introduction

The validation process presented in this paper is performed based on the author visit to approved Maintenance Repair Organisation (MRO) Company in UK. The BAe 146 – 300 is one of the aircraft available for author to study and visualise the aircraft components. Figure 1 show the pictures of BAe 146 – 300 aircraft landing gear used in this study for the purpose of the validation process.

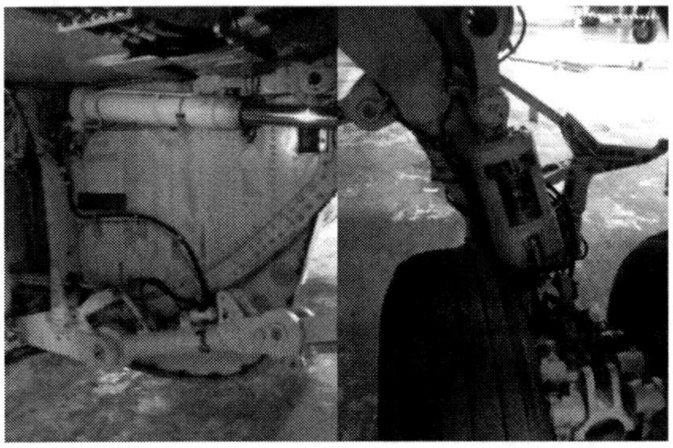

Figure 1 BAe 146 – 300 aircraft landing gear (Source: Author)

2 Service Difficulty Reports (SDR) – Landing Gear

The first process was to understand the SDR historical data related to BAe 146. The historical data and information were analysed for a period of twenty years (1990 – 2009). BAe 146 SDR started from 1992 and increased up to 1998 with various types of part conditions. In total there are 1,688 SDR and the trend is illustrated in Figure 2. The author decided to perform the SDR analysis up to twenty years because the data available for ten years (i.e. 2000 – 2009) are not enough for this research to make appropriate judgements. By extending the SDR analyses period this study expected to be able to understand and undertake necessary action to identify the most suitable aircraft mechanical components. The results show increasing SDR reports in the first nine years (i.e. 1990 – 1998) however, begins to decrease in the following years beginning in 1999. This could be caused by the decrease in demand for this of type aircraft and/or the invention of new types of aircraft that could be more reliable and efficient.

The first part in the analysis study is to identify which part names of the aircraft cause the greatest number of SDR. Figure 3 illustrates the percentage distribution of failed part condition for both BAe 146 – 200 and – 300. Landing gear has contributed the highest percentage of SDR reports with 18.18%, followed by Engine (Turbine/Turboprop) with 15.91%, and Flight Control System with 10.61%. As illustrated in Figure 3 with the total number of SDR for BAe 146 is 132.

Because the landing gear systems show the highest contribution to SDR for BAe 146, the author carried out further analyses within landing gear systems to have a better understanding. There total of 100 SDR reports are collected for BAe 146 Landing gear system. Figure 4 illustrate the percentage distribution for BAe 146

Landing gear system in accordance with part conditions. The most reported part condition include cracked contributed 24.0% SDR reports, failed with 20.0% SDR reports, and corroded with 10.0% of SDR reports. Other types of part condition are faulty with 7.0 SDR reports and followed by damaged and broken which each one contributed 5.0% SDR reports. Table 1 show the detail SDR discrepancy statement which offers and useful to the designer to understand the scenario and condition why the components are failed.

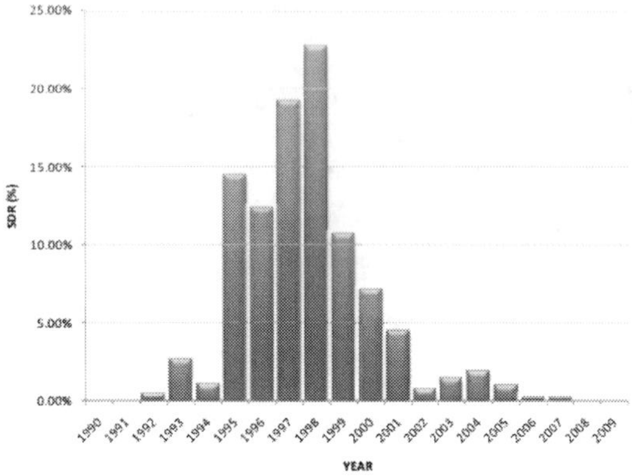

Figure 2 The trend of SDR analyses for all types of BAe 146

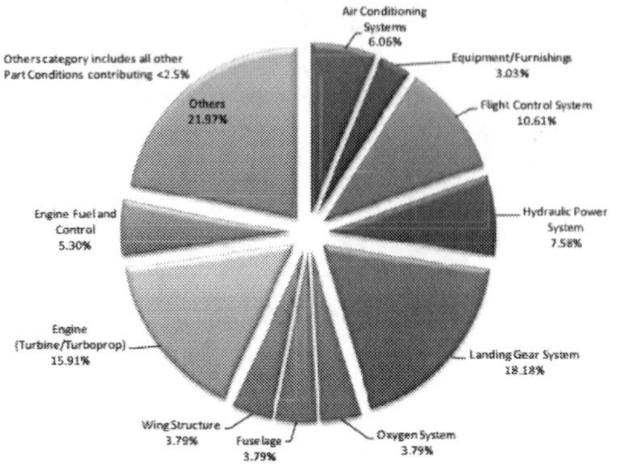

Figure 3 Failed Part Condition percentage distribution for BAe 146 in accordance with JASC Code

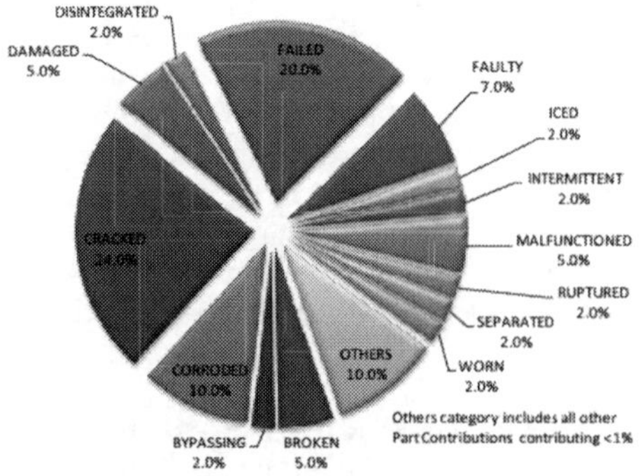

Figure 4 Percentage distributions for BAe 146 in accordance with Part Conditions

Table 1 Landing gear system SDR discrepancy statement in accordance with Part Condition [1]

JASC Code	Part Condition	Part Location	Discrepancy
3260	FAILED	LT MLG	(AUS) LEFT MAIN LANOING GEAR UPIOCK ACTUATOR HOSE ASSEMBLY RUPTURED. HOSE LEAKING FROM END FITTING. LOSS OF GREEN' SYSTEM HYDRAULICS.
3260	FAILED	RT MLG	LANDING GEAR HANOI REQ OVERRIDE ON GEAR RETRACTION AFTER TAKEOFF FROMORD. REMOVED AND REPLACED RIGHT MG WEIGHT OFF HARNESS ASSY.
3230	FAILED	NLG	NOSE LANOING GEAR RETRACTION ACTUATOR FLEXIBLE HYDRAULIC HOSE RUPTURED. LOSS Of GREEN SYSTEM HYDRAULIC FLUID.
3297	FAILED	LT MLG DOOR	UPON GEAR RETRACTION LEFT MAIN GEAR SHOWS RED UNSAFE WITH GEAR HANDLE LT ON. REMOVED AND REPLACED LEFT MLG DOOR UPIOCK SENSOR HARNESS.
3234	FAILED	MLG	LANDING GEAR SELECTOR LEVER WOULD NOT MOVE TO THE UP" POSITION.
3234	FAILED	LANDING GEAR	GEAR FAILED TO EXTEND WHEN SELECTED. WENT TO EMERGENCY AND GEAR EXTENDED. LANDED WITHOUT INCIDENT. SENT REPLACEMENT VALVE AND ACTUATOR TO DEN.

In order to have accurate understanding, the author carried out further analyses to identify which part name mostly affected and received the most SDR reports. Figure 5 illustrate the types of part name contributed the most SDR reports. Pin, actuator, and wheel contributed the most SDR reports which each one of the part names contributed 15.0% for the pin, 8.0% for wheel, and 7.0% for actuator.

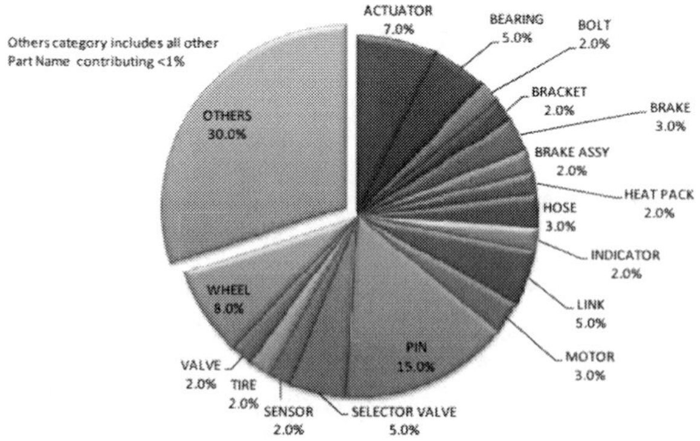

Figure 5 Percentage distribution for BAe 146 in accordance with Part Name

Meanwhile, in Table 2 below shows the detail SDR discrepancy statement for BAe 146 Landing gear system in accordance with the part name. Most of the problems related to SDR is cracked which is also commonly found during Magnetic Partcile Inspection process. Based on this result, the author found that the landing gear system is a part name for the validation process. Table 3 shows the list of information sources of how the validation process was performed.

Table 2 Landing gear system SDR discrepancy statement in accordance with Part Name [1]

JASC Code	Part Name	Part Location	Discrepancy
5230	PIN	NIG	(AUS) NOSE LANDING GEAR RETRACTION JACK ATTAOMNT PIN CRACKED. FOUND DURING MAGNETIC PARTICLE INSPECTION. CRACK LENGTH 19.05 MM (0.75 INCH).
1230	PIN	NLG	(AUS) NOSE LANDING GEAR RETRACTION JACK ATTAOMNT PIN CRACKED FOUND DURING MAGNETIC PARTICLE INSPECTION. CRACK LENGTHS 19.05 MM AND 117 MM.
3230	PIN	NIG	(AUS) NOSE LANDING GEAR RETRACTION JACK ATTACHE NT PIN CRACKED. FOUND DURING MAGNETIC PARTME INSPECTION. CRACK LENGTH 11875 MM <0.625 INCH).
3230	PIN	NIG	(AUS) NOSE LANDINGGEAR RETRACTIONJACK ATTACHMENT PIN CRACKED. FOUND DURING MAGNETIC PARTXIE INSPECTION. CRACK LENGTH 76.2 KM (3 INCHES).
3213	PIN	MjG	(AUS) IT MAIN LANDINGGEAR DIRECTIONALLINK UPPER ATTACHMENT PIN FOUND TO BE MGRATINGFROM THE LUGLOCATEDAT FRAAC 29. LOCKINGPLATE ANO PIN INCORRECTLYFITTED. MAIN LANDUGGEARIS A NEWIYOVERHAUID UNIT AND HAD ONIT JUST BEEN FITTED TO M AIRCRAFT 48CYQES PREVIOUSLY.

Table 3 Summary of information sources for BAe 146 – 300 Landing Gear validation

No	Data and Information	Sources
1	List of Components	Approved Aircraft Maintenance Manual; and Analysed FAA Service Difficulty Reports [2]
2	List of Failure Rate	Non-electronic Part Reliability Data (NPRD) 1995[3]
3	Maintainability Predictions	MIL-HDBK-472[4]; DOD-HDBK-791[5]; CAAIP [6]
4	Maintainability Prediction; Checklist C	Developed approach by author
5	Maintainability Allocation	Improved modules and scores by author
6	Process for Validation	As shown in Integrated Methodology, Chapter 6

3 Maintainability Allocation – BAE 146 – 300 Landing gear

Based on the methodology developed, the authors extended maintainability allocation method was used. The maintainability allocation was the next process to be performed, using extended modules and score values developed by the author. The maintainability allocation was predicted based on Mean Time to Repair (MTTR). The MTTR values can be predicted based on historical data, and/or the decision from top level management. For the purpose of comparison, the author used two different approaches. One of the approaches is to utilise the 1.00 hour as an MTTR value as per advised by industrial experts. The result of the maintainability allocation is shown in Table 4.

The rest of the prediction values were calculated by using the existing methodology developed by Chipchak [7] and by using new module and score improved by the author. There are some listed components requiring more than one module. Component number 4: Brake Assembly for example, required two modules, namely electromechanical equipment and mechanical structures, with mechanisms for which each module offered 3 and 8 score values respectively. Therefore, allocation of the final score values for the brake assembly is by average, i.e. 5.5 hours.

Table 4 Maintainability allocation based on MTTR = 1.00

No	Components	Maintenance Allocation Time MTTR = 1.00 hour
1	Actuator	0.85
2	Antiskid	0.85
3	Brake	1.32
4	Brake Assembly	1.08
5	Connector	0.85
6	Control Unit	0.75
7	Hydraulic	0.85
8	Proximity Sensor	1.13
9	Proximity Switch	1.13
10	Selector Valve	0.85
11	Sensor	0.94
12	Steering	1.13
13	Structure	1.13
14	Strut	1.13
15	Switch	0.85
16	Tire	1.13
17	Uplock Switch	0.85
IS	V-Belt	1.13
19	Warning Light	0.85
20	Warning Systems	1.13
21	Wheel	1.13
22	Wire	0.85
23	Wire Harness	0.85

4 Maintainability Prediction – BAE 146 – Landing Gear

The maintainability full scale prediction was begun with maintainability task time predictions by using MIL-HDBL-472, procedure III. The detailed process of task time predictions has been described in the maintainability prediction chapter. MIL-HDBK-472, procedure III consists of three main elements: Checklist A: Design; Checklist B; Facilities; and Checklist C: Human Factors. Maintainability prediction was performed by using checklists A and B questions. To ensure the score values were accurately identified, additional references were used by the author such as Civil Aircraft Airworthiness Information and Procedure (CAAIP) or also known as CAP 562 and DoD-HDBK-791. The former was used in order to have better illustrations and DoD-HDBK-791 was used to understand the design criteria..

5 Results and Discussions

In this section, the author described the second stage of the maintainability allocation process. In this stage a more accurate MTTR value is used based on the maintainability prediction results to check the allocation method. The second stage utilises the MTTR calculated from the maintainability prediction results above shown in Table 5. In addition, the second stage is utilised for the purpose of comparison as to identify the accuracy of prediction. The result of the maintainability allocation is shown in Table 5.

Table 5 Maintainability Allocation base on MTTR= 0.53

No	Components	Maintenance Allocation Time MTTR = 0.53 hour
1	Actuator	0.45
2	Antiskid	0.45
3	Brake	0.7
4	Brake Assembly	0.57
5	Connector	0.45
6	Control Unit	0.4
7	Hydraulic	0.45
8	Proximity Sensor	0.6
9	Proximity Switch	0.6
10	Selector Valve	0.45
11	Sensor	0.5
12	Steering	0.6
13	Structure	0.6
14	Strut	0.6
15	Switch	0.45
16	Tire	0.6
17	Uplock Switch	0.45
18	V-Belt	0.6
19	Warning Light	0.45
20	Warning Systems	0.6
21	Wheel	0.6
22	Wire	0.45
23	Wire Harness	0.45

The list of components as collected from the SDR analyses. All listed components are received the most reports of SDR. As results, Table 6 show the summary for both MTTR values and both results are illustrated in Figure 6. Both MTTR show almost the same trendlines.

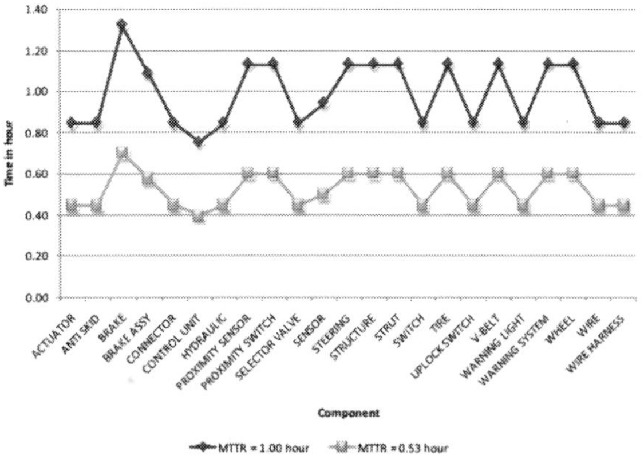

Figure 6 The trend of maintainability allocation times for both MTTR values for BAe 146 – 300 Landing Gear

Table 6 The summary of maintainability allocation for BAe 146 – 300 for both MTTR values.

No	Components	Maintenance Allocation Time in hour	
		MTTR = 1.00 hour	MTTR = 0.53 hour
1	Actuator	0.85	0.45
2	Antiskid	0.85	0.45
3	Brake	1.32	0.7
4	Brake Assembly	1.08	0.57
5	Connector	0.85	0.45
6	Control Unit	0.75	0.4
7	Hydraulic	0.85	0.45
8	Proximity Sensor	1.13	0.6
9	Proximity Switch	1.13	0.6
10	Selector Valve	0.85	0.45
11	Sensor	0.94	0.5
12	Steering	1.13	0.6
13	Structure	1.13	0.6
14	Strut	1.13	0.6
15	Switch	0.85	0.45
16	Tire	1.13	0.6
17	Uplock Switch	0.85	0.45
18	V-Belt	1.13	0.6
19	Warning Light	0.85	0.45
20	Warning Systems	1.13	0.6
21	Wheel	1.13	0.6
22	Wire	0.85	0.45
23	Wire Harness	0.85	0.45

6 Conclusions

The existing maintainability allocation methodology has been improved by the author. The improved methodology is performed by adding more modules and score values related to mechanical aircraft components (i.e. Landing Gear, Mechanical parts). The additional modules and score values have been tested and validated by using several case studies (i.e. Fuel System and Communication Systems).

The additional modules have been assigned to specific score values. The improved methodology, testing and validation results have been described in section 4.3. The linear formula, as shown below, has been chosen as the primary formula to calculate the new score values for new modules and shown in equation (1).

$$y = 1.4364x + 1.769 \quad (1)$$

Furthermore, to ensure the development methodology and improved approach is applicable to industry, the selected approved aircraft has been used for further testing and validation processes. The case studies have been tested and validated in accordance with the approved maintenance manual supplied by a certified aircraft Maintenance and Repair Organisation (MRO). Overall, the results are very successful.

References

1. Federal Aviation Administration (FAA), Service Difficulty Reports, available at: http://av-info.faa.gov/dd_sublevel.asp?Folder=%5CSDRS (accessed April, 13), 2011

2. Federal Aviation Administration, Service Difficulty Reports, available at: http://www.faa.gov/data_research/aviation_data_statistics/data_downloads/ (accessed August, 8), 2010.
3. Mahar, D., Fields, W., Reade, J., Zarubin, P. and McCombie, S., Nonelectronic Parts Reliability Data - 2011, 2011; Reliability Analysis Center (RAC).
4. MIL-HDBK-472, Maintainability Prediction, MIL-HDBK-472, 1966; Department of Defense, Washington, DC.
5. DOD-HDBK-791, Maintainability Design Techniques, 1988; DOD-HDBK-791, Department of Defense, USA.
6. Civil Aviation Authority, Civil Aircraft Airworthiness Information and Procedures (CAAIP), 2008; CAP 562, CAA, UK.
7. Chipchak, J. S., "A Practical Method of Maintainability Allocation", Aerospace and Electronic Systems, IEEE Transactions. 1971, vol. AES-7, no. 4, pp. 585-589.

Numerical Analysis of Flow and Heat Transfer Characteristics of CO_2 at Vapour and Supercritical Phases in Micro-Channels

N.T. Rao[1,a], A.N. Oumer[1], U.K. Jamaludin[1] and A.Y. Adam[1]

[1]*Faculty of Mechanical Engineering, Universiti Malaysia Pahang, 26600 Pekan, Pahang, Malaysia.*

Abstract. Supercritical carbon dioxide (CO_2) has special thermal properties with better heat transfer and flow characteristics. Due to this reason, supercritical CO_2 is being used recently in air-condition and refrigeration systems to replace non environmental friendly refrigerants. Even though many researches have been done, there are not many literatures for heat transfer and flow characteristics of supercritical CO_2. Therefore, the main purpose of this study is to develop flow and heat transfer CFD models on two different phases; vapour and supercritical of CO_2 to investigate the heat transfer characteristics and pressure drop in micro-channels. CO_2 is considered to be in different phases with different flow pressures but at same temperature. For the simulation, the CO_2 flow was assumed to be turbulent, non-isothermal and Newtonian. The numerical results for both phases are compared. From the numerical analysis, for both vapour and supercritical phases, the heat energy from CO_2 gas transferred to water to attain thermal equilibrium. The temperature of CO_2 at vapour phase decreased 1.78% compared to supercritical phase, which decreased for 0.56% from the inlet temperature. There was a drastic increase of 72% for average Nu when the phase changed from vapour to supercritical. The average Nu decreased rapidly about 41% after total pressure of 9.0 MPa. Pressure drop (ΔP) increased together with Reynolds number (*Re*) for vapour and supercritical phases. When the phase changed from vapour to supercritical, ΔP was increased about 26%. The results obtained from this study can provide information for further investigations on supercritical CO_2.

1 Introduction

Carbon dioxide (CO_2) gas, which has zero ozone depleting potential (ODP) and zero effective global warming potential (GWP) was reintroduced as an environmental friendly gas, and used as working fluid in refrigerators and air conditioning systems. Moreover, there are a few advantages of using CO_2, such as it is non-toxic and safe to humans, abundant and non-combustible. As the supercritical CO_2 reaches near to its critical point, the physical properties shows extremely rapid variations with a change in temperature and pressure, which makes it as the most important characteristics [1]. The current refrigerants, such as chlorofluorocarbons (CFCs) and hydro chlorofluorocarbons (HCFCs), which are being used in air conditioning and refrigeration systems, have high ozone depletion and effective global warming potentials. Hence, CO_2 can be the ideal replacement for these non-environmental friendly refrigerants with the thermo-fluid properties and appropriate design, at supercritical phase.

Furthermore, the density and dynamic viscosity of CO_2 at supercritical phase undergo a significant, which is almost vertical within a very narrow temperature range while the enthalpy undergoes a sharp increase near the critical point [2]. As the temperature of supercritical CO_2 was increased in near-critical region, the pressure drop and heat transfer coefficient are increased too [3]. At larger Reynolds number, heat transfer coefficient increased as the heat transfer rate increased [4]. Therefore, to understand the underlying physics, an appropriate implementation of fluid flow and heat transfer correlation is the systems are needed. Even though a few researches had performed studies and investigation on cooling heat transfer and flow of supercritical CO_2 in micro-channels, it still could not solve this issue. The thermophysical properties and variables of supercritical CO_2 are obtained from NIST Refrigerants Database REFPROP [5]. The density (ρ), thermal conductivity (λ), viscosity (μ) and specific heat (C_p) of supercritical CO_2 vary with different pressures along increased temperature [6].

Many researchers have studied the flow and heat transfer characteristics of supercritical CO_2 by using numerical and experimental methods. The geometry often used for mathematical model is the circular tube-in-tube heat exchanger, where supercritical CO_2 flow in the inner tube and water flow in the annular space [7]. Most numerical analysis are done by using Renormalization Group (RNG) k-ε and Low-Reynolds number (LRN) k-ε models as the turbulence model with ANSYS FLUENT CFD codes [8-11]. Besides, the flow domains are divided into two; CO_2 and water for cooling process [12, 13].

[a] Corresponding author: thiwaan@hotmail.com

The main purpose of this study is to develop flow and heat transfer mathematical models for CO_2 at vapour and supercritical phases in micro-channels, and to compare both phases to prove which phase is the best for heat transfer and flow characteristics of CO_2. This study is expected to provide better knowledge on reducing the ozone depletion effect and global warming potentials by replacing the existing non-environmental friendly refrigerant with supercritical CO_2.

2 Mathematical formulations

2.1 Governing equations

In this study, the flow field is assumed to be incompressible, steady, non-isothermal and two-dimensional (2D) flow. Therefore, the governing equations for the continuity, momentum and energy can be expressed as [14]:

Continuity equation for incompressible flow:

$$\frac{\partial u}{\partial x}+\frac{\partial v}{\partial y}+\frac{\partial w}{\partial z}=0 \quad (1)$$

$$\vec{\nabla}\cdot\vec{V}=0 \quad (2)$$

where u, v and w are velocity components in the x, y and z directions, respectively, and $\vec{\nabla}$ is divergence operator.

Momentum equation:

$$\rho\frac{D\vec{V}}{Dt}=\rho\left[\frac{\partial\vec{V}}{\partial t}+\left(\vec{V}\cdot\vec{\nabla}\right)\vec{V}\right]=-\vec{\nabla}P+\rho\vec{g}+\mu\nabla^2\vec{V} \quad (3)$$

where ρ is density of the fluid (kg/m^3), V is velocity vector of the fluid (m/s), t is time (seconds), g is gravitational acceleration (m/s^2) and μ in the fluid viscosity (kg/m.s)

Conservation of energy equation:

$$\rho\frac{DE}{Dt}=\frac{\partial(u\sigma_{xx})}{\partial x}+\frac{\partial(v\sigma_{yy})}{\partial y}+\frac{\partial(w\sigma_{zz})}{\partial z}+\frac{\partial(u\tau_{yx})}{\partial y}+\frac{\partial(v\tau_{zx})}{\partial z}+\frac{\partial(v\tau_{xy})}{\partial x}+\frac{\partial(v\tau_{zy})}{\partial z}+\frac{\partial(w\tau_{xz})}{\partial x}+\frac{\partial(w\tau_{yz})}{\partial y}-\frac{\partial q_x}{\partial x}-\frac{\partial q_y}{\partial y}-\frac{\partial q_z}{\partial z} \quad (4)$$

where σ is stress and τ is shear stress.

2.2 Pressure drop equations

Pressure drop (ΔP) takes place due to pressure loss in a system due to friction in the system. The ΔP equation represents the relationship between friction factor, length to diameter of tube ratio, and density and velocity of the fluid [14]. The general equation of ΔP is:

$$\Delta P=f\cdot\frac{L}{D}\cdot\frac{\rho V^2}{2} \quad (5)$$

where ΔP is pressure drop (MPa): f is friction factor, L is length the tube (m) and D is diameter of the tube (m).

Meanwhile, Reynolds number also is calculated with following formula:

$$\mathrm{Re}=\frac{\rho VD}{\mu} \quad (6)$$

where Re is Reynolds number and ρ is density of the fluid (kg/m^3). The friction factor for equation (1) will be calculated with Reynolds number obtained from equation (6). For laminar flow, the friction factor was:

$$f=\frac{64}{\mathrm{Re}} \quad (7)$$

Meanwhile, for turbulent flow, Colebrook equation was used to calculate friction factor:

$$\frac{1}{\sqrt{f}}=-2.0\log\left(\frac{\varepsilon D}{3.7}+\frac{2.51}{\mathrm{Re}\sqrt{f}}\right) \quad (8)$$

where ε is pipe roughness.

2.2 Heat transfer rate equations

Convective heat transfer rate ($\dot{Q}_{conv}$) is the amount of heat transferred per unit time. The rate of convection heat transfer was expressed by Newton's Law of Cooling [15]:

$$\dot{Q}_{conv}=hA_S\left(T_S-T_\infty\right) \quad (9)$$

where $\dot{Q}_{conv}$ is convection heat transfer rate (W), h is convection heat transfer coefficient (W/m^2 °C), A_S is heat transfer surface area (m^2) , T_S is temperature of surface (°C) and T_∞ is temperature of the fluid sufficiently far from the surface (°C).

Besides, Nusselt number (Nu) was calculated using the following equation:

$$Nu=\frac{hL_C}{\lambda} \quad (10)$$

where λ is thermal conductivity (W/m.K) and L_C is characteristics length (m).

3 Numerical analysis

3.1 Computational domain

The flow domain of CO_2 and water was designed in Design Modeler software. It was designed according to the tube-in-tube heat exchanger concept. The diameter of the inner tube was measured from the smallest tube available in market and the cooling length was the minimum value obtained from previous studies. The model was designed in 2D, as shown in Figure 1. The pipe-in-pipe flow domain was designed with the dimensions stated in Table 1. The 2D model is the half of the whole tube-in-tube heat exchanger.

Table 1. Dimensions of flow domain.

Parameter	m
Inner tube height (H_i)	0.0025
Outer tube height (H_o)	0.01
Cooling length (L)	1

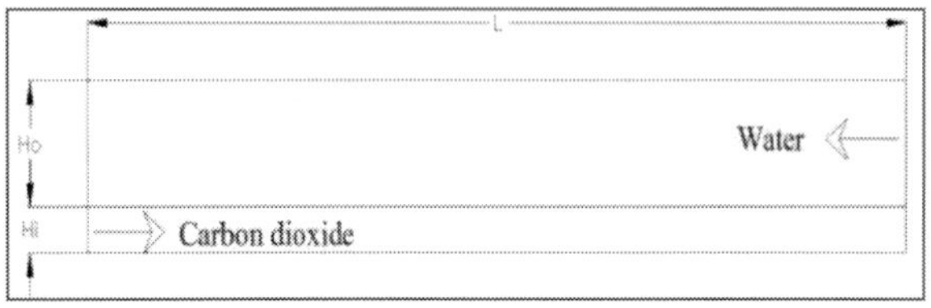

Figure 1. 2D pipe-in-pipe heat exchange design in Design Modeler.

As mentioned above, the fluids used for the numerical analysis are water as the cooling fluid and CO_2 as the main fluid. Since the thermophysical properties of water are at room temperature, the details are available in ANSYS FLUENT material database. However, the thermophysical properties of CO_2 available in ANSYS FLUENT database are only at room temperature. Therefore, the thermophysical process of CO_2 at vapor and supercritical were obtained from National Institute of Standards and Technology (NIST) web book [6]. The density, thermal conductivity, viscosity and specific heat of supercritical CO_2 vary with different pressures along the increased temperature. The thermophysical properties of CO_2 in both vapour and supercritical phases from 3.5 MPa to 10.0 MPa pressures respectively with 313 K inlet temperature are obtained, and are tabulated in Table 2.

3.2 Meshing

The mesh size is setup to be fine. Meanwhile, the mesh size at the interface between inner tube and annular part was setup to be finer to obtain accurate heat transfer data between CO_2 and water. Moreover, mesh independent test was conducted to make sure that the numerical analysis results were same for all mesh sizes at 3.5 MPa pressure only. The inlet and outlet faces for CO_2 and water, interfaces, symmetries and walls were renamed. For both vapour and supercritical phases' models, the number of meshes used is same, which are 15,000 elements.

Table 2. Thermophysical properties of CO_2 [6].

Pressure, P (MPa)	Density, ρ (kg/m^3)	Thermal conductivity, λ (W/m.K)	Viscosity, μ (Pa.s)	Specific heat, C_p (J/g.K)
3.5	70.871	0.020268	1.6298×10^{-5}	1.1735
4.0	83.758	0.020901	1.6494×10^{-5}	1.2527
5.0	113.05	0.022622	1.7029×10^{-5}	1.4697
6.0	149.26	0.025392	1.7860×10^{-5}	1.8335
7.5	231.53	0.034950	2.0452×10^{-5}	3.3553
8.0	277.90	0.042376	2.2345×10^{-5}	4.9501
9.0	485.50	0.070590	3.4806×10^{-5}	12.833
10.0	628.61	0.071410	4.7825×10^{-5}	5.6575

3.3 Boundary conditions

For this study, the CO_2 was assumed as incompressible flow for both vapour and supercritical phases. The CO_2 and water flow in the inner tube and outer tube respectively. The parameters, which were used in the numerical analysis, are tabulate in Table 3.

Table 3. Boundary conditions.

Boundaries	Variables	Vapour phase	Supercritical phase
CO_2 inlet	Pressure-inlet with constant temperature	3.5 MPa	7.5 MPa
CO_2 outlet	Constant	313 K	313 K
Water inlet	Velocity-inlet with constant temperature	10 m/s	10 m/s

4 Results and discussions

The heat transfer and pressure drop are obtained from dynamic pressure, velocity, temperature, Nusselt number and Reynolds number data from the numerical analysis. These data are analysed and compared. The convective heat transfer coefficients, Nusselt number, Reynolds number and pressure drop are recorded, and calculated.

4.1 Heat transfer

The temperature of CO_2 for both vapour and supercritical phases decreased along the 1 m tube, as shown in Figure 2. The temperature of CO_2 is constant up to 0.2 m and started reducing subsequently. However, the temperature of CO_2 at vapour phase decreased 1.78% compared to supercritical phase, which decreased for 0.56% from the inlet temperature. This is because the tube length was not enough for CO_2 at supercritical phase to transfer heat energy to attain thermal equilibrium at high total pressure. For vapour phase, due to low total pressure, it attained thermal equilibrium faster.

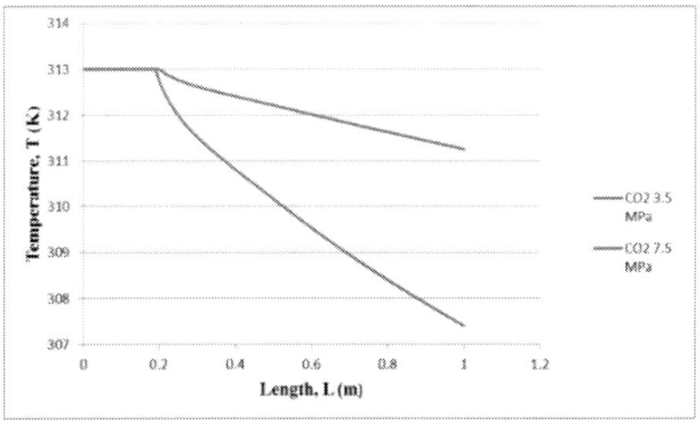

Figure 2. Temperature versus length of carbon dioxide and water for both vapour and supercritical phases.

In Figure 3, the average Nusselt number (Nu) for CO_2 for both vapour and supercritical phases were varying according to total pressures. At vapour phases, Nu of CO_2 increased slowly as the total pressure increased from 3.5 MPa to 6.0 MPa. Meanwhile, at supercritical phase, average Nu also increased together with the total pressure from 7.5 MPa to 9.0 MPa. There was a drastic increase of 72% for average Nu when the phase changed from vapour to supercritical. However, after 9.0 MPa total pressure, average Nu decreased rapidly for 41% because the tube length is not enough to transfer the thermal energy from CO_2 to water.

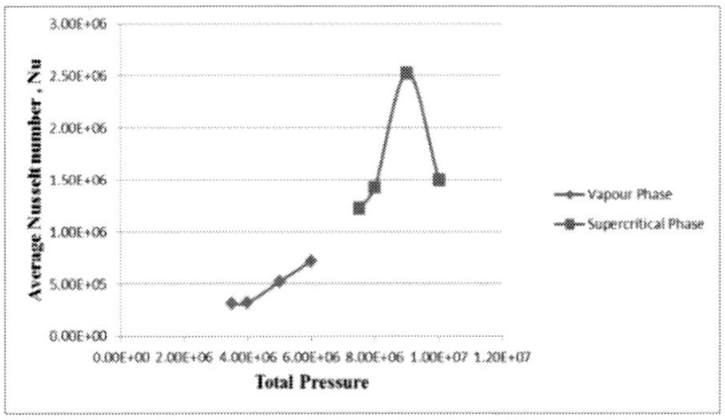

Figure 3. Average Nusselt number (Nu) versus total pressure for both vapour and supercritical pressure phases.

Besides, as shown in Figure 4, the temperature changes of CO_2 along the tube at supercritical pressures were decreasing. The temperature of CO_2 decreased linearly for all supercritical pressure. At total pressure of 7.5 MPa, 8.0 MPa, 9.0 MPa and 10.0 MPa, the temperature decreased 0.56%, 0.4%, 0.44% and 0.53%, respectively from the inlet temperature. At 7.5 MPa, which is closer to critical pressure, the amount of heat loss is highest. Increase on pressure made the heat transfer coefficient decreased closely at pseudocritical temperature [16]. As the total pressure increased from critical pressure, the heat transfer rate reduced. Besides, the length of the tube is not enough for supercritical CO_2 to heat up the water.

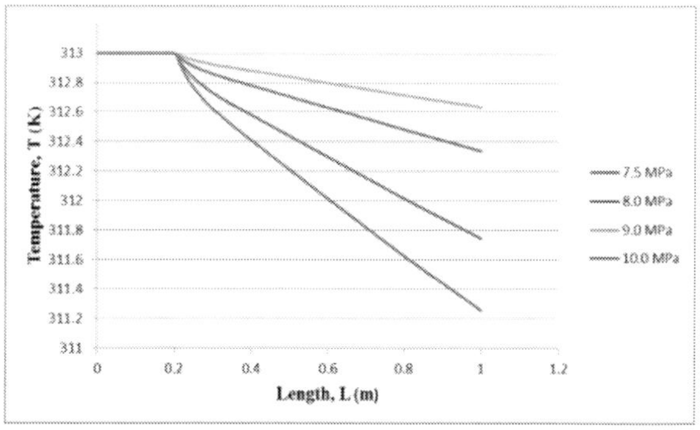

Figure 4. Temperature versus length of CO_2 at supercritical pressures.

From the numerical analysis, for both vapour and supercritical phases, the heat energy from CO_2 gas transferred to water to attain thermal equilibrium, as shown is Figures 2 and 4. The thermal energy from CO_2, which was at 313 K temperature, was transferred to water, which was at 300 K. It was proved that as the total pressure increased, the heat transfer coefficient decreased [3], which is same with the situation in this study. Moreover, for supercritical phase, the tube length was not enough for the heat energy from CO_2 to be transferred fully to water. The tube length should be increased to obtained better results on heat transfer.

4.2 Pressure drop

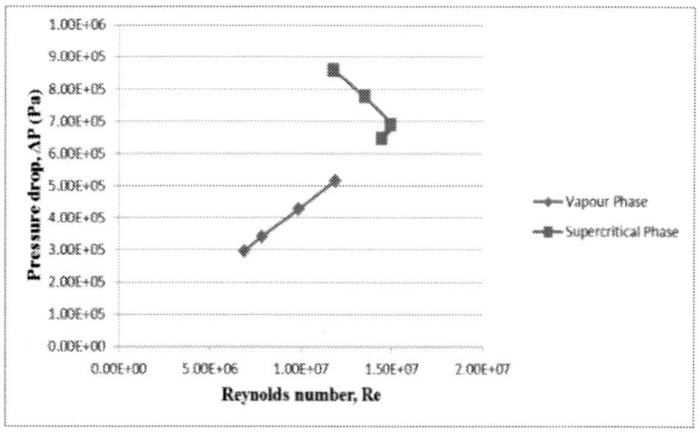

Figure 5. Pressure drop versus Reynolds number of CO_2 at both vapour and supercritical phases.

The dynamic pressure and velocity were analyzed, and compared for both vapour and supercritical phases along the 1 m cooling length. For all total pressures, the dynamic pressure and velocity were directly proportional to each other. As the total pressure of CO_2 flowing in the tube decreased due to heat loss to water, the velocity of CO_2 increased along the tube.

Moreover, the pressure drop, ΔP of CO_2 was calculated from the dynamic pressure data from the numerical analysis. The ΔP increased linearly as the total pressure of increased from 3.5 MPa to 10.0 MPa. As shown in Figure 5, the ΔP increased together with Re for vapour phase, as the pressure is approaching the critical point. When the phase changed from vapour to supercritical, ΔP increased about 26%. For supercritical phase, the ΔP increased along with Re up to 8.0 MPa. Even though the ΔP increased after 8.0 MPa, the value of Re keep decreasing up to 10.0 MPa. Hence, the pressure drop of CO_2 at both vapour and supercritical phases do increased as it was proven. The decrease in Re is due to drastic increase in density and viscosity of CO_2 at total pressure from 8 MPa to 10 MPa. C.H. Son and S.J. Park [17] investigated that variation in the density of CO_2 resulted to decrease in ΔP with increased inlet pressure of gas cooler at supercritical phase, which is against the results in Figure 5.

5 Conclusion

In this study, the mathematical models to investigate the flow and heat transfer characteristics of CO_2 at vapour and supercritical phases in micro-channels were developed. The flow domain of CO_2 and water was designed in 2D by using Design Modeler software, according to the pipe-in-pipe heat exchanger concept, with length of 1000 mm. The thermophysical properties of CO_2 in both vapour and supercritical phases from 3.5 MPa to10.0 MPa pressures respectively with 313 K inlet temperature were obtained from NIST. The temperature of water was 300 K at velocity of 10 m/s.

From the numerical analysis, for both vapour and supercritical phases, the heat energy from CO_2 gas transferred to water to attain thermal equilibrium. The temperature of CO_2 at vapour phase decreased 1.78% compared to supercritical phase, which decreased for 0.56% from the inlet temperature. According to average Nusselt number (*Nu*), the heat transfer rate of CO_2 increased as the total pressure is increased. It was proved that as the total pressure increased, the heat transfer coefficient decreased, which is same with the situation in this study. There was a drastic increase of 72% for average Nu when the phase changed from vapour to supercritical. However, after 9.0 MPa total pressure, average Nu decreased rapidly for 41% because the tube length is not enough to transfer the thermal energy from CO_2 to water.

Pressure drop (ΔP) increased together with Reynolds number (*Re*) for vapour and supercritical phases. When the phase changed from vapour to supercritical, ΔP increased about 26%. There is a rapid heat transfer between CO_2 and water at low pressure compared to high pressure. The results obtained from this study can provide information for further investigations on supercritical CO_2

Acknowledgement

The authors would like to be thanking Universiti Malaysia Pahang for providing laboratory facilities and Ministry of Higher Education Malaysia for financial assistance under FRGS project no. RDU130132.

References

1. Bolaji, B.O. and Z. Huan, *Ozone depletion and global warming: Case for the use of natural refrigerant - a review.* Renewable and Sustainable Energy Reviews`, 2013. **18**: p. 49-54.
2. Liao, S.M. and T.S. Zhao, *An experimental investigation of convection heat transfer to supercritical carbon dioxide in miniature tubes.* International Journal of Heat and Mass Transfer, 2002. **45**: p. 5025-5034.
3. Huai, X.L., S. Koyama, and T.S. Zhao, *An experimental study of flow and heat transfer of supercritical carbon dioxide in multi-port channels under cooling conditions.* Chemical Engineering Science, 2005. **60**: p. 3337-3345.
4. Hsieh, J.C., et al., *Experimental study of heat transfer for supercritical carbon dioxide with upward flow in vertical tube.* International Journal of Advanced Science and Technology, 2014. **7**: p. 66-71.
5. Cao, X.L., Z.H. Rao, and S.M. Liao, *Laminar convective heat transfer of supercritical CO2 in horizontal miniature circular and triangular tubes.* Applied Thermal Engineering, 2011. **31**: p. 2374-2384.
6. Lemmon, E.W., M.O. McLinden, and D.G. Friend. *"Thermophysical Properties of Fluid Systems" in NIST Chemistry WebBook, NIST Standard Reference Database Number 69, Eds. P.J. Linstrom and W.G. Mallard.* 2015; Available from: http://webbook.nist.gov.
7. Pilta, S.S., E.A. Groll, and S. Ramadhyani, *New correlation to predict the heat transfer coefficient during in-tube cooling of turbulent supercritical CO2.* International Journal of Refrigeration, 2002. **25**: p. 887-895.
8. Xu, J., et al., *Turbulent convective heat transfer of CO2 in a helical tube at near-critical pressure.* International Journal of Heat and Mass Transfer, 2015. **80**: p. 748-758.
9. Mohseni, M. and M. Bazargan, *Modification of low Reynolds number k-e turbulence models for applications in supercritical fluid flows.* International Journal of Thermal Sciences, 2012. **51**: p. 51-62.
10. Lisboa, P.F., et al., *Computational-fluid-dynamics study of a Kenics static mixer as a heat exchanger for supercritical carbon dioxide.* Journal of Supercritical Fluids, 2010. **55**: p. 107-155.

11. Yadav, A.K., M.R. Gopal, and S. Bhattacharyya, *Transient analysis of subcritical/supercritical carbon dioxide based natural circulation loops with end heat exchangers: Numerical studies.* International Journal of Heat and Mass Transfer, 2014. **79**: p. 24-33.
12. Yadav, A.K., M.R. Gopal, and S. Bhattacharyya, *CFD analysis of a CO2 based natural circulation loop with end heat exchangers.* Applied Thermal Engineering, 2012. **36**: p. 288-295.
13. Jiang, P.X., et al., *Experimental and numerical study of convection heat transfer of CO2 at super-critical pressures during cooling in small vertical tube.* International Journal of Heat and Mass Transfer, 2009. **52**: p. 4748-4756.
14. Cengel, Y.A. and J.M. Cimbala, *Fluid Mechanics: Fundamentals and Applications*. 2013, New York: McGraw-Hill.
15. Cengel, Y.A. and A.J. Ghajar, *Heat and Mass Transfer: Fundamentals and Applications*. 4th Edition ed. 2011, New York: McGraw-Hill Higher Education.

 DOI: 10.1051/matecconf/20167400012

Ultrasonic Guided Wave Method For Crack Detection In Buried Plastic Pipe

Wan Sofian Wan Hamat[1,a], Mohd Fairusham Ghazali[1] and Gigih Priyandoko[1]

[1]Faculty of Mechanical Engineering, Universiti Malaysia Pahang, 26600 Pekan, Pahang, Malaysia

Abstract. Plastic pipe are widely used in many fields for the fluid or gaseous product conveyance but basic components of a plastic material made it very sensitive to damage, which requires techniques for detecting damage reliable and efficient. Ultrasonic guided wave is a sensitive method based on propagation of low-frequency excitation in solid structures for damage detection. Ultrasonic guided wave method are performed to investigate the effect of crack to the frequency signal using Fast Fourier Transform (FFT) analysis. This paper researched to determine performance of ultrasonic guided wave method in order to detect crack in buried pipeline. It was found that for an uncrack pipe, FFT analysis shows one peak which is the operating frequency by the piezoelectric actuator itself while the FFT analysis for single cracked pipe shows two peak which is the operating frequency by the piezoelectric actuator itself and the resultant frequency from the crack. For multi cracked pipe, the frequency signal shows more than two peak depend the number of crack. The results presented here may facilitate improvements in the accuracy and precision of pipeline crack detection.

1 Introduction

Recent developments in pipelines systems have heightened the need for study about the pipe defect detection using ultrasonic guided wave method. In recent years, several studies have focused on defect detection in pipeline systems because it is very important used extensively all over the world to transport and distribute natural gas, crude oil, water, and other easy-flowing products. Indeed, in the process of continuous transportation, gas pipeline leakage cannot be avoided because of the corrosion of pipe walls [1], third-party interference [2], aging of the pipes [3], and so on. Therefore, leakage detection and localization is one of the paramount concerns of pipeline operators and researchers all over the world.

Many researchers have argued that Non-destructive testing (NDT) is a wide group of analysis techniques used in science and industry to evaluate the properties of a material, component or system without causing damage. NDT is a highly valuable technique that can save both money and time in product evaluation, troubleshooting, and research. Common NDT methods include ultrasonic, magnetic-particle, liquid penetrant, radiographic, remote visual inspection (RVI), eddy-current testing, and low coherence interferometry. According to K Haller (2007), damage can be monitored in many ways [4].

The choice of composite materials is no longer restricted to specific applications. In other fields, using these new materials as innovation tool and performance rising of their products such civil engineering. It is important first to have the best possible knowledge about these materials in somehow to monitor any changes in these properties and consequently their influence on the structure behavior [5, 6].

Plastic pipe is a good material because high resistance, light, cheap and longer life span. Therefore, plastic pipes usually are used to transport gas, water and chemicals. In general, plastic pipes are widely used over metal or concrete pipes. Due to the material being immune to water corrosion and highly resistant to fouling, it is also being considered as a replacement for stainless steel in safety-critical applications in nuclear power stations [7].

According to Carvalho et al. (2008), researchers have spent an ample amount of time, investigations and researched on non-destructive test techniques. As much as during the setting up phase during their useful life, these techniques are used as a method to asses engineering structures and system [8]. Crack inspection is a valuable non-destructive test (NDT) method in many industries. Crack inspection method is useful, which often exhibit complex geometries, are exposed to high thermal and mechanical loads and are expected to satisfy

[a] Corresponding author: w_sofian45@yahoo.com

strict safety requirements. Lately, there is a need to expand a powerful crack defect detection methods [9].

Early detection of damage is not necessarily visible from the external face and even when it is the case of visual inspection does not realize its case state. In order of reliable and efficient use, these new materials, the development of a measurement system that can answer the following two questions of primary importance is required whether it could be measured directly in NDT and reliable mechanical properties of a composite structure. Is it possible to detect damage as quick as possible in order to monitor, evaluate and repair if necessary the structure [10]. Piezoelectric materials constitute another important research topic with respect to Structural Health Monitoring (SHM) applications. Piezoelectric transducers are often used to excite and receive ultrasonic guided waves due to their low costs and an easy integration into existing structures. This research area offers a wide variety of contributions [11, 12].

Although there were many researches about ultrasonic guided wave method to detect pipe damage, few of them focused on result obtained only for this method. So it is necessary to do deep research on the design and develop an experimental test rig for pipe defect detection to run easily, faster, efficient and cost-effective.

2 Experimental Set Up

Design of Experiment

The aim of this paper is to determine the performance of ultrasonic guided wave method in detecting cracks in the pipeline system using an experiment. An experimental test rig design and develop for buried plastic pipe defect detection to prevent the leakage in the piping system. All the experiments were conducted on plastic pipe specimens with piezoelectric wafer actuator (PSI-5A4E) and accelerometer mounted on the surface of each pipe. Piezoelectric wafer actuator as a transmitter and the accelerometer as a receiver. A National Instruments Data Acquisition system (NI USB-4431) was used to drive the transmitter and to record data and waveform at the receiver. The specifications of the experimental specimens are given in Table 1.

Table 1. Specification of the experimental specimens

Exp. No.	Buried Plastic Pipe Specimens				Sensors & Trans-ducers
	Dimensions [mm]			Specificati-ons	Dist.[c] [mm]
	Length	OD[a]	WT[b]		
1.1	600	40	2	Difference distance between the transmitter and the receiver.	300
1.2	600	40	2		500
1.3	1300	40	2		600
1.4	1300	40	2		900
1.5	1300	40	2		1200
2.0	600	40	2	1 lateral saw cut.	500
3.0	600	40	2	2 lateral saw cut.	500

([a]OD: outside diameter; [b]WT: wall thickness; [c]Dist: distance between the transmitter and the receiver).

Experimental Procedure

This experiment carried out a few times to make sure a more accurate result is obtained. First, the experiment will be carried out on an uncrack specimen. Data will be collected and analyzed. Next, the experiment will be carried out on a specimen with a single crack. Lastly, the experiment will be carried out on a specimen with multi cracks. All the data were collected and analyzed. From the data collected, we can compare the result of all the three specimen.

Piezoelectric wafer actuator (PSI-5A4E) was used to excite the specimen. By placing the specimen on the foams and using a significantly smaller size of wafer actuator, this is the best compromise method for diagnosing defects. Nonlinear effects and noise effects from the attachment of the mechanical shaker or rigid supports can be prevented by using this method. A comparison was made between this method and the former method that using a mechanical shaker for the excitation. Although the maximum excitation level from the piezoelectric actuator is relatively lower than with the mechanical shaker, results from the latter method also showed a good signature of the sidebands effect against the defects [10, 13].

The experiment for uncrack specimen begin with distance between piezoelectric wafer actuator as a transmitter and accelerometer as a receiver is 300 mm. The frequencies applied to the specimen are 500 Hz, 3000 Hz, 6000 Hz, and 9000 Hz. After the data is collected and recorded, the experiment is repeated with different value of distance between transmitter and receiver. This process is repeated with 600 mm, 900 mm and 1200 mm distance between transmitter and receiver. The relationship between the distance and amplitude for frequency signal can be identify from results obtained for this experiments.

The experiment for single crack specimen and multi cracks specimen was performed on a pipe with length 600 mm which the piezoelectric wafer actuator and accelerometer are placed on the surface of the pipe where the distance between each other is 500 mm. Figure 1 illustrates the specimen for (a) uncrack specimen, (b) single crack specimen and (c) multi cracks specimen. Lateral saw cut made as a crack shown at Figure 1(b) and Figure 1(c). The crack made at the middle of this distance (250 mm) for single crack specimen; see Figure 1(b). For multi cracks specimen, two crack made at distance 165 mm from piezoelectric stack actuator and 165 mm from accelerometer; see Figure 1(c).

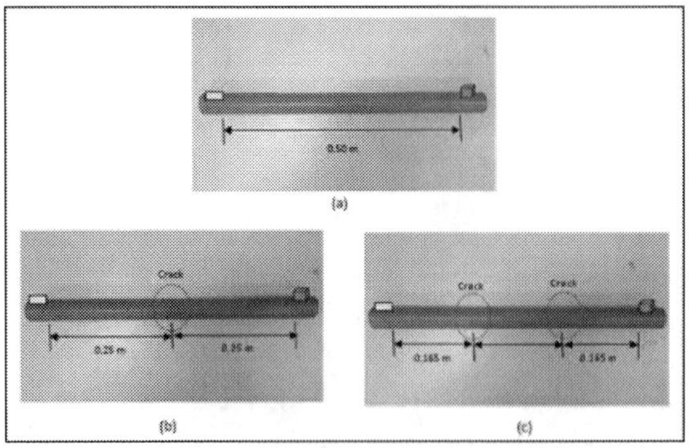

Figure 1. Piezoelectric wafer actuator and accelerometer placed on the surface of three pipe specimens. (a) Uncrack specimen; (b) Single crack specimen; (c) Multi cracks specimen.

The data collected will undergo analysis using the Dasylab and Matlab software. The data will also undergo amplification and filtering module. The amplification module is to amplify the data collected and filtering of the data is extremely vital in obtaining an accurate time base response graph. The plastic pipe will be attached the sensors and then it will be connected to Data Acquisition (DAQ) system. The Data Acquisition system will be synchronized to the Dasylab software using the NI-MAX software. The system will be configured to analyze the voltage as the sensor medium. The accelerometer will be controlled using the Dasylab software. After collecting the raw data, analyzing data were done by using Matlab software. The experiments setup for this study are illustrated in Figure 2(a) and Figure 2(b).

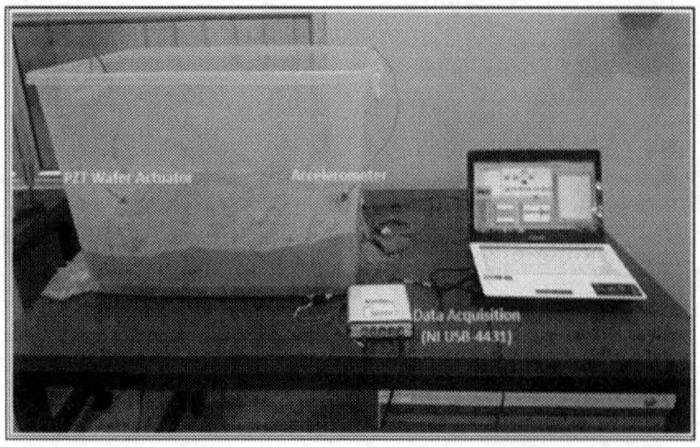

Figure 2(a). Full view for the experimental setup

Figure 2(b). Buried pipe specimen

3 Results and Discussion

The relationship between distance and amplitude

The experiment results will be shown in the form of graph for Fast Fourier Transform (FFT) analysis which is produced by using Matlab software. Table 2 shows the comparison of amplitude value for each of the spectrum at different distance between sensor and actuator for uncrack specimen. The value of the amplitude decreases as the distance between actuator and sensor increases. This can be deduced that the waves tend to dissipate more when the distance traveled by the wave is increased. So, the waves will become weaker when the distance is increased and this explained the reason for the value of amplitude decreases when the distance increases. The amplitude value is inversely proportional with distance between sensor and actuator.

Table 2. Comparison of amplitude value for each spectrum at different distance and frequency applied

Frequency (Hz)	Amplitude			
	300 mm	**600 mm**	**900 mm**	**1200 mm**
3000	0.13267	0.06291	0.05385	0.02730
6000	0.09017	0.04210	0.02834	0.01482
9000	0.08762	0.00753	0.00376	0.00135

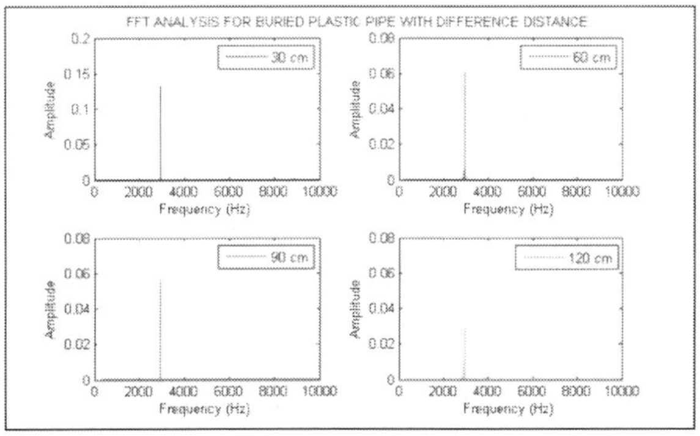

Figure 3. Results obtained for 3000 Hz frequency applied to the 300 mm, 600 mm, 900 mm 1200 mm of uncrack specimen.

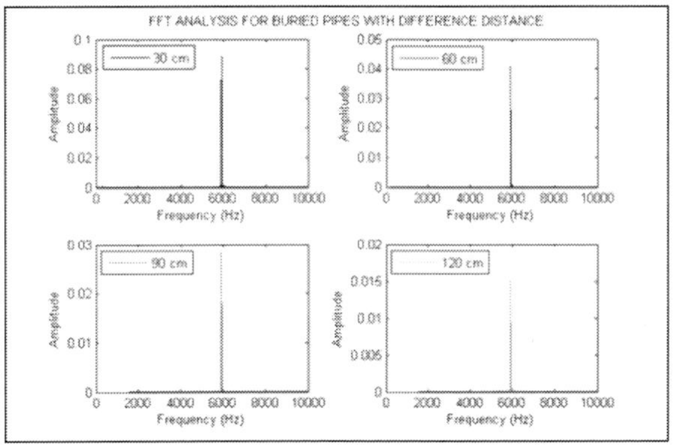

Figure 4. Results obtained for 6000 Hz frequency applied to the 300 mm, 600 mm, 900 mm 1200 mm of uncrack specimen.

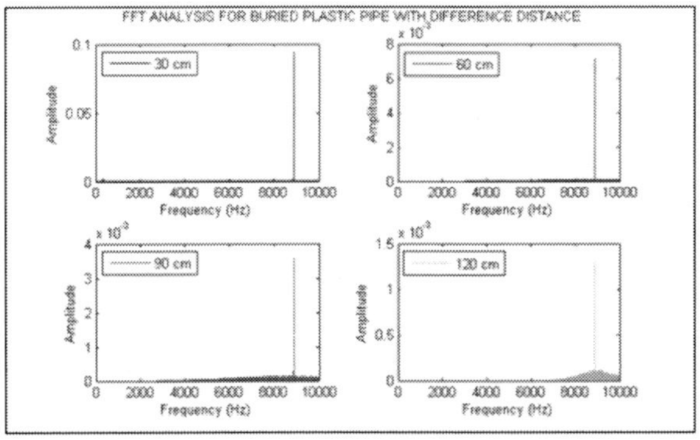

Figure 5. Results obtained for 9000 Hz frequency applied to the 300 mm, 600 mm, 900 mm 1200 mm of uncrack specimen.

From the FFT analysis graph in Figure 3, Figure 4, and Figure 5, it could be clearly seen that the amplitude of the frequency response signal decreases as the distance between the piezoelectric wafer actuator and accelerometer increases. For 3000 Hz, the frequency signal response amplitude is 0.13267 at 300 mm followed by 0.06291 at 600 mm, 0.05385 at 900 mm and 0.02730 at 1200 mm; see Figure 3. For 6000 Hz, the frequency signal response amplitude is 0.09017 at 300 mm followed by 0.04210 at 600 mm, 0.02834 at 900 mm and 0.01482 at 1200 mm; see Figure 4. For 9000 Hz, , the frequency signal response amplitude is 0.08762 at 300 mm followed by 0.00753 at 600 mm, 0.00376 at 900 mm and 0.00135 at 1200 mm; see Figure 5. All the frequencies show that that the frequency signal amplitude are the highest at the shortest distance. This shows that it is capable of travelling at a short distance.

In general, the frequency signal amplitude decreases in all samples due to the attenuation inside the testing medium as the distance from the source is increased too. It is clear that the attenuation increases with increasing distance. The transmitted signal is attenuated by the transmission medium which is the time required by the signal to reach the sensor point. This transmission makes some turbulence in the transmission medium and such turbulence may remain exist over this period of time. When transmission frequency increases, the transmission medium will contain effects of consecutive signals and constructive and destructive interferences would take place. Accordingly, the accelerometer would receive transmitted signals at a higher frequency and if the accelerometer needs longer time to distinguish the received signals from each other, then higher frequency may cause destructive interference at the sensor, which in turn affects the final recorded signal amplitude.

Comparing the results for uncrack and crack specimen.

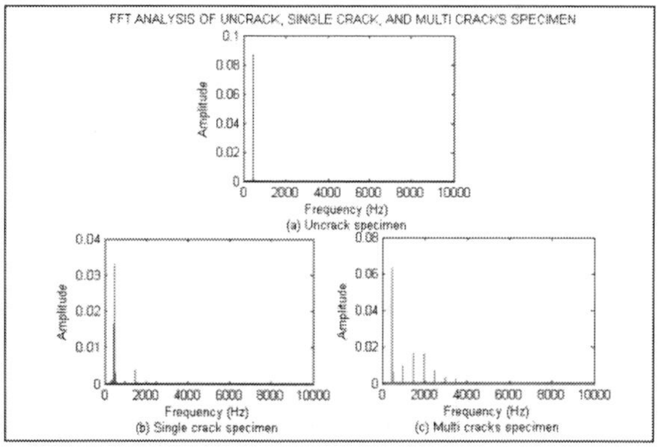

Figure 6. FFT analysis graph for (a) uncrack specimen, (b) single crack specimen, (c) multi cracks specimen.

Figure 6(a) shows the FFT analysis graph of the results obtained for an uncrack specimen. From the results obtained, the FFT analysis graph shows one peak which is the operating frequency by the piezoelectric actuator itself. Figure 6(b) shows the FFT analysis graph for single crack specimen. The graph FFT analysis shows two peak which is the operating frequency by the piezoelectric wafer actuator itself and the resultant frequency from the crack. Figure 6(c) shows FFT analysis graph for multi cracks specimen. From the frequency signal obtained, the data shows more than two peaks which one peak is the operating frequency by the piezoelectric actuator itself and other peaks is the resultant frequency from the cracks.

Accelerometer has its own area of influence characterized by a sensing radius and the corresponding circle. Damage within the area of influence of the sensor is highly sensitive as it creates higher disturbances than the damage outside the area of influence. An exciting vibration at creates appreciable stress in the material affecting interfaces at all scales of damage. An ultrasound probe wave traveling through these interfaces is modulated due to vibration-induced changes in the interface condition and, as a result, additional spectral components appear at the frequencies. These new frequency components indicate that a flaw or crack is present.

4 Conclusions

This study set out to determine the performance of ultrasonic guided wave method in detecting cracks in the pipeline system using an experiment. An experimental test rig design and develop for buried plastic pipe defect detection to prevent the leakage in the piping system. The findings of this study suggest that detecting the defects in buried pipe using ultrasonic guided wave method is faster, efficient and cost-effective. From the data collected and analyzed, uncrack and crack specimen can be distinguished more obviously in a short time. The FFT analysis graph clearly shows only one peak for the uncrack specimen and two peak for the single crack specimen. There are more than two peak for the multi cracks specimen. Further investigation and experimentation for defect detection in pipe is strongly recommended. A number of possible future studies using the same experimental set up are apparent. It would be interesting to assess the effects of ultrasonic guided wave detecting the defect in pipe.

Acknowledgements

The work presented in this paper was supported by Universiti Malaysia Pahang (UMP) through the RDU Grant research project (RDU130111) lead by Dr. Mohd Fairusham Bin Ghazali. The authors are grateful to the Universiti Malaysia Pahang and Education Sponsorship Division, Ministry Of Education Malaysia for supporting the present study.

References

1. Edalati K, R.N., Kermani A, Seidi M, Movafeghi A, *The use of radiography for thickness measurement and corrosion monitoring in pipes.* International Journal of Pressure Vessels and Piping, 2006. **83**: p. 736-741.
2. Liang, W., et al., *Assessing and classifying risk of pipeline third-party interference based on fault tree and SOM.* Engineering Applications of Artificial Intelligence, 2012. **25**(3): p. 594-608.
3. Vargas-Arista, B., J.M. Hallen, and A. Albiter, *Effect of artificial aging on the microstructure of weldment on API 5L X-52 steel pipe.* Materials Characterization, 2007. **58**(8–9): p. 721-729.
4. Haller, K., *Nonlinear Acoustics Applied to NonDestructive Testing.* 1st ed. 2007, Department of Mechanical Engineering, School of Engineering Blekinge Institute of Technology, Sweden: Blekinge Institute of Technology.
5. Lakhdar, M., et al., *Damages Detection in a Composite Structure by Vibration Analysis.* Energy Procedia, 2013. **36**: p. 888-897.
6. Xu, H.L.H., *Damage Detection for Structural Health Monitoring Using Ultrasonic Guided Waves*, in *School of Energy, Power and Mechanical Engineering.* 2013, North China Electric Power University: Beijing, China.
7. Assunção E, C.L., Hagglund F, Troughton M, & Spicer M, *Advanced NDT techniques for plastic pipeline inspection.* 2013, European Union under the Framework-7 TestPEP: EWF, Oeiras, Portugal.
8. Zenzinger, G., et al., *Crack Detection Using EddyTherm.* AIP Conference Proceedings, 2005. **760**(1): p. 1646-1653.
9. Carvalho, A.A., et al., *Reliability of non-destructive test techniques in the inspection of pipelines used in the oil industry.* International Journal of Pressure Vessels and Piping, 2008. **85**(11): p. 745-751.
10. Jenal, R., *Fatigue crack detection using nonlinear acoustic – Analysis of vibro–acoustic modulations.* , in *Department of Mechanical Engineering.* 2010, The University of Sheffield, United Kingdom.
11. C. Willberg, S.D., J.M. Vivar-Perez, Z.A.B. Ahmad, *Simulation Methods for Guided Wave-Based Structural Health Monitoring: A Review. Applied Mechanics Reviews.* January 2015. **Vol. 67/010803-1**.
12. Ahmed, M.N., *A Study of Guided Ultrasonic Wave Propagation Characteristics in Thin Aluminum Plate for Damage Detection*, in *Graduate Faculty.* 2014, The University of Toledo: United States.
13. Shi Yan, J.Q., Nai-Zhi Zhao, Yang Cheng and Sheng-Wenjun Qi, *Multiple Crack Detection of Pipes Using PZT-based Guided Waves*, in *School of Civil Engineering.* 2014, Shenyang Jianzhu University, Shenyang Liaoning 110168, China.

Experimental Investigation of an Automobile Air-Conditioning System using Integrated Brushless Direct Current Motor Rotary Compressor

M.F. Sukri[1,4,a], M.N. Musa[2], M.Y. Senawi[1] and H. Nasution[3]

[1]Faculty of Mechanical Engineering, Universiti Teknologi Malaysia,81310 Johor Bahru, Johor, Malaysia
[2]UTM-Ocean Thermal Energy Center (UTM-OTEC), Universiti Teknologi Malaysia, Wilayah Persekutuan Kuala Lumpur, Jalan Semarak, 54100 Kuala Lumpur, Malaysia
[3]Automotive Development Centre, Universiti Teknologi Malaysia, 81310 Johor Bahru, Johor, Malaysia
[4]Efficient Energy & Thermal Management Research Group, Universiti Teknikal Malaysia Melaka, Hang Tuah Jaya, 76100 Durian Tunggal, Melaka

Abstract. The present study presents an experimental investigation on the effect of condenser air inlet temperature and dimensionless parameter of X on the performance of automobile air-conditioning (AAC) system using integrated brushless direct current motor-rotary compressor and electronic expansion valve. The other components of AAC system are from original component of AAC system used for medium size passenger car. The experimental results showed that the increment of the condenser air inlet temperature and X caused an increase in condensing temperature, cooling capacity and compressor work, while decreasing the coefficient of performance (COP). Meanwhile, the evaporating temperature increase with the increment of condenser air inlet temperature, but decrease with decrement of X. In general, AAC system have to work at higher value of X in order to produce more cooling capacity, thereby increment in compressor work also occurs due to energy balance. However, at higher value of X, the COP of the system dropped due to dominant increase in compressor power, as opposed to a rise in cooling capacity. Due to this reason, the best operation of this compressor occurs at $X = 4.96$ for constant T_5 (35°C), or at $T_5 = 30$°C for constant X (4.96).

1 Introduction

An automotive air-conditioner is one of the most important accessories in typical land vehicles, which provides thermal comfort to the driver and passenger. Thermal comfort especially for drivers is crucial since road safety also improves with the comfort of the driver and a pleasant environment reduces driver fatigue [1]. However, automotive air-conditioning (AAC) system consumes second largest of energy after power train [2], thereby becoming a critical issue because of increasing concerns on energy cost, thermal comfort, greenhouse gas emissions, and vehicle performance.

Since the compressor consumes the highest amount of energy (65%) in a typical AAC system [3], a small improvement in compressor efficiency will leads to a larger reduction in power consumption. As a result, energy is consumed efficiently, and a significant improvement in the system performance is achieved. In the other aspect, reduction in energy consumption with more efficient compressor can decreases greenhouse gas emissions. This low-carbon technology is highly relevant in the real world and an important research interest related to energy-efficient systems. Many countries, particularly developed ones, have policies to reduce greenhouse gas emissions, especially at the transportation industry. Due to these reasons, numerous studies have been conducted to improve the AAC system performance using alternative compressor of conventional belt-driven fixed capacity compressor.

Tian et al. [4] highlighted that the compressor capacity control technique by using variable capacity compressor (VCC) has been used popularly in the AAC system due to its continuous operation, better thermal comfort inside car, and lower fuel consumption. Due to these reasons, Park et al. [3], Nasution and Wan Hassan [5], Qi et al. [6], Tian et al. [7] and Alkan and Hosoz [8] conducted few studies concerning AAC system equipped with VCC, particularly in the aspect of performance as compared to AAC system with fixed capacity compressor. In general, they found that AAC system with VCC yields better performance, less in exergy destruction and good thermal comfort sensation in the rapid changing environment with less energy usage/improved fuel consumption.

In different aspect, Ooi [9] optimized rotary compressor design and theoretically reduces mechanical losses by approximately 50% and improves the COP of the compressor by more than 14%. Later, a new gas compression mechanism called the "Revolving Vane" (RV) compressor was invented and theoretically shown to achieve improved mechanical, volumetric, and

[a] Corresponding author: mohdfirdaus@utem.edu.my

compression efficiencies compared with other existing compressor designs [10]. Further theoretical studies on this compressor have shown that friction, leakage, and discharge loss were substantially reduced in the new design, leading to significant improvements in efficiencies[10-12]. An experimental study at a shaft speed of 600–1200 rpm on the prototype of a RV compressor with air as the working fluid proved the reliability of the mechanism, achieving pressure ratios higher than 8:1 and more than 30 h of operation without any failure [13]. Further design improvements of the RV compressor with a fixed-vane significantly reduced the frictional losses by 18% to 41% and mechanical efficiency by 96% compared with the older RV compressor design [14].

Recently, Ekren et al. [15] demonstrated that direct current (DC) compressors have the potential to be used in energy-efficient refrigeration systems because these compressors do not require additional components such as a power inverter that an alternative current compressor would require. By utilizing this compressor, any problem related to the interaction of the refrigerant cycle components and the rapidly-changing operating conditions of the vehicle (speed, revolutions per minute, and so on) is solved. All types of ground vehicles are equipped with direct current batteries; thus, the authors agreed that efficient DC compressors are also among the best options to be used in an energy-efficient AAC system in future. One way of achieving an efficient DC refrigerant compressor is by integrating compressor pump with the latest brushless DC (BLDC) motor technology. The BLDC motors are microprocessor-controlled to keep the stator current in phase with the permanent magnets of the rotor, requiring less current for the same torque and therefore resulting in greater efficiency. Brushless DC motor has an excellent mechanical efficiency ranging from 85 to 90%. In addition, the BLDC motor-rotary compressor can be designed for low voltage (particularly 12V) up to high voltage of up to 300V. Hence, it is a better choice for powering an electric refrigerant compressor.

So far, however, research of AAC system coupled with BLDC motor-rotary compressor is very limited in the publication, at least to the author's knowledge. Therefore, it is the intent of this paper to investigate the performance of BLDC motor-rotary compressor available in the market, integrated with typical vapor compression refrigeration, AAC system.

2 Experimental set up

2.1 Design of experiment

The present study aims to investigate the correlation between the condenser air inlet temperature, condenser air face velocity, compressor speed and evaporator air volumetric flowrate to the performance of AAC system equipped with BLDC motor-rotary compressor and electronic expansion valve. The actual components of heat exchangers (evaporator and condenser), internal and external fans of 1.6L Proton Wira Aaeroback passenger car are used. A SIERRA06-0982Y3 high voltage BLDC variable speed hermetic compressor and electronic expansion valve (EEV) from Danfoss, model ETS 6 – 14 are used. The selection of these two components are based on vehicle compartment cooling load model proposed by Sukri et al. [16] for typical thermophysical data of 1.6L Proton Wira Aaeroback passenger car. Details of the selection process can be found in Sukri et al. [17]. The experimental test rig is mounted on an actual 1.6L Proton Wira Aeroback. Figure 1 shows the schematic diagram of the experimental test rig. Table 1 shows the model and accuracies of all test instruments.

Table 1. The accuracies of measuring instruments

Instruments	Measurements	Accuracies
T-Type Thermocouple with Pico TC-08 Data Logger	Refrigerant and air temperatures	±0.01°C
AKS 32 Danfoss Pressure Transducer	Refrigerant pressure	±0.01 bar
923 Fluke Air Velocity Meter	Air velocity	±0.01 m/s
Platon NGX Glass Flowmeter	Refrigerant mass flow rate	±1 g/s

The design of condenser ductwork is an open tunnel, rectangular shape and made of 20 mm thick insulation form (type insoform) with an overall length of 2870 mm. In order to provide a uniform air flow through the face of the condenser, air mixer and honey comb type air flow straightener were installed upstream of the condenser coil. The air sampling tube is constructed based on BS 5141-1:1975, Specification for Air Heating and Cooling Coils – Part 1: Method of Testing for Rating of Cooling Coils [18]. The frequency converter is used to control the condenser air face velocity. Meanwhile, the original evaporator ductwork of 1.6L Proton Wira Aeroback passenger car as in Figure 2 is utilized as experimental evaporator ductwork. The evaporator air volumetric flowrate is controlled by controlling the percentage of energy input supplied to the internal fan.

The condenser air face velocity profile is measured in the condenser ductwork with cross-sectional area of 500 mm x 500 mm at point 5 as in Figure 1. It is measured according to SAE International Surface Vehicle Standard 2008, Procedure for Measuring System COP of a Mobile Air Conditioning System on a Test Bench [19]. To obtain an averaged velocity, the readings have been taken by using a 923 Fluke hot-wire air velocity meter, placed successively at nine measuring points (at the same measuring plane). For each point, data is recorded for 5 minutes with averaged 10 seconds of sampling time. In total, 30 data is recorded for each point. The evaporator air volumetric flowrate is measured at point 7 as in Figure 1. At this point, the cross-sectional area of the original ductwork is 168 mm x 196 mm. The methodology adopted in measuring the averaged condenser air face velocity profile is also used in measuring averaged evaporator air velocity. The averaged evaporator air volumetric flowrate, V_7 is then calculated by multiplying

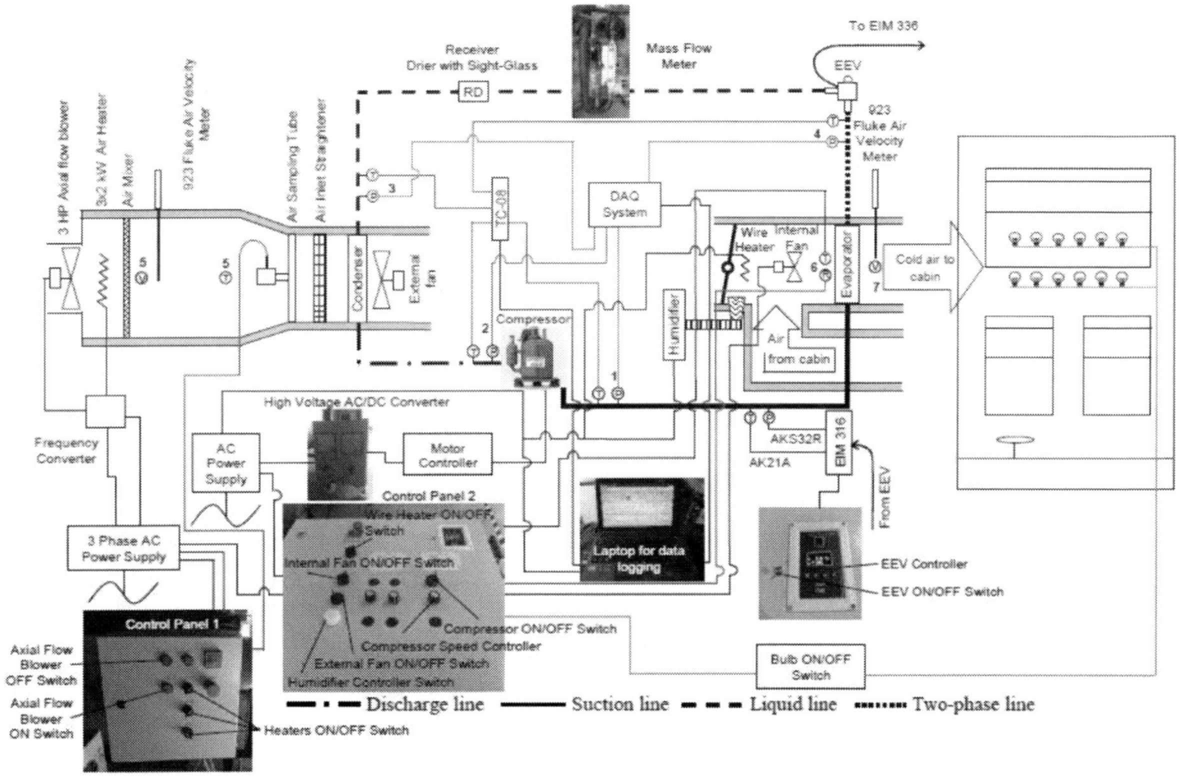

Figure 1. Schematic diagram of the experimental setup

the averaged evaporator air velocity, v_7 at point 7 with its cross-sectional area, thus:

$$V_7 = 0.032928 v_7 \quad (\mathrm{m^3/s}) \qquad (1)$$

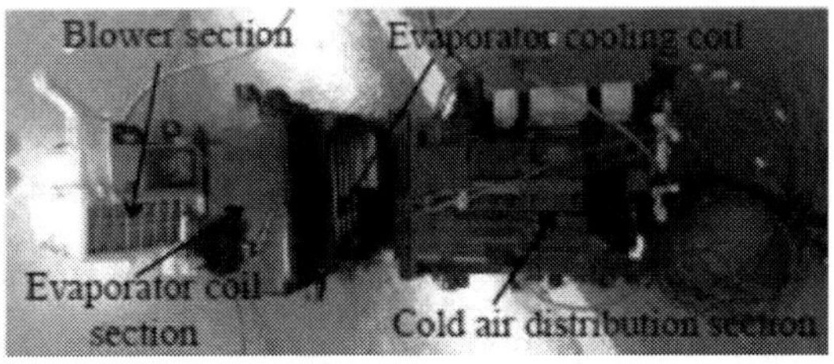

Figure 2. Evaporator ductwork

Since point sampling for each set of experimental work is time consuming, relationships are obtained between averaged condenser air face velocity and frequency of the inverter motor drive, and evaporator air volumetric flowrate with percentage of energy input. Simple linear regression are created using Microsoft Excel yields correlations as in Figure 3(a) and 3(b) for condenser and evaporator ductworks respectively. The subsequent experiments are conducted by setting an input frequency of the inverter motor drive and percentage of energy input, and the averaged air face velocity and averaged evaporator air volumetric flowrate are obtained from equations 2 and 3 respectively.

$$v_5 = 0.0896 f - 0.17 \,(\mathrm{m/s}) \qquad (2)$$

$$V_7 = 5.9449 c + 7.6091 \,(\mathrm{m^3/h}) \qquad (3)$$

2.2 Experimental Procedure

The experimental work is conducted in steady state condition. The tests is conducted by varying the value of dimensionless parameter, $X\left(V_7 N_c^{\,2} / v_5^{\,3}\right)$ and condenser air inlet temperature, T_5 at constant value of evaporator air inlet temperature, T_6 (30°C) and evaporator air inlet humidity ratio, ϕ_6 (50%). To stabilize the air-conditioning system, the system is run at least 20 minutes prior to each test before measured output variables are recorded as proposed by SAE International Surface Vehicle Standard 2008, Procedure for Measuring System COP of a Mobile Air Conditioning System on a Test Bench [19]. The output variables of refrigerant mass flow rate, $\dot{m}_r$, temperature and pressure at point 1 to 4 as in Figure 1 are recorded for ten minutes as proposed by SAE International Surface Vehicle Standard 2008, Procedure for Measuring System COP of a Mobile Air Conditioning System on a Test Bench [19]. The sampling time is 30 seconds. In total 20 data points is collected for each setting.

DOI: 10.1051/matecconf/20167400013

At each experimental setting, X is maintained at 4.96, 11.74 and 22.92 by controlling the value of N_c, v_5 and V_7 as in Table 2. At each value of X, T_5 is kept at 35°C. Then, the experimental is repeated at T_5 of 30, 35 and 40°C, but at constant value of X (11.74). During the entire experimental investigation, the ϕ_6 is maintained at 50% by controlling the humidifier manually. The EEV is also manually controlled at 100% of opening degree, and the input direct current voltage to the compressor is kept at 300 VDC.

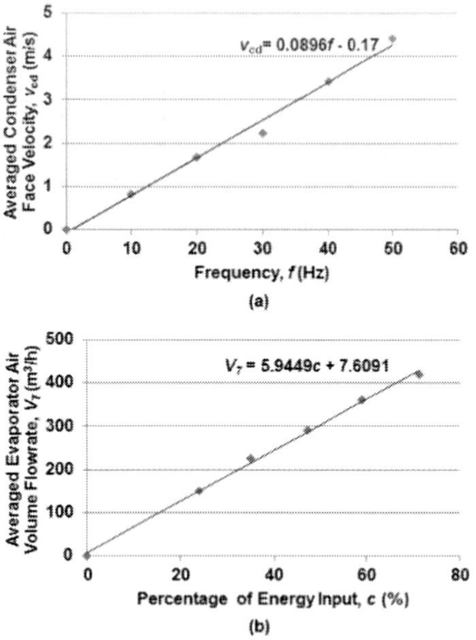

Figure 3. Liner regression between (a) averaged condenser air face velocity and frequency and (b) averaged evaporator air volumetric flowrate and percentage of energy input.

Table 2. The setting input variables during experimental work.

$X = \frac{V_7 N_c^2}{v_5^3}$	N_c (rpm)	V_7 (m³/h)	v_5 (m/s)
4.96	2300	190	2.5
11.74	3700	300	3.0
22.92	6500	450	4.0

The cooling capacity, Q_e and compressor work, W_c are calculated based on refrigerant enthalpy, h difference between inlet and outlet of the evaporator and compressor respectively, where:

$$Q_e = \dot{m}_r (h_1 - h_4) \tag{4}$$

$$W_c = \dot{m}_r (h_2 - h_1) \tag{5}$$

with , = 1, 2, 3, 4.

The coefficient of performance, COP is then determined as:

$$COP = Q_e / W_c \tag{6}$$

3 Results and discussion

The condensing temperature is higher at increased condenser air inlet temperature (Figure 4(a)) and X (Figure 4(b)). When the condenser air inlet temperature and the value of X are increased, the temperature of refrigerant discharged from the compressor will be higher cause an increased in condensing temperature. Simultaneously, the higher condensing temperature cause an increased in compressor work, as shown in Figure 5(a) and 5(b).

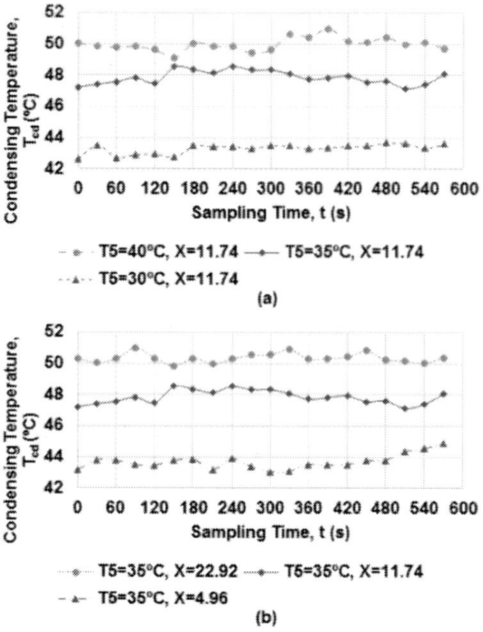

Figure 4. Condensing temperature over time (a) at constant X = 11.74 and (b) at constant condenser air inlet temperature, T_5 = 35°C.

Figure 6(a) and 6(b) shows the effect of cooling capacity at different condenser air inlet temperature and X respectively. The decreases in the cooling capacity at lower condenser air inlet temperature as in Figure 6(a) could be due to decrease in the enthalpy difference in the evaporator caused by drop in evaporating temperature as in Figure 7(a). Therefore at constant value of X, the refrigeration capacity decreased with the decrement of evaporating temperature. However, Figure 6(b) and 7(b) show that at constant condenser air inlet temperature, the cooling capacity decreased with the increment of evaporating temperature. It can be clearly seen that for energy balance, at lower operating conditions of X (lower of compressor speed, evaporator air volumetric flowrate and condenser air face velocity), the system produces lower cooling capacity at high evaporating temperature.

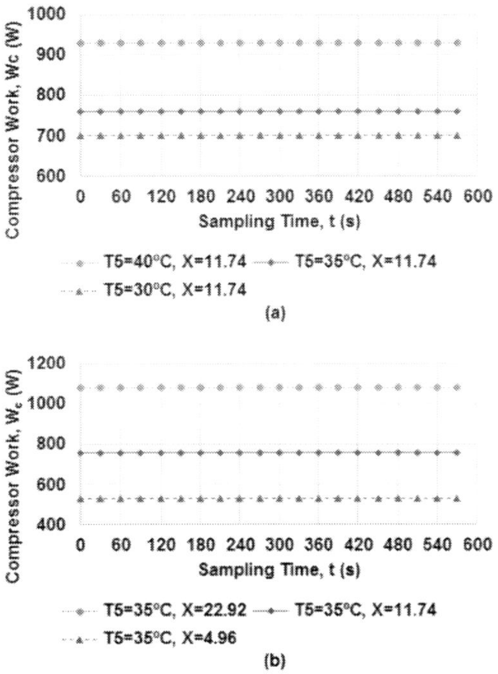

Figure 5. Compressor work over time (a) at constant X = 11.74 and (b) at different constant condenser air inlet temperature, T_5 = 35°C.

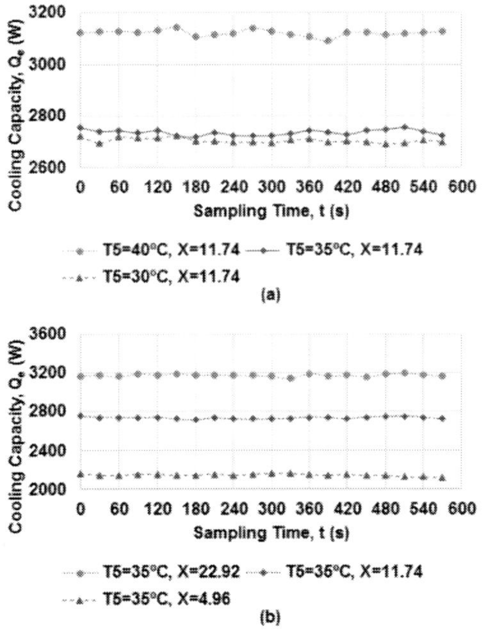

Figure 6. Cooling capacity over time (a) at constant X = 11.74 and (b) at different constant condenser air inlet temperature, T_5 = 35°C.

Figure 8(a) and 8(b) show the relationship between the *COP* and condenser air inlet temperature at X of 11.74, and *COP* with X at condenser air inlet temperature of 35°C respectively. It shows that the COP decrement is due to the increase in the condenser air inlet temperature and also with the increase of X. According to Eq. (6), the *COP* is influenced by the cooling capacity and compressor work. An increment in condenser air inlet temperature (Figure 5(a) and 6(a)) as well as in X (Figure 5(b) and 6(b)) causing increase in both, cooling capacity and compressor work. Hence, a dominant increase in compressor power, as opposed to a rise in cooling capacity lead to lower *COP* in both cases.

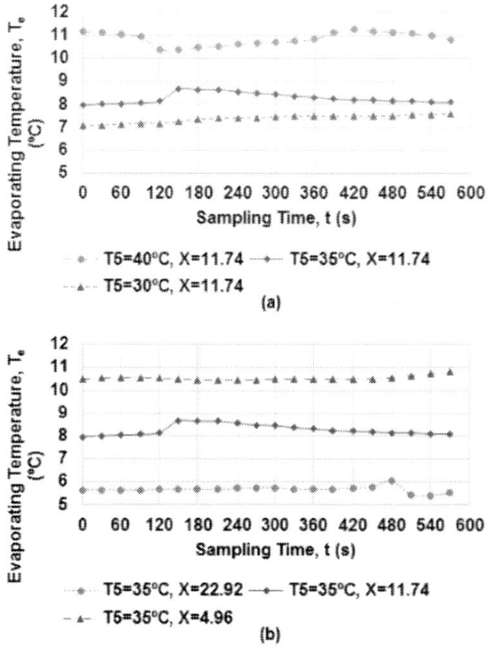

Figure 7. Evaporating temperature over time (a) at constant X = 11.74 and (b) at constant condenser air inlet temperature, T_5 = 35°C.

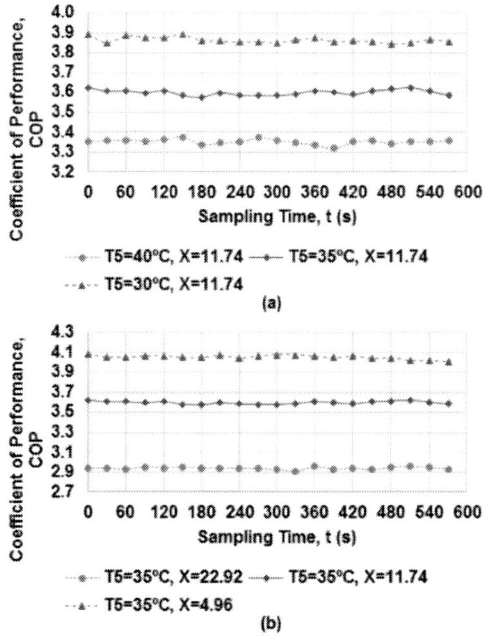

Figure 8. Coefficient of performance over time (a) at constant X = 11.74 and (b) at constant condenser air inlet temperature, T_5 = 35°C.

4 Conclusions

The present study provides experimental result of an automobile air-conditioning system equipped with integrated BLDC motor-rotary compressor and EEV. The results of this study showed that the increment of the condenser air inlet temperature and value of X respectively caused an increase in condensing temperature, cooling capacity and compressor work, while decreasing the COP. However, the evaporating temperature increases with the increment of condenser air inlet temperature, but decrease with decrement of X. In general, AAC system have to work at higher value of X, particularly at high compressor speed in order to produce more cooling capacity, thereby increment in compressor work also occurs due to energy balance. However, at higher value of X, the COP of the system dropped due to dominant increase in compressor power, as opposed to a rise in cooling capacity. As a result, the best operation of this compressor occurs at $X = 4.96$ for constant T_5 (35°C) or at $T_5 = 30$°C for constant X (4.96).

Acknowledgements

The authors would like to be obliged to Universiti Teknologi Malaysia, Universiti Teknikal Malaysia Melaka and Ministry of Education Malaysia for providing laboratory facilities and all financial assistances, especially under project no. Q.J130000.2424.00G41.

References

1. M. Konz, *A generic simulation of energy consumption of automobile air conditioning systems* (M Tech Eng Thesis. Nelson Mandela Metropolitan University, 2007).
2. M.A. Roscher, W. Leidholdt, J. Trepte, Int J Elec Power **37**(1), 126-130 (2012).
3. S.K. Park, H. Kim, H. Ahn, H.S. Park, *SAE Technical Paper*, 2006-01-0164 (2006).
4. C. Tian, H. Xu, L. Zhang, X. Li, Appl Therm Eng **29**, 2824–2831 (2009).
5. H. Nasution, M.N. Wan Hassan, Clean Techn Environ Policy **8**, 105–111 (2006).
6. Z. Qi, J. Chen, Z. Chen, W. Hu, B. He, Appl Therm Eng. **27**, 927–933 (2007).
7. C. Tian, H. Xu, X. Li, Y. Liao Y. Appl Therm Eng. **27**, 1868–1875 (2007).
8. A. Alkan, M. Hosoz. Int J Refrig **33**(3), 487-495 (2010).
9. K.T. Ooi, Appl Therm Eng. **25**(5-6), 813–829 (2005).
10. Y.L. Teh, K.T. Ooi, Int J of Refrig **32**(5), 1092-1102 (2009).
11. Y.L. Teh, K.T. Ooi, Int J Refrig **32**(5), 1103-1111 (2009).
12. Y.L. Teh, K.T. Ooi, Int J Refrig **32**, 945-920 (2009)
13. Y.L. Teh, K.T. Ooi, Appl Therm Eng **29**(14-15), 3235–3245 (2009).
14. K.M. Tan, K.T. Ooi, Int J Refrig **34**(8), 1980-1988 (2011).
15. O. Ekren, S. Celik, B. Noble, R. Krauss, Int J Refrig **36**(3), 745-757 (2013).
16. M.F. Sukri, M.N. Musa, M.Y. Senawi, H. Nasution, *7th International Meeting on Advances In Thermofluids (7th IMAT)* (Swiss-Garden Hotel and Residences, Kuala Lumpur Malaysia, Nov 2014).
17. M.F. Sukri, M.N. Musa, K. Sumeru, S. Sodiya, JAMT **8**(2), 39-49 (2014).
18. BS 5141-1:1975, *Specification for Air Heating and Cooling Coils – Part 1: Method of Testing for Rating of Cooling Coils* (1975).
19. SAE International Surface Vehicle Standard, *Procedure for Measuring System COP of a Mobile Air Conditioning System on a Test Bench* (2008).

A Study of Postural Loading in Malaysian Mining Industry using Rapid Entire Body Assessment

Norhidayah M.S.[1,a], Nik Mohd Zuki Nik Mohamed[1], Muhamad Arifpin Mansor[2] and Ahmad Rasdan Ismail[3]

[1] *Faculty of Mechanical Engineering, Universiti Malaysia Pahang, 26600 Pekan, Pahang, Malaysia*
[2] *Faculty of Engineering Technology, Universiti Malaysia Pahang, Lebuhraya Tun Razak, 26300 Gambang, Pahang, Malaysia*
[3] *Faculty of Creative Technology and Heritage, Universiti Malaysia Kelantan, Locked Bag 01. 16300 Bachok, Kelantan, Malaysia*

Abstract. The ergonomics and environment factors have been the core issue for the mining industry for many years, and its profiles are rising. To ensure an ergonomics work environment, it is possible to require specific attention especially in this industries sector. It is becoming increasingly difficult to ignore the essential issue in Malaysia due to lack of ergonomics knowledge and low awareness among the engineers in the mining sector. The focus of this study is to evaluate and validate the physical risk factor associated with work-related musculoskeletal disorder (WMSDs) by using Rapid Entire Body Assessment (REBA) among mining industry workers. All the physical risk factors involved the main body regions such as upper arm, lower arm, wrist, trunk, neck and leg that has been identified associated with WMSDs. There were 18 subjects were selected to involve in this study. Those subjects were chosen according to their job task. To increase the reliability of the result, each subject was evaluated thrice in the trials. From the analysis, the average of final score of the REBA is 8.24 indicates high risk and calls for engineering/or work method changes to reduce or eliminate muscular disorder risk. The results of the analysis were used to improve the process of work, design of workstation and also improving the work posture to enhance the comfort level of operators. This study is crucial among the mining industry that is a lack of the information and research about the ergonomics issues in the industry. The overall finding indicated that the whole process of selected work task will contribute to musculoskeletal disorder either for a short or long time exposure.

1 Introduction

In recent years, there has been an increasing interest in quality, health and safety requirement in several occupations. Researchers have increased interest in occupational safety and health issue for mining industries in Malaysia. It is becoming gradually more difficult to ignore the issue related to occupational workplace ergonomics risk assessment due the recent studies and statistics. According to statistic report by the Department of Occupational Safety and Health, Malaysia (DOSH) 3 of victims were reported for the category of Death until June 2014, 17 of victims were reported for Non-Permanent Disability (NPD) and 3 of victims were reported for Permanent Disability for mining industries [1]. Furthermore, according to statistics report on the number of accidents by industry for the past year 2012 conducted by Social Security Organisation, (SOCSO), 9 cases were reported for fatal accident and 417 cases were reported for disability in mine and quarry industry [2].

However, those accidents can be minimized throughout the application of engineering and administrative controls [3]. Lack of information about ergonomics is observed in the various industries in which task are carried out. Musculoskeletal disorders are observed in the welding process where workers are working in kneeling posture, and it shows that there is need to adjust the body postures [4]. The purpose of ergonomic principles would help to enhance machine performance and productivity, help workers to be comfortable and safe [5]. Some workers work under harsh conditions to carry out the required task. These tough circumstances normally increase to various MSD within the workers. These disorders emerge within the workers body due to repetitive lifting, differential lifting height, ambient conditions, etc. [6]. The significance of ergonomics should be product design, working environment, and industrial workstation design, in order to increase productivity and reduce MSD among the workers.

The study exposed that there had been numerous gaps in the work environment, tools, and equipment that affect the health and safety of workers at the work site [8]. Ergonomics related to the design of methods and processes can help reduce or decrease works related risks, as well as advance the company's quality and productivity [9]. Awkward posture, lifting, forceful movement and physical work at rapid rate contribute to the musculoskeletal disorder. The Current study is focused on assessing the work posture of worker

[a] Corresponding author: hidayahsout@gmail.com

occupied in different activities of casting [10]. The application of ergonomic principles not only help to increase machine performance and efficiency, but also help the human operator to be comfortable and protected [11]. It was found that: there was the lack of ergonomics development and methods in small scale forging industry. A significant proportion of the workers were working in very terrible postures. It is recommended that the implementation of ergonomics intervention with accurate awareness among worker [12]. The Study recommended that an ergonomic workstation design can contribute widely to improve the physiological performance of the operators [13].

Therefore, MSD is a regular disorder characterized by ergonomics. Whereby, 448 cases were reported by, SOCSO [2]. The increasing cases reported can be the major issues for the workers are at high risk of developing WMSDs that are associated with exposure factors in this work environment. Despite the high prevalence of work-related musculoskeletal disorder (WMSDs) in mining industries, therefore the aims of this study is to investigate the physical risk factor among the workers in selected job task by using Rapid Entire Assessment (REBA) introduced by Hignett & McAtamney[14]. The aim for this study is to evaluate and validate the physical risk factor associated with work-related musculoskeletal disorder (WMSDs) by using Rapid Entire Body Assessment (REBA) among mining industry workers.

2 Methodology

2.1 Subjects and selected job task

Selected mining companies in Pahang and Kelantan states were randomly selected as a field study in this research. From the three workplace of the mining industry, 18 workers in the selected job were randomly selected as subjects. The selected job task was wet screening that related to screening the raw material by using manual handling hose.

The work task was based on the main procedure of selecting the best raw material before the next process continues. Those workers handle the hose manually to screening all the raw material from 8.00 am in the morning until 5.00 pm afternoon with 30 to 45 minutes break at 1.00pm. The hose weight more than 20kg under strong pressure, and their body was excessively exposed to repetitive motion throughout the working hours. All subjects were exposed to standing position while handling the hose manually. The investigation was taken three times, in the morning, at noon and after the noon session.

2.2 REBA method

The concepts of assessing risk in REBA methods can be used to develop solutions leading to a comprehensive method appropriate for all work tasks and all parts of the body [15]. Basically the idea of REBA has been developed to fill a perceived need for a practitioner's field tool, specifically designed to be sensitive to the type of unpredictable working posture found in health care and other service industries [16]. There were several researches that used REBA methods as their ergonomics tools in several industries. The user of the tools needs to observe the worst posture and time aspect not considered. Janowitz et al. (2006) had measured the physical demands of work in hospital setting by using REBA and the result showed that the inter-rater reliability kappa was 0.54 for UB and 0.66 for LB. The scoring algorithm demonstrated significant variation (ANOVA $p<0.05$) by occupation in anticipated directions (administrators ranked lowest; support staff ranked highest on both scores). A supplemental self-assessment measure of spinal loading correlated with high strain LB scores ($r = 0.30$; $p<0.001$) [17].

All risk assessment methodologies examined by using REBA with agreed a level of risk was associated with performance of the saw-filer job and compare the results of 5 ergonomic risk assessment methods calculated with quantified physical exposure information [18]. REBA is sensitive to musculoskeletal risk in variety of task such as the assessment of working postures in health care and service industries. It concentrates on upper arm, lower arms, wrists, trunk, neck and legs[16] which is suitable method for postural loading tools in mining industry. In a user study (N=18 participants) a group with the REBA feedback was compared to a control group. The levels of MSD demonstrate that the real-time ergonomics feedback significantly decreased the risk of MSDs at global and segmental levels. The real-time ergonomics tool presented in this study could be used directly to reduce the risk of MSDs in an industry and to optimize the long-term performance of workers. The results of REBA score involve the neck, legs, trunk, arm lower, wrist, upper arm, force and coupling to get the final score to determine the level of MSD for the selected job task. The main value to be considered were the mean, maximum, minimum and standard deviation (SD) to validate and evaluate the significant of WMSDs and the physical risk factor involve.

3 Results and discussion

3.1 Description of the subjects

From the wet screening job, out of 18 workers, the age range from 19 to 36 years (mean 2.44±0.86). The working experience ranges was from less than a year to 5 years (mean 1.89±0.47). Table 1 shows the demographics of the workers in wet screening.

Table 1. Demographics of the workers in wet screening job

Job	Age (year)			Working experience (year)		
	Mean	SD	Range	Mean	SD	Range
Wet Screening	2.44	0.86	19-36	1.89	0.47	< 1-5

3.2 Results and analysis

The Figure 1 shows the REBA assessment during the morning session. The neck score was 1.06±0.24 as the highest score was 2 as the neck position located at the range of 20° + from the natural position of the neck. Legs score was 2.17±0.38 as the highest score was 3 as the operators can freely move throughout the REBA assessment with the adjustable leg between 30° to 60°. Trunk score was 2.11±0.32 and had the same highest score with legs score was 3. Most of the operators were repetition movement with the range of the trunk between 0° to 20°. The operators' lower arm positions were always at the range of 0° until to 100° from the natural of the lower arm position, hence the score was 1.78±0.43. The wrist mean score was 1.0±0.0 as all of the operators had locate wrist position as same with each other because they had most the same of wrist position throughout the REBA assessment. Upper arm score was 3.0±0.0 as all the operators had the same position at the range 20° to 45° in handling the equipment throughout the REBA assessment. The score for force was 2.0±0.0 as the loading was more than 10kg for each operator. Coupling score was 1.0±0.0 as the acceptable coupling with another body part of the operators along the working job performance. Final mean score was 8.11±1.78 and it is indicates high risk and calls for further investigation and engineering or method changes to reduce or eliminate musculoskeletal risk for the selected work task.

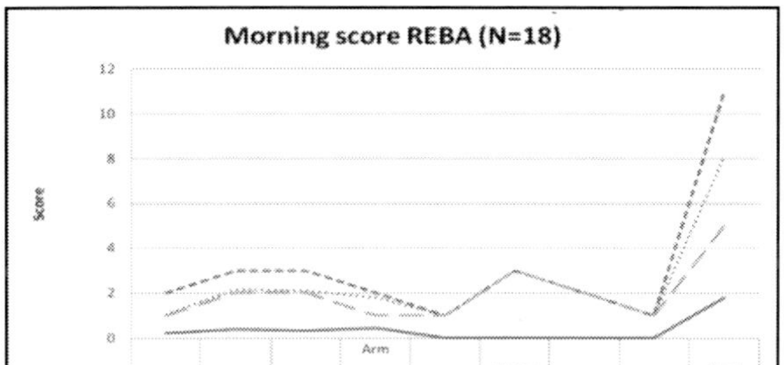

	Neck	Legs	Trunk	Arm Lower	Wrist	Upper Arm	Force	Coupling	Final Score
Mean	1.06	2.17	2.11	1.78	1	3	2	1	8.11
Max	2	3	3	2	1	3	2	1	11
Min	1	2	2	1	1	3	2	1	5
(SD)	0.24	0.38	0.32	0.43	0	0	0	0	1.78

Figure 1. REBA Morning Score

The next Figure 2 was explained the REBA assessment for noon session. From the data, neck score was 1.06±0.24 as the most of the operators were locate natural neck position between the ranges of 0° to 20°. Legs score was 2.17±0.38 which was most of the operators can move the legs and adjust to the range of 30° to 60° during the working job task. Trunk score was 2.11±0.32 as the highest score was 3. There was some of the operators had trunk position at the range 20° to 60° from the natural position as the position of their workstation cause the operators need to bent forward during the working job task. The score for lower arm was 1.78±0.43 which was caused of the range of work task was around 0° until 100° during handling the equipment at the workstation. As the score for lower arm was 1.0±0.0 obtain the wrist score also had no changes with REBA assessment compare to morning session. Upper arm score was remain same as morning session at 3.0±0.0 because the upper arm position does not have major changes between the morning and noon session throughout the working REBA assessment. The force score still had the same score as previous REBA assessment at 2.0±0.0 with the same loading with more than 10kg involved in the working job task. Coupling score also remain same as before with score at 1.0±0.0 as all the operators had the same coupling handling in the working task. For the final score was 7.94±1.86 which is indicates medium risk with further investigation and need to be change the type of working station soon.

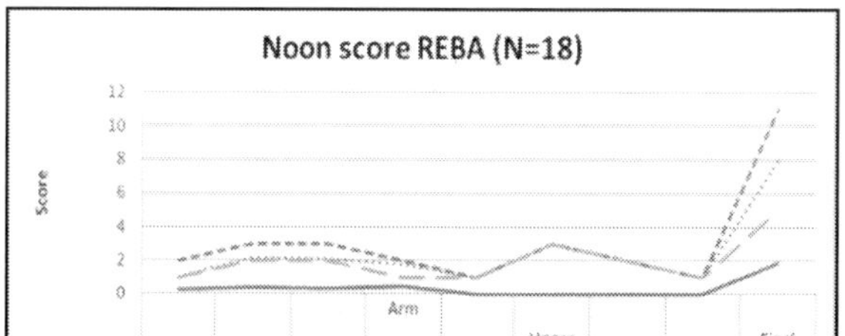

	Neck	Legs	Trunk	Arm Lower	Wrist	Upper Arm	Force	Coupling	Final Score
Mean	1.06	2.17	2.11	1.78	1	3	2	1	7.94
Max	2	3	3	2	1	3	2	1	11
Min	1	2	2	1	1	3	2	1	5
(SD)	0.24	0.38	0.32	0.43	0	0	0	0	1.86

Figure 2. REBA Noon Score

Figure 3 below shows the slight different score between the physical risk factor of REBA assessment after the noon session. As the neck score had slight lower score than noon session at 1.0±0.0. Legs score showed slightly higher than noon session at 2.22±2.3, because of there were few operators that had the same posture for the whole day working task. The trunk score was 2.22±0.0 had slightly increase score compare to noon session assessment. Lower arm score was 2.0±0.0 which means all the operators had the same position of lower arm for the rest of REBA assessment after the noon session. The wrist score does not had any different changes for the whole day at 1.0±0.0. Same goes with upper arm score with same score as morning and noon session at 3.0±0.0 because the operators were had same position during the REBA assessment from morning till after noon. The work task had same loading force which was more than 10kg involved. The coupling score was 1.0±0.0 same as other session and acceptable with another body part toward the coupling work task. Final score was 8.67±1.14 indicates the high risk and calls for further investigation and engineering with method changes to reduce or eliminate musculoskeletal disorder risk.

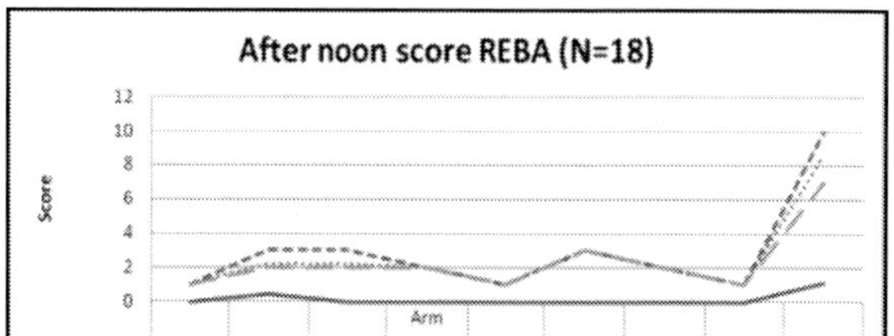

	Neck	Legs	Trunk	Arm Lower	Wrist	Upper Arm	Force	Coupling	Final Score
Mean	1	2.22	2.22	2	1	3	2	1	8.67
Max	1	3	3	2	1	3	2	1	10
Min	1	2	2	2	1	3	2	1	7
(SD)	0	0.43	0	0	0	0	0	0	1.14

Figure 3. REBA After noon Score

4 Conclusions

The result of this assessment shows that the subjects were exposed to all the physical risk factors such as neck, trunk, and upper extremities. As a conclusion, we can say that the wet screening task was in very high-risk level and need to be change immediately. The task needs to be revamped change for the health concern to the workers involve. The present study was designed to determine the level of physical MSD risk among the workers in mining. The further study needs to be done to ensure the cause of significant risk factors to the WMSDs among the mining workers.

References

1. DOSH, Statistic Report of Occupational accidents for the category of death until August 2010.Department of Occupational Safety and Health, Malaysia, 2010a.
2. SOCSO, Statistic report about the numbers of accidents by industry year 2007 Social Security Organization, Malaysia, 2007.
3. Isa Halim & Abdul Rahman Omar, "A Review on Health Effects Associated with Prolonged Standing in The Industrial Workplaces," International Journal Of Research And Review In Applied Science, vol. 8, issue 1, July 2011.
4. D. N. Agrawal et al.," Study and Validation of Body Postures Of Workers Working In Small Scale Industry through Rula," International Journal of Engineering Science and Technology, Vol. 3 No.10 October 2011
5. Aide Maldonado-Macias et al., "Ergonomic Evaluation of Work Stations Related With the Operation of Advanced Manufacturing Technology Equipment: Two Cases of Study", XV CONGRESO INTERNACIONAL DE ERGONOMIA SEMAC, 2009.
6. Aman Sachdeva et al., "Minimizing Musculoskeletal Disorders in Lathe Machine Workers," International Journal of Ergonomics, Volume 1 Issue 2 2011.
7. Mohammad Iqbal et al., "Ergonomics and Design," International Conference on Industrial Engineering and Operations Management Kuala Lumpur, Malaysia, January 22 – 24, 2011.
8. Parimalam P. et al., "Ergonomic interventions to improve work environment in garment manufacturing units," Indian Journal of Occupational and Environmental Medicine, August 2006 -Volume 10 - Issue 2.
9. M. C.I. Javier Antonio Lom Holguín et al., "Ergonomics and its relationship to the design of production processes," Ponencia Conergo 2008.
10. Lakhwinder Pal singh., "An Investigation Work Posture Of Workers Engaged In Casting Industry: A Study In India,"Asian Journal of Managerial Science, Volume 1 Number 1, January – June 2012.
11. Asim Zaheer et al., "Ergonomics: A work place realities in Pakistan," International Posture Journal of Science & Technology, Volume 2, Issue 1, April, 2012.
12. L. P. Singh., "Work Posture Assessment in Forging Industry: An Exploratory Study in INDIA," International Journal of Advanced Engineering Technology, vol. 1 issue III Oct.-Dec. 2010.
13. Isa Halim et al., "Posture, muscle activity and oxygen consumption Evaluations among metal stamping operators: A pilot study in Malaysian small and medium Industries," Journal - The Institution of Engineers, Malaysia, Vol. 72, No. 4, December 2011.
14. S. Hignett and L. McAtamney, "Rapid entire body assessment (REBA)," *Appl. Ergon.*, vol. 31, pp. 201–205, 2000.
15. Roman-Liu, "Comparison of concepts in easy-to-use methods for MSD risk assessment.," *Appl. Ergon.*, vol. 45, no. 3, pp. 420–7, May 2014.
16. J. Singh and K. Rula, "A Literature Review On MSDs Using Ergonomic Body Assessment Tools : RULA And REBA Himanshu Chaudhary," pp. 147–149, 2013.
17. I. L. Janowitz, M. Gillen, G. Ryan, D. Rempel, L. Trupin, L. Swig, K. Mullen, R. Rugulies, and P. D. Blanc, "Measuring the physical demands of work in hospital settings: design and implementation of an ergonomics assessment.," *Appl. Ergon.*, vol. 37, no. 5, pp. 641–58, Sep. 2006.
18. T. Ã. Jones and S. Kumar, "Comparison of ergonomic risk assessments in a repetitive high-risk sawmill occupation : Saw-filer," vol. 37, pp. 744–753, 2007.

MATEC Web of Conferences 74, 00015 (2016) DOI: 10.1051/matecconf/20167400015
ICMER 2015

Characteristics of the Biodiesel Palm Oil Methyl Ester Pool Fire

Mohd Rosdzimin Abdul Rahman[1,a], Raj Mohan[2], Kin Yuen Leong[1] and Siti Noor Aliah Baharom[2]

[1]*Dept. of Mechanical, Faculty of Engineering, Universiti Pertahanan Nasional Malaysia, Kem Sg. Besi, 57000, Kuala Lumpur, Malaysia*
[2]*Malaysia Agricultural Research and Development Institute, Serdang, Selangor, Malaysia*

Abstract. The characteristics of biodiesel pool fires with palm oil methyl ester concentrations of 7% by volume in various pan diameter sizes (0.05 m, 0.10 m and 0.15 m) were studied. Initially, biodiesel was heated on a hot plate to increase the temperature close to the flash point temperature. Later fuel was ignited and stable pool fire formed. Flame temperature along the centerline pool fire from flame base to flame tip were measured. Flame temperature shows that the temperature at flame base is higher than that the flames tip. This is due to the complete burning occurs near flame base and at flame tip incomplete burning exists. Incomplete burning produces yellow flame appearance. Moreover, the mass burn rate varies linearly with pan size. Small diameter pan produces low mass burn rate compared to the large diameter pan. Low and high mass burn rate contributes to the flame height. Flame with higher mass burn rate is longer than that the low mass burn rate. Thus, the characteristic of the biodiesel palm oil methyl ester is successfully studied in the present study.

1 Introduction

Alternative fuels are developed to replace petroleum-based fuels due to the gas exhaust emission issues and to promote energy security. Biodiesel is one category of the alternative fuels that is environmental friendly. It is used widely in Malaysia since 2010. Started with B5 and the recent B7 is in the market since November 2014. Biodiesel in Malaysia is made from palm oil methyl ester. Methyl ester is derived from vegetable oil (biodiesel). Conversion of crude palm oil to palm oil methyl ester (POME) was successfully done by The Palm Oil Research Institute of Malaysia (PORIM) through trans-esterification. Studies on the engine performance using POME were done. Moreover, due to low sulphur content (1.112 wt %) of the POME, clean emission was produced. Performance studies on the engine using biodiesel and POME were reported by various researchers [1-5] . However, the fire safety precautions also must be taken into consideration to ensure safe storage and fueling. Studies on the fire characteristic of other biodiesel [6-8] have been carried out. However, there is no scientific study on the POME fire behavior. Knowledge of the fire behaviors will lead to better safety and fire's control when it happens as the use of biodiesel becomes widespread and increase the possibility of fire occurrence.

Pool fire is used to imitate the fire scenario including fires ranging in size from a 0.005 m to fire diameter as large as 10^5 [9]. Lots of studies on the pool fire using different types of fuel have been done. The important parameters in analyzing pool fire are flame characteristic and shape, pulsation frequency, flame height, detailed structure of the flame, enthalpy balance in the flame, enthalpy balance at the fuel surface, mass vaporization rate and radiative emission [9]. These parameters are required in modeling and combating fires.

Biodiesel fire research received less attention in fire research community. Only Refs. [6-8] conducting biodiesel pool fire research recently. In 2011, Ref. [6] studied on the biodiesel pool fires of canola, soy and blends with petroleum diesel. Ref. [6] observed that the fuel regression rate of the surface in all pans size was comparable. It is also revealed that burning rate of the fuel mass as well as the surface regression rate varied non-monotonically with the concentration of biodiesel in the blend. Later, Ref. [7] blended canola methyl ester, soy methyl ester with Jet A fuel. They found that the fuel mass burning rate in the small pan was less than that of larger pan for all type of flames. Moreover, the fuel surface regression rate was reduced when the biodiesel concentration in the blends was greater than 75% in volume. Since there is no study on the POME fire behavior, which is important for safety, therefore, the objective of the study is to investigate the pool fires characteristics of palm oil methyl ester in various pan diameters.

2 Experiment Set Up

An experimental setup diagram is shown in Fig. 1.The fuel pan was placed on a top of hot plate in this experiment. Three cylindrical aluminum pans were used to fire the biodiesel; The pan's inner diameter were 0.05 m, 0.10 m and 0.15 m while its height is 0.1 m. The pans

[a] Corresponding author: rosdzimin@gmail.com

were filled with the fuel at various volumes. A 0.5 cm gap was left between the biodiesel surface and the container lid. Gap between the lid and biodiesel surface serves two purposes. There are: to allow volumetric expansion of biodiesel upon heating and also to prevent spilling. Six K-type thermocouples were spaced at 0.02 m vertically started at 0.01 m above the container lid. Flames temperatures were recorded by the thermocouples as shown in Fig. 1. Initially the hot plate was turned on and once the fuel reached flash point temperature, an igniter was used to ignite the biodiesel. The hot plate was turned off and the temperatures were recorded when the fire was sustained. When a fire was sustained, the hot plate was turned off and all the data were recorded. The flame height is measured by investigate the flame with scale on it side.

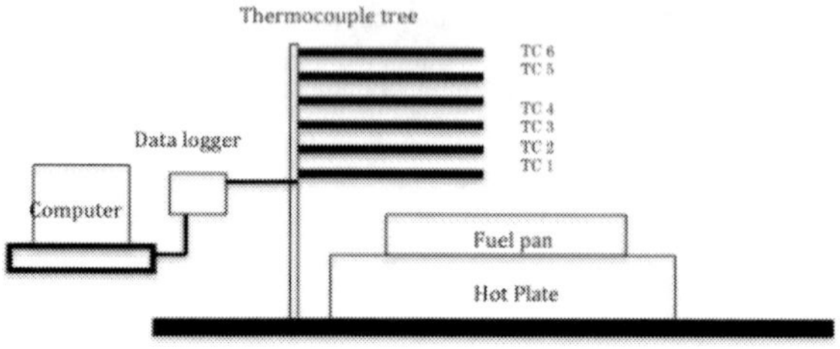

Figure 1. Schematic diagram of the experimental setup.

3 Results and discussion

Figure 2 shows photographs of the biodiesel flames at various pan's diameter. The flames were sensitive to the ambient wind condition and yellow color is attributed to the incomplete combustion. Moreover, 0.05 m pan produces laminar flame. On the other hand 0.1 m and 0.15 m produces flame in a transition regime. This result agreed with Ref. [9]. The flame height was 18 cm for the 0.05 m diameter pan (Fig. 3) and increase monotonically with increase in pan diameter. The increase of pan diameter increases flame height is due to increase in mass burn rate. Moreover, high fuel mass burn rate leads to higher heat release rate, thus increases flame height. It can be seen in Fig. 4, the fuel mass burn rate of 0.05 m diameter pan was significantly smaller than that of the 0.15 m diameter pan flames (Fig. 4). Higher mass burn rate for large diameter pan is because of more heat being transferred by conduction to the fuel. This can be explained by theory of heat transfer in pool fire.

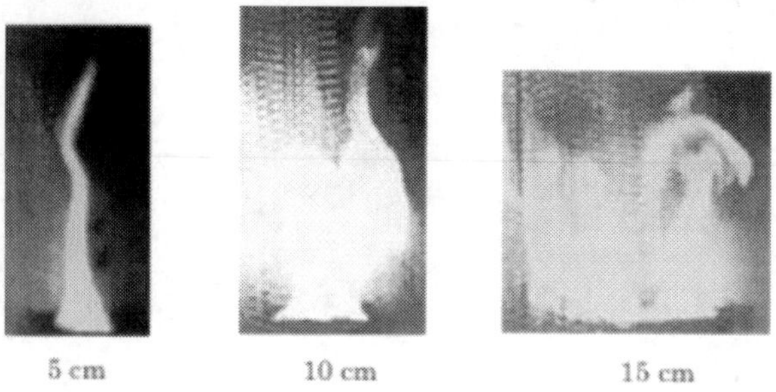

Figure 2. Photograph of the biodiesel palm oil methyl ester pool fire at various pan diameters.

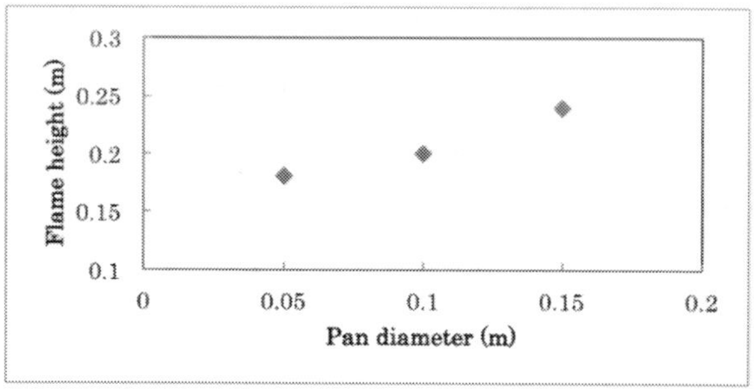

Figure 3. Flame height of the biodiesel palm oil methyl ester pool fire at various pan diameters.

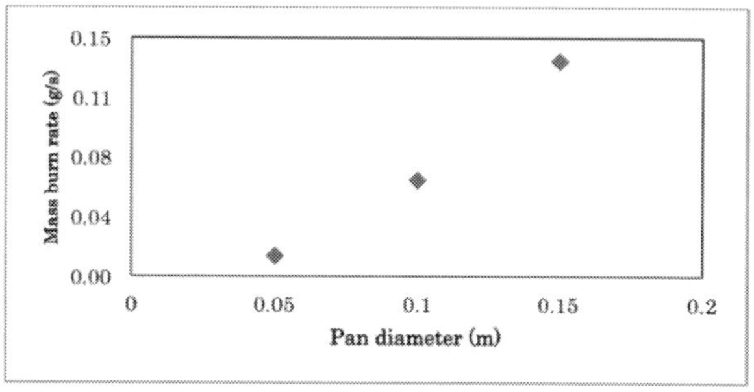

Figure 4. Mass burn rate of the biodiesel palm oil methyl ester pool fire at various pan diameters.

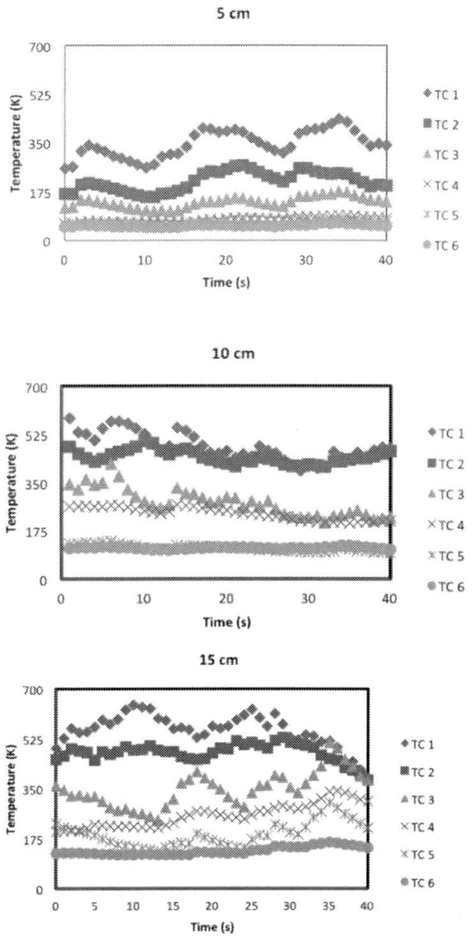

Figure 5. Flame temperature profile of the biodiesel palm oil methyl ester pool fire at various pan diameters.

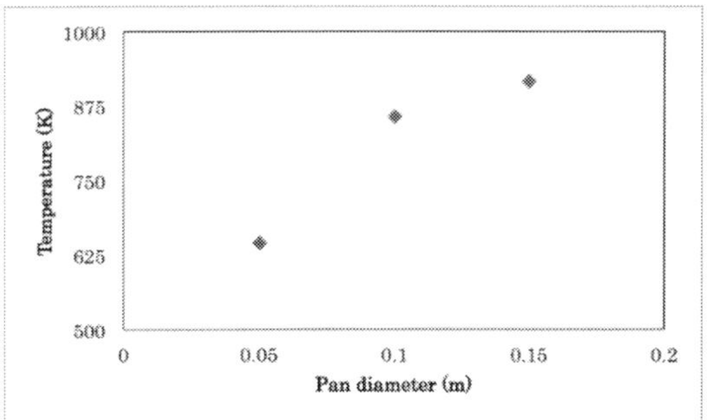

Figure 6. Highest temperature of the biodiesel palm oil methyl ester pool fire at various pan diameters.

4 Conclusions

The present study investigate combustion characteristics of pool fires biodiesel palm oil methyl ester fuel in various pan diameters. The flame appearance, fuel mass burn rate, temperature profile were measured. From this study, it can be concluded that:

(a) More heat is being transferred by conduction in large pan resulted smaller mass burning rate from small container flames compared to that of large container flame.

(b) Smaller pan produced shorter flame due to the lower rate of heat release

(d) Similar temperature profiles were observed for all type of flames. The maximum temperature was recorded at the flame base attributed to complete combustion.

Acknowledgements

The authors would like to be obliged to Universiti Pertahanan Nasional Malaysia for providing laboratory facilities and financial assistance under project no. UPNM/2014/GPJP/TK/6.

References

1. A. Murugesan, C. Umarani, R. Subramanian and N. Nedunchezhian , *Renewable and Sustainable Energy Reviews*, **13**, 653–662 (2009)
2. E Sadeghinezhad, S. N. Kazia, A. Badarudin, C. S. Oon N.M. Zubir and Mohamad Mehrali, *Renewable and Sustainable Energy Reviews*, **28**, 410–42 (2013)
3. Mofijur M, Masjuki HH, Kalam MA, Atabani AE, Fattah IMR, Mobarak HM., *Industrial Crops and Products*, **53**, 78-84 (2014)
4. A. Sanjid, H.H. Masjuki, M.A. Kalam, S.M.A. Rahman, M.J. Abedin, S.M. Palash, *Journal of Cleaner Production*, **65**, 295- 303 (2014)
5. Ahman, S. M. A., Masjuki, H. H., Kalam, M. A., Abedin, M. J., Sanjid, A., & Rahman, M. M.. *Renewable Energy*, **68**, 644-650 (2014)
6. Tran V., Morton C., Parthasarathy R. N. and Gollahalli S. R., *9th Annual International Energy Conversion Engineering Conference USA*, (AIAA 2011-6013)
7. Tran V., Morton C., Parthasarathy R. N. and Gollahalli S. R., *10th Annual International Energy Conversion Engineering Conference USA*, (AIAA 2012-3722)
8. Tran V., Morton C., Parthasarathy R. N. and Gollahalli S. R., *International Journal of Green Energy*, **11**, 595-610 (2014)
9. Hamins A., Kashiwagi T., and Buch R., Fire Resistance of Industrial Fluids, ASTM STP 1284, American Society for Testing and Materials, 1995.

The study of mechanical properties of pineapple leaf fibre reinforced tapioca based bioplastic resin composite

D. Mathivanan [1,a], H. Norfazilah[1], J.P. Siregar [1], M.R.M. Rejab [1], D. Bachtiar[1], and T. Cionita[2]

[1]Faculty of Mechanical Engineering, Universiti Malaysia Pahang, 26600 Pekan, Pahang, Malaysia,
[2] Department of Mechanical Engineering, Faculty of Science, Technology, Engineering and Mathematics, INTI International University, 71800 Nilai, Negeri Sembilan, Malaysia

Abstract. Natural fibre reinforced composite has brought the material engineering to a high new level of research. Natural fibres are compatible with matrices like polypropylene and can be used as reinforcement material to reduce the composition of plastic in a material. Natural fibres such as kenaf, pineapple leaf, and coir already found its importance in reducing the dependence of petroleum based products. However the biodegradability of the product at the end of the intended lifespan is still questionable. This has led many researches to look for a suitable replacement for synthetic fibres and achieve better adhesion between fibre and matrix. In this study, fiber and matrix which are hydrophilic in nature was used and the mixture was extruded and hot compressed to acquire better mechanical properties. The specimens were fabricated and tested according to ASTM D638. The 30% composition illustrates the best average modulus value among other composition and from this result it can be concluded that the increase of PALF fibre in TBR composite increases the modulus strength of the composite.

1 Introduction

The depletion of petroleum, landfill issues and degradability of product has urged variety of studies to be carried out worldwide to find a suitable solution for the problem stated[1]. New materials have been tested and studied to replace the current synthetic fibres which are made from petroleum and difficult to be degraded naturally by enzyme or bacteria[2]. Natural fibres which are found abundantly and a renewable resource has caught the attention of researches all over the world[3]. There are three types of natural fibres which are plant fibre, animal fibre and mineral fibres which are already being studied[4]. Plant fibres which are mostly from agricultural waste can be found in most of the countries in the world[5]. Every year, tropical countries like Thailand, Malaysia, and Indonesia records tonnes of waste from agricultural crops[6].

Natural plant fibres' usage in mass industries such as automotives, construction, sports and medicine has already seen rapid increase due to its good properties such as high tensile strength, acoustic insulate, light weight, low cost, less health risk and non abrasiveness towards equipment[7]. Plant fibres has good mechanical properties which almost equivalent with glass fibres.

These properties are results of high cellulose content in plant fibre which has high number of hydroxyl group bonded in intramolecular pattern among macrocellulose[8].

In past studies done by researchers on PALF with synthetic fibres such as polypropylene and polyvinyl acid has shown drastic improvement in terms of strength[9], however these composite are found hard to be decomposed naturally by microorganisms. Moreover the adhesion between PALF and synthetic matrices are low due to the difference in polarity between fibre and matrix which makes the bonding to be weak[10]. In this study we will use tapioca based bioplastic resin as matrix because of the chemical similarity between starch in tapioca flour and cellulose in PALF where both are hydrophilic in nature[11].

We will use extrusion[12] followed by hot compression moulding[13] method to fabricate the specimen as this techniques are used widely and shown better results than most of the fabrication methods. Fabrication technique plays an important role in enhancing the properties of the composite as the temperature and the method used determines the properties of the final product.

In this study we will use pineapple leaf fibre (PALF) as reinforcement material to reinforce tapioca based bioplastic resin (TBR). Pineapple leaf is one of the main agricultural wastes in Malaysia as this country is one of the major export countries of this fruit which enables PALF to be found abundantly and easily accessible for the project. The main objective of this research was to study the mechanical properties of PALF reinforced TBR composite.

[a] Corresponding author:davindrabrabu@gmail.com

2 Methodology

In this research, we will study the adhesion and the best compatibility between PALF and TBR through several tests by manipulating the composition of the composite according to Table 1. The temperature for extrusion and hot compress process were set to 160°c at maximum to avoid PALF fibre from damaging due to excess heat. The mould was prepared according to ASTM D638 specimen size to achieve the best result. The parameters such as span length and crosshead speed for the testing also was set constant as stated in ASTM D638. Altogether 28 experimental test run was conducted and 5 best results from each composition will be taken for average value calculation.

2.1 Experimental Procedure

PALF is made through scrapping the pineapple leaf with a sauce plate. The fibres are then cut into small sizes and grind. The grinded fibres were then sieved to get the highest amount of fibre length available. Then the fibres are placed in oven for 24 hours before the experiment.

The composites were made through mixing the PALF and TBR according to the composition in Table 1. The mixture was first extruded into pellet using extrusion, followed by pelletizer. The extrusion temperature for all zone were set at 160°c. The pellets were then placed in the mould and hot pressed at 160°c for 5 minute at 8 Mpa, and then cold pressed at room temperature for 5 minute at 8 Mpa. The composites were then cut into ASTM D638 specimen size using a hand saw. The specimens were tested for tensile strength using INSTRON machine according to ASTM D638 parameters. The top five results were taken from the INSTRON machine for analysis and average value calculation.

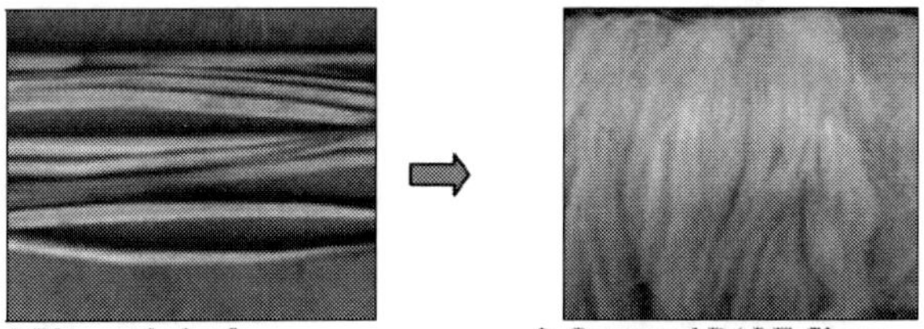

1.Pineapple leaf 2. Scrapped PALF fibres

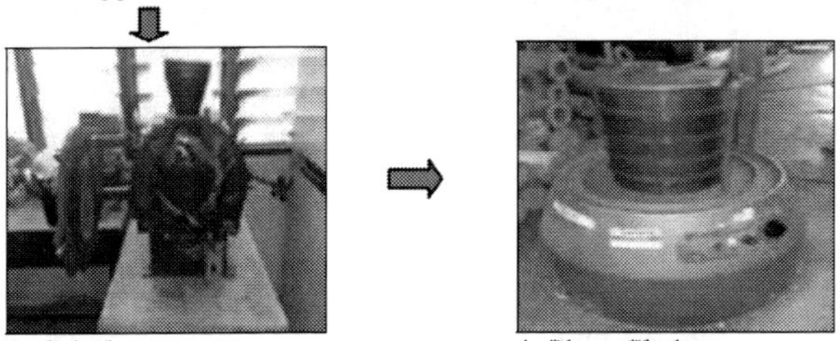

3. Grinder 4. Sieve Shaker

6. Vacuum Oven 5. Sieved PALF fibres

Figure 1-6. Preparation of materials

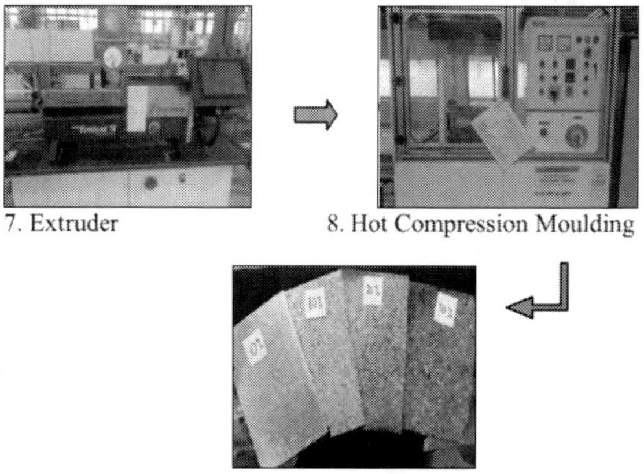

7. Extruder 8. Hot Compression Moulding

9. PALF/TBR Composites

Figure 7-9. Preparation of Composites

Table 1. PALF and TBR mixture by weight fraction.

TBR/PALF Mixture	PALF	TBR
Pure TBR plate	0%	100%
10% PALF reinforced TBR composite plate	10%	90%
20% PALF reinforced TBR composite plate	20%	80%
30% PALF reinforced TBR composite plate	30%	70%

3 Results and discussion

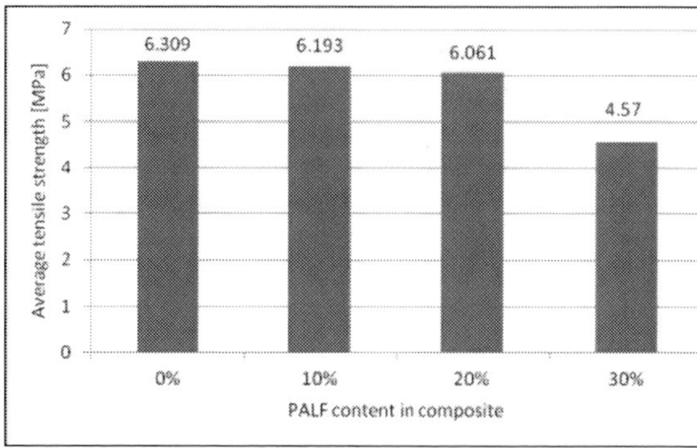

Figure 10. Average Tensile strength of PALF/TBR Composite

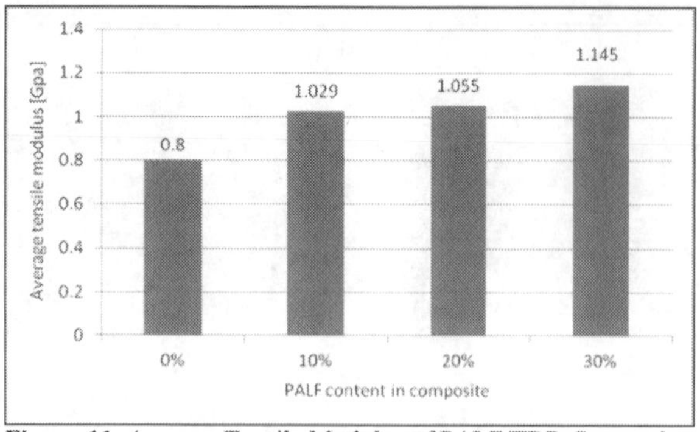

Figure 11. Average Tensile Modulus of PALF/TBR Composite

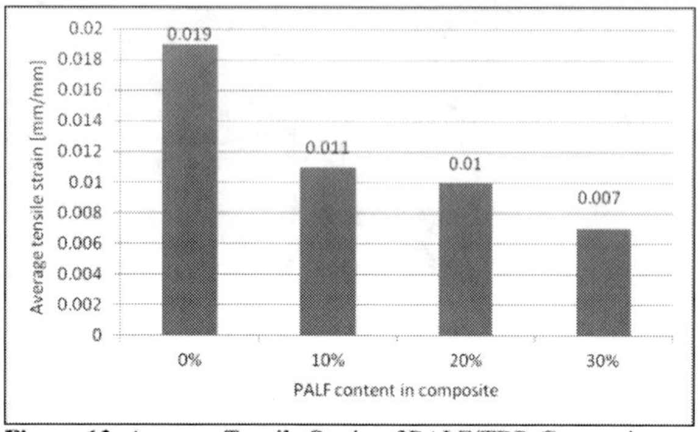

Figure 12. Average Tensile Strain of PALF/TBR Composite

Figure 10 and 11 shows that the average modulus and average tensile strength obtained from the experiment. It can be observed that the average modulus increases as the fibre content increases in the composition and the highest were 1.145GPa at 30% PALF content. This is due to the availability to transfer stress through fibre which enables the composite to hold the load longer and better as the fibre loading increases. However the average tensile strength presents decrease as the fibre content and the highest average tensile strength were 6.3088MPa at 0% PALF content, this is caused by the increase of fibre in the composites causes' fibre agglomeration or fibre overlapping which causes weak bond between TBR and PALF.

Another reason for this result is the fabrication method where the composites were cut into specimens using hand saw which also weakens the bond between matrix and fibre. Furthermore, in Figure 12 it was noted that the graph showed decrease as the fibre loading increased. This is because the change in characteristic of the composite in term of ductility and brittleness where the composite becomes more brittle as the fibre composition increases in the composite making the specimens easier to break.

4 Conclusions

From this research, it can be concluded that the effort to reinforce tapioca based bioplastic resin with PALF has shown positive result where the average modulus of the composite show an improvement at 30% PALF/TBR composite which is higher value than the pure tapioca based bioplastic resin although the average tensile strength of 10% PALF/TBR composite shown a slight decrease than the pure TBR. The best compatibility and adhesion between PALF and TBR were at 30% because of the increase in PALF fibre. All the objective of the studies has been achieved in this experiment.

Acknowledgements

The authors wish to thank the Ministry of Higher Education, Malaysia, for funding the research through the Fundamental Research Grant Scheme (FRGS) grant number 140120. The authors also would like to be obliged to Universiti Malaysia Pahang for providing laboratory facilities.

References

1. Kengkhetkit, N. and T. Amornsakchai, *A new approach to "Greening" plastic composites using pineapple leaf waste for performance and cost effectiveness.* Materials & Design, 2014. **55**(0): p. 292-299.
2. Leao, A.L., et al., *Agro-Based Biocomposites for Industrial Applications.* Molecular Crystals and Liquid Crystals, 2010. **522**(1): p. 18/[318]-27/[327].
3. Davindrabrabu, M., et al. *Effect of Fibre Loading on the Flexural Properties of Natural Fibre Reinforced Polymer Composites.* in *Applied Mechanics and Materials.* 2015. 695(0): p.85-88.
4. Satyanarayana, K.G., G.G.C. Arizaga, and F. Wypych, *Biodegradable composites based on lignocellulosic fibers—An overview.* Progress in Polymer Science, 2009. **34**(9): p. 982-1021.
5. Arib, R.M.N., et al., *Mechanical properties of pineapple leaf fibre reinforced polypropylene composites.* Materials & Design, 2006. **27**(5): p. 391-396.
6. Nayan, N.H.M., W.A.W.A. Rahman, and R.A. Majid, *The effect of mercerization process on the structural and morphological properties of Pineapple Leaf Fiber (PALF) pulp.* Malaysian Journal of Fundamental and Applied Sciences, 2013. **10**(1).
7. Vinod B and D.S.L. J, *Effect of Fiber length on the Tensile Properties of PALF Reinforced Bisphenol Composites.* International Journal of Engineering, Business and Enterprise Applications (IJEBEA), 2013. **5**(2): p. 158-162.
8. Kabir, M.M., et al., *Chemical treatments on plant-based natural fibre reinforced polymer composites: An overview.* Composites Part B: Engineering, 2012. **43**(7): p. 2883-2892.
9. Leao, A.L., et al., *Pineapple Leaf Fibers for Composites and Cellulose.* Molecular Crystals and Liquid Crystals, 2010. **522**(1): p. 36/[336]-41/[341].
10. Mishra, S., et al., *A Review on Pineapple Leaf Fibers, Sisal Fibers and Their Biocomposites.* Macromolecular Materials and Engineering, 2004. **289**(11): p. 955-974.
11. Ozemoya, P.O., E.S.A. AJISEGIRI, and P.A. IDAH, *Production of Adhesives from Cassava Starch.*

Leonardo Electronic Journal of Practices and Technologies, 2007(10): p. 93-98.
12. Gunning, M.A., et al., *Effect of Compatibilizer Content on the Mechanical Properties of Bioplastic Composites via Hot Melt Extrusion.* Polymer-Plastics Technology and Engineering, 2014. **53**(12): p. 1223-1235.
13. Rett, H.T., *Thermal and microscopic analysis of biodegradable laminates madefrom cassava flour, sorbitol and poly (butylene adipate-coterephthalate) PBAT.* Acta Scientiarum. Technology, 2013. **35**(4): p. 765-770.

Analysis of optimum wire rope configuration for equal unidirectional torsional stiffness for flexible steering shaft

Najaf Hussain[1], Masri Bin Baharom[1,a] and Mui'nuddin Maharun[1]

[1]*Department of Mechanical Engineering, Universiti Teknologi PETRONAS Malaysia*

Abstract. The design and modeling of Low Stiffness Resilience Shaft (LSRS) for the Semi-Active Steering (SAS) system using wire ropes is discussed in this paper, along with the static structural torsion test simulation of the wire ropes in order to determine the best possible configuration which serves the purpose of an LSRS. The importance of this study arises due to the unidirectional torsional properties of a wire rope. For an effective operational LSRS, the wire ropes need to have similar angular deflection in both the clockwise and anti-clockwise direction. LSRS, an integral component of the SAS is a flexible shaft that can replace the conventional rigid shaft of the steering system and allows active control to be performed. 3D solid models of the simple strand and the 4 strand wire ropes used in finite element analysis were generated in CAD software SolidWorksTM. The single strand and the different configuration of wire ropes required to function the LSRS effectively were then analyzed using Finite element simulation in ANSYS$_{TM}$. A single wire rope could not be used because its construction has inconsistency in the torsional stiffness in clockwise and anti-clockwise direction. The single-strand right-direction lay wire rope is found to have 16.05% angular deflection percentage difference in the clockwise and anticlockwise directions which indicates that using a single strand wire rope for the LSRS will cause the vehicle to have a variable response in the clockwise and anti clockwise direction upon turning the steering wheel. Due to this inconsistency, two variations namely Variation 1 and Variation 2 with arrangement of 4 strand wire rope were devised so that the angular deflection percentage difference would be negligible. Simulation results indicated that Variation 1 of the two variations with an angular deflection percentage difference of 0.34% in the clockwise and anti-clockwise direction respectively is best suited for the use in LSRS as it has almost negligible angular deflection percentage difference and will allow the vehicle to have similar steering response in the clockwise and anti-clockwise direction.

Keywords: Steering; Automotive; flexible shaft; wire rope; Finite Element simulation

1 Introduction

It has been more than 100 years since the introduction of the first automobile and the engineers have been able to make amazing technological advancements in the automotive field since then. For the need of drivers or passengers, engineers have been constantly improving the design of various systems of vehicles such as improved ride comfort, different suspension set ups and various tire profiles. With the increasing speeds of vehicles, safety became one of the important issues and therefore measures were taken to increase the safety of the drivers and passengers of moving vehicles, such as new and improved chassis designs. One of the examples is the case with the development of the steering system[1].

Drivers of vehicles were having difficulty to steer vehicles during parking at low speed maneuvers 50 years after the invention of automobile. With the increase in travelling speeds of automobiles, drivers were also looking for increased handling stability at high speeds. Furthermore, the need for higher power and speed led to the creation of heavier engines and chassis of the vehicles which meant that the rack loads increased and the drivers needed a lot of strength to turn the steering wheels [2]. In order to solve these problems, automotive engineers have come up with different types of steering systems such as Steer-by-Wire (SBW) steering system or the SAS system. SBW steering systems allow the amount of steering wheel operation to be transmitted in the form of electric signals to the vehicle wheels. These systems help improve control performance for vehicle safety while increasing vehicle design freedom [3]. One another design concept in pursuit of solving these problems is the Semi-Active-Steering (SAS) system technology [4] which takes care of all these problems, and provides advantages such as the Active control, ease of packaging and improved safety to the drivers.

The SAS system is like a SBW steering system but instead of a backup system there is a permanent connection between the steering wheel and the rack and pinion. The problem arises because current steering shafts cannot be used in this system as they are torsionally rigid

[a] Corresponding author: masrib@petronas.com.my

and will affect the SBW steering capabilities when the SBW system is active and will not allow active steering to be performed on the vehicle. Furthermore the conventional steering shaft is rigid along the axis, which may cause injury to the driver in case of a front end collision. Therefore in the SAS instead of a rigid shaft, there is a Low-stiffness-resilient-shaft (LSRS) which introduces active control to the vehicle [4]. The LSRS is the integral component of the SAS system and is always connected to the system unlike the (SBW) where the clutch mechanism is only engaged when the electric system fails. The properties of the LSRS should be as such that if the SBW system is still engaged, the LSRS doesn't affect the steering capability, but if the SBW system fails the LSRS should be torsionally rigid enough to provide the vehicle steering capabilities so that the vehicle can be maneuvered to safety. Furthermore, the LSRS should be flexible along its axis so that in case of an accident or a front end collision, the LSRS buckles and the driver isn't harmed. A suitable arrangement of wire ropes supported by bearings can give the properties needed to make a LSRS for an SAS system.

Wire rope is a versatile, flexible, high-strength member that is used in many mechanical systems to provide excellent tensile strength while it remains flexible. It is also used in power transmission application components which are physically separated or not co-linear. Wire rope is typically classified by four parameters: (1) nominal diameter; (2) rope construction; (3) core type; and (4) breaking load. Variations in the make of the wire rope affect physical properties, such as flexibility, rupture strength, and service life [5].

The mechanical analysis of a wire rope structure is made complex by contact friction interactions taking place between elementary wires [6]. Similarly the analytical solutions for the torsional or tensile stresses of the wire rope normally neglect the frictional and contact effects. With the use of solid modeling and finite element analysis in a computer environment, it is possible to consider frictional, contact and other working conditions [7]. Finite element predictions of rope properties started in the early 1970s but it was restricted at the time. The developments in finite element software as well as in computer technology in respect to contact and model size in the recent years have made it much more feasible not only from a scientific but also from a practical point of view [8].

Because of the lay pattern of a single strand wire rope, it cannot be used for the construction of the LSRS as it will have varying angular deflection in the clockwise and anti-clockwise direction. The author models and performs simulation on the single strand wire rope to find out the angular deflection upon application of torque in both directions. Two different arrangements of wire ropes are then modeled so that the angular deflection in both directions is similar. FEA simulations are performed on the 3D solid models in ANSYS^{TM} and the angular deflection is noted.

2 System Concept Design

The system setup of the SAS is shown in Figure 1. The LSRS, Power motor, Reaction motor, sensors and controllers are the parts and parcels of the system. The presence of the LSRS makes this system different from any other steering system in the automotive industry. The LSRS is basically used instead of a rigid steering shaft because firstly it offers a packaging advantage as it can allow the steering wheel to be placed on either side of the dashboard. Secondly, in case of a head on collision it will buckle. The LSRS can be made out of a cable which is flexible and resilient to the torque acting along its length. The main concern is the rigidity and stiffness of the cable so that when the main electric system fails, the driver should be able to steer the vehicle safely using just the LSRS.

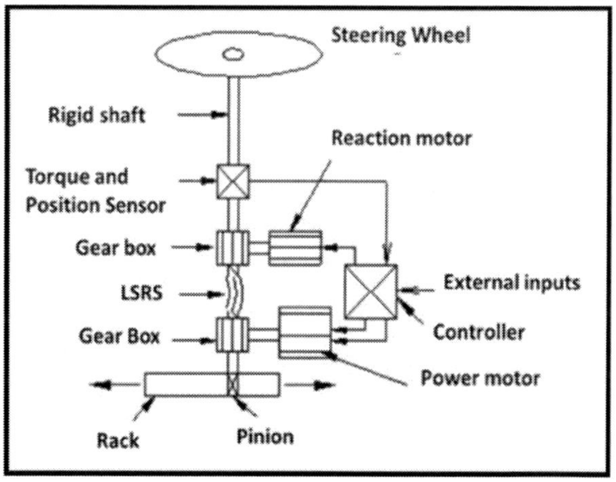

Figure 1. Schematic diagram of the SAS [4]

2.1 Basic Configuration of the LSRS

The proposed LSRS design mainly constitutes of two important components which are the bearings and wire ropes as shown in Figure 2(a).

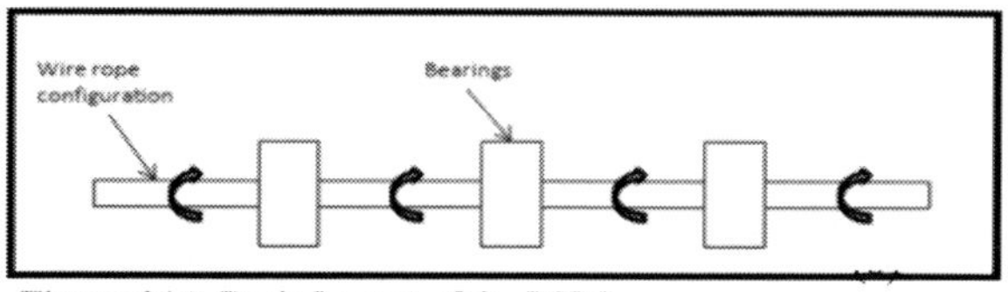

Figure 2(a). Basic Layout of the LSRS

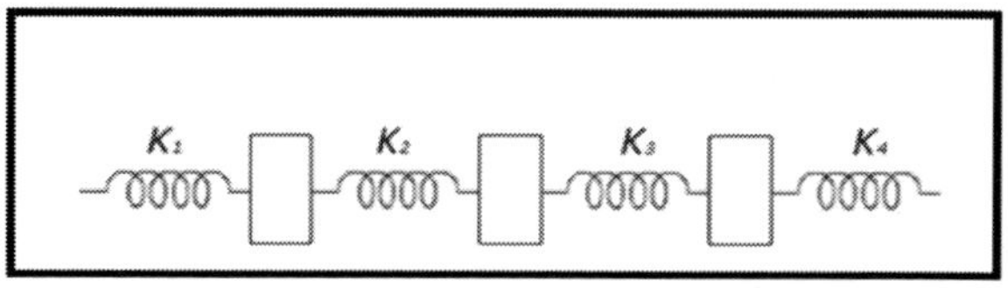

Figure 2(b). Wire rope sections denoted by their spring constants

Figure 2(a) shows the basic layout of the LSRS while Figure 2(b) denotes the wire rope sections as their spring constant. The main properties concerned with LSRS that affect the SAS are its torsional stiffness and axial flexibility which are determined by two factors namely:

- The distance between the bearings.
- The configuration/arrangement of the wire rope.

The LSRS should be designed in such a way that a vehicle should be steered comfortably to safety in the event of a failure of the SBW system, meaning that the LSRS should be torsionally rigid enough to provide control to the car in the event of failure of the SBW but flexible so that active steering can be performed when the SBW system is engaged. The torsional stiffness is characterized by the spring constant or the torsion coefficient. The formula for the equivalent spring constant depends on the number of wire rope sections and can be derived from the Figure 2(b).

$$\frac{1}{K_{eq}} = \frac{1}{K_1} + \frac{1}{K_2} + \frac{1}{K_3} + \ldots\ldots \quad (1)$$

The unit for the Keq is Nm/rad. Hence the equivalent Torque on the LSRS can be calculated using the following equation:

$$\tau = K_{eq}\theta \quad (2)$$

Where τ, is the torque exerted on the LSRS and θ is the angular deflection made by the LSRS in rad. The negative sign indicates that the direction of torque is opposite to the direction of twist.

2.2 Configuration of the Wire Ropes in the LSRS

The configuration of the wire rope is also of grave importance in designing the LSRS. The individual wires in a wire rope are organized in a circular pattern but are wound in a certain direction, either clockwise or anti clockwise. The wire ropes used to transmit torque are categorized under uni-directional or bi-directional. The uni-directional wire ropes are wound for one direction only and they can transmit torque in just one direction efficiently. Although bi-directional ropes can transmit torque in both directions, they have varying torsional stiffness because of the variable laying patterns of the wire ropes. The varying torsional stiffness values of the wire ropes is the reason why the configuration of the wire ropes is very important for the LSRS.

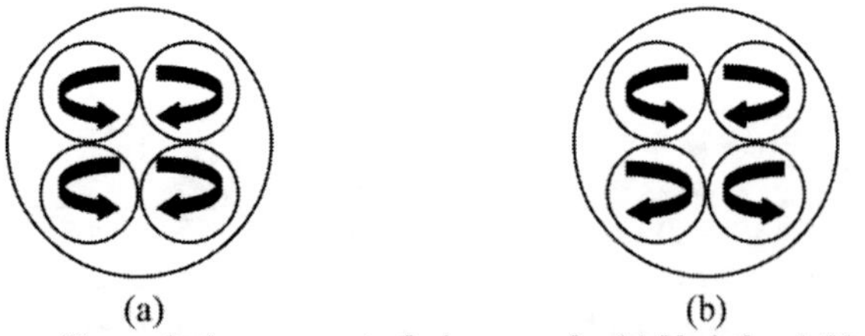

(a) (b)

Figure 3. Arrangement of wire ropes for **(a)** Variation 1 **(b)** Variation 2

Two different configurations of the wire ropes were modeled and analyzed using finite element analyses (FEA) in ANSYS Workbench 15.0 in order to determine the best possible configuration which has similar torsional stiffness in clockwise and anti-clockwise directions. Figure 3(a) shows Variation 1, in which a simple arrangement of 4 strand wire ropes has been chosen. In this variation, two of the adjacent wire ropes are right-hand lay and the other two are left-hand lay so that the overall arrangement has similar stiffness, Keq in both directions. Variation 2 on the other hand, as shown in Figure 3(b), has two alternate wire ropes of right-hand lay and the other two alternate wire ropes of left-hand lay for similar purposes.

3 MODELING OF THE WIRE ROPE

This section discusses the modeling of the 1x7 wire rope strand and the complete wire rope model. The 1 x 7 is the basic strand construction which contains 7 individual wires wrapped in a right or left lay direction. Since there are many parameters which can be varied in the construction of a rope strand, it is important to have a general and accurate model which can predict the effects of possible variations of these parameters on the performance of the strand [9]. The solid models of the right and left lay wire ropes were developed and used to perform the simulations in ANSYSTM, so that the behavior of the wire rope at the application of torque could be studied.

The selected wire rope was a 1x7 strand made out of Galvanized steel in a nominal diametric size of 3.56 mm. Table 1 shows the geometrical parameters of the simple strand.

Table 1. Geometric parameters of the 1x7 strand wire rope

Geometrical parameters	Value
Strand Diameter *(d)*	3.56mm
Center wire diameter $(2R_1)$	1.20mm
Outer wire diameter $(2R_2)$	1.18mm
Pitch length or Lay length *(p)*	54.3mm

For the current 1x7 strand, the outer wire helix radius and angle from Equation (3) and Equation (4) can be found out to be,

$$R = R_1 + R_2 = 1.19mm \quad (3)$$

$$\tan\alpha = \frac{p}{2\pi R} = \frac{54.3}{2\pi 1.19} = 7.26 \quad (4)$$

$$\alpha = 82.15^o$$

The simple strand model of the above mentioned geometrical parameters is shown in Figure 4(a). Meanwhile Figure 4(b) shows the solid model with torque handles in order to make it easier to apply boundary conditions to the wire rope model. The two different types of variations were also modeled in order to analyze which variation is better in terms of equality of torsional stiffness in both directions, which can be seen in Figure 4(c) and Figure 4(d).

 DOI: 10.1051/matecconf/20167400017

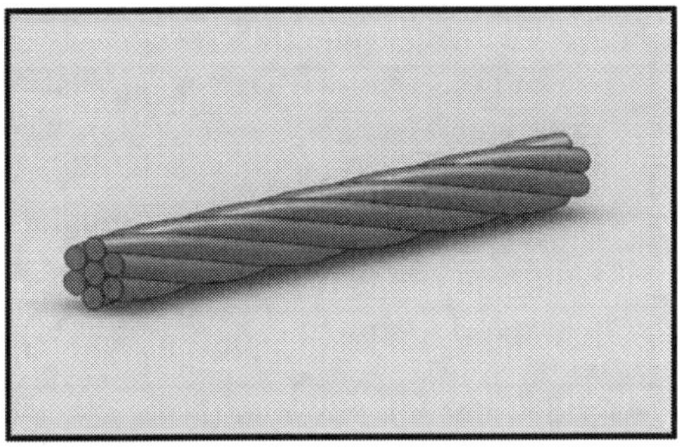

Figure 4(a). Solid model of the 1-strand wire rope

Figure 4(b). Solid model of the 1-strand wire rope with torque handles

Figure 4(c). Top view of Variation 1 – adjacent wire ropes in same lay direction

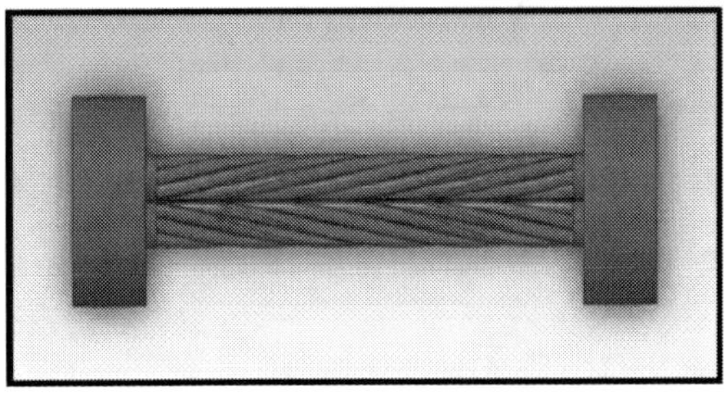

Figure 4(d). Top view of Variation 2 – alternate wire ropes in the same lay direction

4 SIMULATION AND RESULTS

4.1 Simulation of the 1-strand wire rope

In order to find the angle of deformation made by the wire rope at the application of a certain torque in both directions, static structural torsion test was performed on the 1-strand wire rope. The only available wire rope for this experiment is the right-lay direction. For the analysis, the torque was applied both in clockwise and anticlockwise directions in order to determine the difference in behavior of the wire rope in both directions. The frictional contact coefficient between individual wires was 0.115 [10]. Figure 5(a) shows the frictional contact between two individual wires of the wire rope.

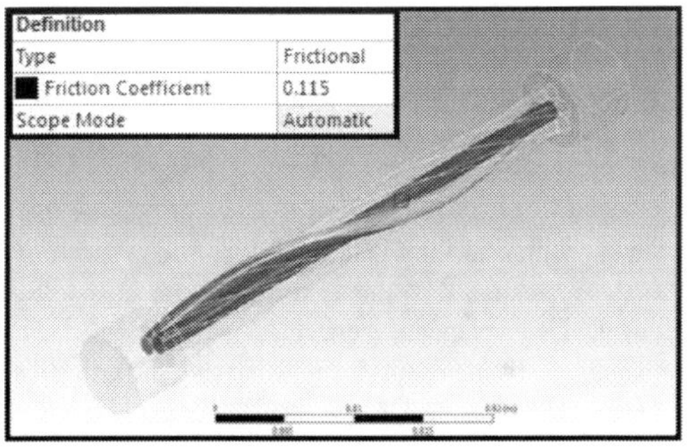

Figure 5(a). Frictional contact between the core and outer wire

Face Sizing was applied to every individual wire, in order to obtain a uniform mesh. The material assigned to the model was Galvanized steel as the purchased wire rope for the experiment was made of the same material. For the boundary conditions, one end of the wire rope was fixed and not allowed to displace in the x, y or z directions and on the other end a torque of 0.05 N.m magnitudes was applied. The results were then analyzed and compared. The total deformation for the anti-clockwise and clockwise directions as a result of the applied torque can be seen in Figure 5(b) and 5(c) respectively.

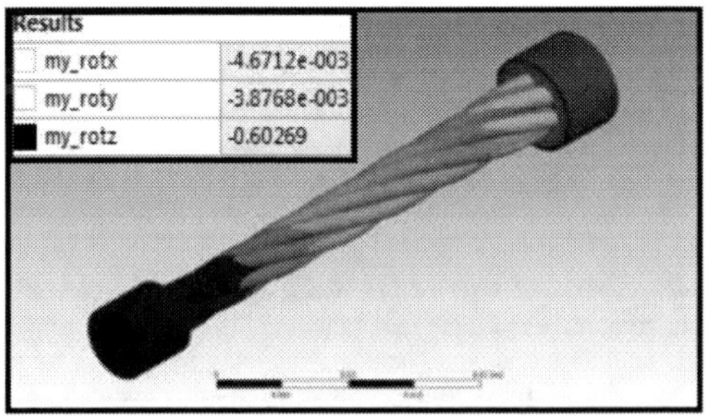

Figure 5(b). Angular deformation in degrees by application of Anti-clockwise torque

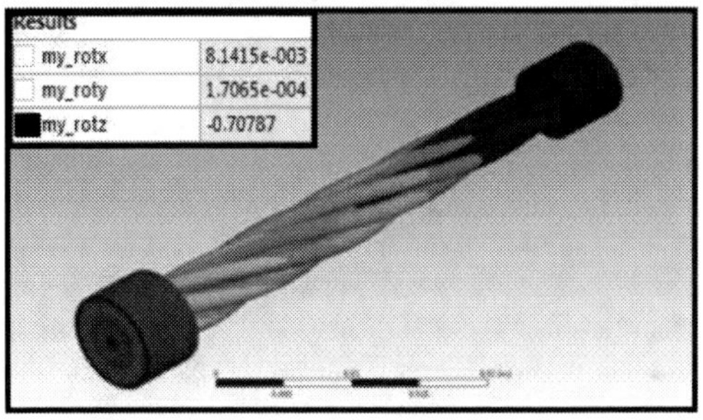

Figure 5(c). Angular deformation in degrees by application of Clockwise torque

A vast difference in the angular deflection is observed at the application of torque in the clockwise and anti-clockwise direction. Table 2 elucidates the angular deflections in degrees which are observed through the simulation results.

Table 2. Angular deflection and percentage difference of 1 strand

Angular Deflection	
Clockwise deflection	0.708°
	0.012 rads
Anticlockwise deflection	0.603°
	0.010 rads
Percentage difference	16.051 %

An angular deflection of 0.708 degrees or 0.012 rad is found in the clockwise direction and a deflection of 0.603 degrees or 0.010 rad is noted in the anti- clockwise direction. The observed percentage difference is 16.05% which proves that the torsional stiffness in both directions varies hugely. The torsional stiffness can be calculated using Equation (5):

$$K = \tau/\theta \tag{5}$$

For clockwise direction:

$K = 4.04\ N.m/rad$

For anti-clockwise direction:

$K = 4.75\ N.m/rad$

The reason for the huge difference in the torsional stiffness as discussed earlier was due to the lay direction of the wire rope. The wire rope under consideration is right-hand lay and the application of the torque in the anti-clockwise direction will make it tighter and hence it will have higher stiffness in that direction. If the torque is applied in the clockwise direction, the torsional stiffness will be lower as the application of torque in that direction will cause the wire rope to un-wind. Because of these reasons, Variation 1 and Variation 2 were subjected to static structural torsion simulation.

4.2 Simulation of the 4-strand wire rope arrangement

To counter the torsional behavioral inconsistency of a single wire rope, two different variations of different arrangement of wire rope were simulated in ANSYSTM. The simulation procedure for the different variations was similar to the ones conducted on the 1-strand wire rope i.e. one end was fixed and the other end was allowed to rotate at a fixed torque. The torque applied on one end of both variations was 0.1 N.m.

The simulation results indicate a vast improvement in the angular deflection for both Variation 1 and 2 (see Table 3). The observed angular deflection in clockwise and anticlockwise direction is much more consistent and similar in values. Table 3 shows the Angular deflection in degrees and the percentage differences for both Variation 1 and Variation 2.

Table 3. Angular deflection and percentage difference of Variation 1 and Variation 2

Angular Deflection		
Variation 1 Clockwise deflection	0.358o	0.00625 rad
Variation 1 Anticlockwise deflection	0.357o	0.00623 rad
Percentage difference	0.34%	
Variation 2 Clockwise deflection	0.346o	0.00603 rad
Variation 2 Anticlockwise deflection	0.359o	0.00627 rad
Percentage difference	3.90%	

For Variation 1, the clockwise angular deflection is found to be 0.358 degrees and the anticlockwise deflection is 0.357 degrees. The percentage difference between the two was calculated to be 0.34%. Similarly, for the Variation 2, the Clockwise deflection is 0.346 degrees and the Anti-clockwise deflection is 0.359 degrees with a percentage difference of 3.90%. The torsional stiffness for the different Variations is calculated below using Equation (5).

- Variation 1

  For clockwise direction:

  $K = 15.99\ N.m/rad$

  For anti-clockwise direction:

  $K = 16.04\ N.m/rad$

- Variation 2

  For clockwise direction:

  $K = 16.58\ N.m/rad$

  For anti-clockwise direction:

  $K = 15.94\ N.m/rad$

The values of the coefficient of torsional stiffness indicate that these two variations have a higher value and indeed much stiffer than the 1-strand wire rope due to the introduction of 3 more wire ropes. The lower values of percentage difference indicate that the torsional inconsistency problem for the 1-strand can be overcome using the two suggested variations, especially Variation 1 with such a low percentage difference of 0.34% compared to the percentage difference of 1-strand wire rope which was found out to be 16.05%.

5 CONCLUSION

The construction of the LSRS and the best configuration of the wire ropes required for its construction have been discussed in this paper. The use of wire ropes for the LSRS of the SAS system serves the purpose of allowing Active steering to be performed on the system because of its torsional flexibility, as well as making it safe in terms of a head on collision or beneficial in terms of packaging due to the property of it being axially flexible.

The simulations performed on the single strand wire rope indicated that there is a vast torsional inconsistency due to its construction as it is laid in the right lay direction. An angular deflection percentage difference of 16.05% was calculated in the clockwise and anti-clockwise direction which indicates that a single wire rope cannot be used for the construction of the LSRS as it will cause the vehicle to have a variable response in the clockwise and anti-clockwise direction when the steering wheel is turned. Using this as the background, two different configurations of wire rope arrangements namely Variation 1 and Variation 2 were also simulated using similar boundary conditions. Variation 1, with an angular deflection percentage difference of 0.34% compared to the 3.90% of Variation 2 is considered suitable to be used for the construction of the LSRS.

REFERENCES

[1] Zhai P. Bilateral control of steer-by-wire vehicles. School of Electrical, Computer and Telecommunications Engineering, University of Wollongong. 2013.
[2] Holt DJ. Electric steering. a revolution in steering technology. Society of Automotive Engineers. 2002.
[3] Mogi K, Sugai T, Sakurai R, Suzuki N. Development of a new steer-by-wire system. NTN Technical Review No.79. 2011.
[4] Baharom MB, Hussain K, Day AJ. Design concepts and analysis of a semi- active-steering system for a passenger car. Proc. IMechE. 2009;223:283-292.
[5] Reid JD, Lechtenberg KA, Stolle CS. Development of advanced finite element material models for cable barrier wire rope. University of Nebraska, Lincoln. 2010.
[6] Otano N, Durville D, Usabiaga H. Finite element simulation of the mechanical behaviour of wire ropes, comparison with analytical models and experimental tests. 5th European Conference on Computational Mechanics (ECCM V). Barcelona, Spain. 2014.
[7] Gerdemeli I, Kurt S, Anıl AS. Fatigue life analysis of wire rope strands with finite element method. Key Engineering Materials. 2014;572: 513-516.
[8] Bechtold M. Modeling of steel ropes. 2009 SIMULIA Customer Conference. 2009.
[9] Jiang WG, Yao MS, Walton JM. A concise finite element model for simple straight wire rope strand. International Journal of Mechanical Sciences 41. 1999; 41:143-161.
[10] Jiang WG, Henshall JL. The analysis of termination effects in wire strand using finite element method. Journal of Strain Analysis. 1999;34:31-38.

Notations

K	Spring constant/ Torsion constant (Nm/rad)
Keq	Equivalent torsion constant of the LSRS (Nm/rad)
τ	Torque exerted by the LSRS (Nm)
θ	Angle of twist from the equilibrium position (Rad)
d	Strand diameter (mm)
R	Radius of the helix on the outer wire (mm)
R_1	Radius of the center wire (mm)
R_2	Radius of the outer wire (mm)
p	Pitch or Lay length of the wire rope (mm)
α	Helix angle of the outer wire (degrees)

DOI: 10.1051/matecconf/20167400018

Review on Finite Element Material Modelling Of Brain Tissue for Surgical Simulation

Faris Tarlochan , Siva Mahesh Tangutooru

[a] *Department of Mechanical and Industrial Engineering, Qatar University, Qatar*

Abstract. Surgical simulation is a growing field of interest as it serves as a good training for the medical practitioners as well as planning and preparing for surgery. For a reliable and a good outcome from the simulation, the challenge is to model the material properties of tissues as accurate as possible. For a real time surgical simulation the numerical model has to be simple and computationally inexpensive. One of the preferred numerical models is the finite element model. Hence, the objective of this paper is to review finite element modelling of the brain tissue. The human brain tissue is an interesting organ to be studied because of their use in understanding trauma related injuries and tissue injury mechanism during brain surgery.

1 Introduction

Interactive and real time surgical simulation is a very important tool in today's health care industry [1 - 5]. Latest technology has reduced to less invasive surgeries by the surgeons. This is accomplished by smaller cuts during surgery. This results in loss of direct contact and tactile information to the surgeons. As a result, surgeons need to adapt themselves to improve their hand-eye coordination. Here simulation tools such as finite element models can to some extend accurately predict tissue injuries. This is also useful for surgeons to practice and plan their surgeries [6-8].

The motivation for this review paper is related to an ongoing research onto addressing the high incidence of blindness in eastern Arabia. Glaucoma is the major ophthalmological challenge for the eastern Arabian population. Glaucoma destroys retinal ganglion cells, the neurons which connect the eye to the visual brain. Without these cells visual signals cannot reach those parts of the central nervous system that give rise to perceptions. Since human ganglion cells do not grown anew and once lost are gone forever, the only practical means of restoring vision to those blinded through glaucoma is a neuroprostheses that interfaces visual neurons of the brain to stimulating electrodes driven by an image capture system and electronic circuitry. This involves navigating the electrodes and penetrating the brain tissue to reach the target side. Hence one of the objectives is to study the level of tissue injury by means of the penetrating electrode. This will help in planning the trajectory for the invasion. Hence a finite element model that best represents the brain tissue will be needed. This paper will share all the developed brain tissue material models for finite element analysis along with their limitations.

2 Material Models

The main key in developing brain tissue finite element material model is the level of accuracy needed. There is a paradox here where the surgical simulation of tissues has to be simple to be used for a broad range of medical conditions but with sufficient complexity to realistically describe the accurate physical response due to variety of applied loading conditions. The review of various material models will help in identifying a solution for this paradox.

<u>Linear Elastic Models</u>
Linear elastic models are commonly used in most finite element analysis as it is the easiest to be implemented in terms of computational time and setting the model up. Several researchers have developed linear elastic model for measuring brain deformation [9 – 12]. From [12], it was found that for a strain rate of 0.01 s-1 (quasi static testing) of actual brain tissue, the tissue can be approximated with a linear elastic model provided the strain does not exit 10%. Higher strain values will significantly differ between the linear model and the observe experimentation data. Even though the brain tissue can be approximated with linear elastic model, however from a clinical point of view this is not really good since most surgical procedures involve deformations of more than 10% [12].
<u>Hyperelastic Models</u>

Hyperelastic material model is used to describe the non-linear stress-strain behaviour of complex materials such as tissues. The model assumes that the material behaviour can be described by means of a strain energy density function, from which the stress–strain relationships can be derived. Hyperelastic materials can be considered to be isotropic, incompressible and strain rate independent. Due to the large deformations seen on tissues during surgeries, these models are used extensively for all kinds of tissues such as brain [13, 18], liver tissue [14], skin [15], soft tissues [16] and for general laparoscopic simulation [17]. The relevant equations of the strain energy function can be explained as follows [18]:

Hyperfoam function:

$$U = \sum_{i=1}^{N} \left[\frac{2\mu_i}{\alpha_i^2} \left(\varphi_1^{\alpha_i} + \varphi_2^{\alpha_i} + \varphi_3^{\alpha_i} - 3 \right) + \frac{1}{\beta_i} \left(J^{-\alpha_i \beta_i} - 1 \right) \right] \tag{1}$$

Where φi are the principal stretches: J is the determinant of the deformation gradient tensor; N is the number of terms; and μ, α, β are material parameters.

Ogden Strain Energy Potential:

$$U = \sum_{i=1}^{N} \frac{2\mu_i}{\alpha_i^2} \left(\bar{\varphi}_1^{\alpha_i} + \bar{\varphi}_2^{\alpha_i} + \bar{\varphi}_3^{\alpha_i} - 3 \right) + \sum_{i=1}^{N} \frac{1}{D_i} (J - 1)^{2i} \tag{2}$$

Where φi are the deviatoric principal stretches, D are material parameters.

Polynomial Strain Energy Potential:

$$U = \sum_{i+j=1}^{N} C_{ij} (I_1 - 3)^i (I_2 - 3)^j + \sum_{i=1}^{N} \frac{1}{D_i} (J - 1)^{2i} \tag{3}$$

Where Cij are material parameters and I_1 and I_2 are the first and second deviatoric strain invariants. From the study conducted by Moran et al [18], it was found that the Hyperfoam, Ogden and polynomial functions have relatively good fit to the experimental data. The coefficient of determination for the Hyperfoam, Ogden and Polynomial were 0.97, 0.92 and 0.80 respectively. In order to achieve a good fit, a value of N = 2 had to be used to avoid strain softening. For each of the above functions the Poisson's ratio were 0.38, 0.42 and 0.44 respectively. Hyperfoam model was the best amongst the three models for the compression test. In terms of shear test, the polynomial function did not fare as well. Tables 1-3 show the parameters for all the three models obtained from [18].

Viscoelastic Models

The coupling of elasticity and viscosity creates a viscoelastic material with specific mechanical responses and it needs specific experimental investigations and modelling approaches. Viscoelastic materials are materials where the stress and strain depends on time or, in the frequency domain, on frequency. Polymers are good example of viscoelastic materials. Several researchers have used linear viscoelastic models to model the brain tissue [19]. This model has also been used by other researchers to model other tissues such as liver [20 - 21] and joints [22]. Some have used this for surgical simulation purpose [23, 24].

For the brain tissue, it is not very accurately to represent them as linear viscoelastic. To some extend brain tissues are non-linear viscoelastic [12]. Several researchers have worked in this aspect to develop non-linear viscoelastic material model for finite element application [25, 26]. Even though the nonlinear viscoelastic model is a good representation of the brain tissue, it is difficult to estimate the parameter values for the model and can be computationally expensive [27].

Some researchers tried to improve the linear viscoelastic model by introducing quasi-linear viscoelastic viscoelasticity [12]. Here the tissue is assumed to have two material behaviour: a time dependent elastic response and a linear viscoelastic stress relaxation response. It was found that this model fits experimental behaviour accurately up to strains of 15%, a much better improvement than the linear elastic models.

The finite element (FE) model for the brain has vast applications. For specific applications, it is important to specific the domain of the brain where a particular response is being computed. This requires discretization of the brain itself. The discretization depends on the nature of the surgery if it requires patient specific information. This leads to the development of a computational grid based on CT and MRI images. Due to computational time efficiency, low order elements are preferred such as linear hexahedron elements to discretize the computational grid. However, due to complexity that arrives due to the presence of tumors, mixed meshes having both hexahedral and linear tetrahedral elements are the most convenient [28]. Figure 1 depicts the patient specific discretization using finite elements. Variations of brain FE models exist with element densities varying from 9000 elements to 50 000 elements [29]. Hence this indicating the element size depends on the accuracy details, the model of the brain itself and the type of response required.

Table 1: Parameters for Hyperfoam [18]

Brain Tissue Region	Mu 1 (Pa)	Alpha 1	Mu 2 (Pa)	Alpha 2	β_1, β_2	G (Pa)	R^2 Compression	R^2 Tension	R^2 Shear
White matter	67.19	10.65	981.67	-3.45	9.15, 1.57	1048	0.97	0.98	0.98
Gray matter	76.46	10.96	1267.70	-3.20	11.70, 0.89	1344	0.97	0.98	1.00
Corona radiata	75.50	11.63	1359.40	-3.84	12.00, 0.72	1434	0.97	0.98	1.00

Table 2: Parameters for Ogden [18]

Brain Tissue Region	Mu 1 (Pa)	Alpha 1	Mu 2 (Pa)	Alpha 2	D_1, D_2	G (Pa)	R^2 Compression	R^2 Tension	R^2 Shear
White matter	71.33	11.61	826.35	-6.07	3.13E-04, 4.98E-04	897	0.99	0.98	0.99
Gray matter	116.43	7.49	1058.9	-6.75	3.19E-04, 3.89E-04	1175	0.98	0.96	0.99
Corona radiata	90.94	6.97	1150.6	-7.42	3.18E-04, 2.62E-04	1241	0.98	0.92	0.99

Table 3: Parameters for polynomial function [18]

Brain Tissue Region	C10 (Pa)	C01 (Pa)	C20	C11	C02	D_1, D_2	G (Pa)	R^2 Comp.	R^2 Tension	R^2 Shear
White matter	-1455	2186	3235	-5276	1875	3.13E-04, 0.00	1462	1.00	0.95	0.98
Gray matter	-1526	2291	3226	-5276	1958	3.13E-04, 0.00	1530	0.99	0.95	0.89
Corona radiata	-1545	2372	3123	-5146	2052	3.13E-04, 0.00	1653	0.99	0.91	0.80

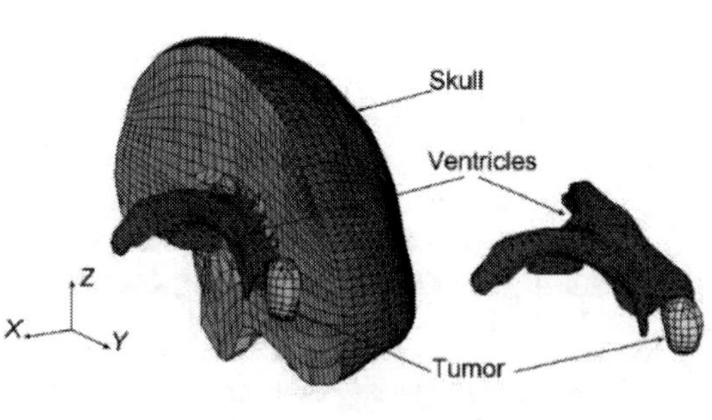

Figure 1. Patient-specific hexahedron-dominant brain mesh [28]

3 Conclusion and Challenges

From published experimental data, the mechanical response of brain tissue to external loading is very complex. The response is non-linear where brain tissue has a higher stiffness in compression than in tension. Three types of common brain tissue material models used in finite element analysis were discussed; namely the linear elastic, hyperelastic and viscoelastic. The linear elastic model is the simplest and computationally fast. However, this model is only accurate for strains up to 10%. From these three models, the hyperelastic model, especially the Hyperfoam model represents accurately the tissue behavior up to strains of 50% [12]. Quasi-linear viscoelastic models are good for strains up to 15%. Hence, it can be concluded here that defining the elastic response of the brain tissue is best done with the hyperelastic material model. There is however some challenges that need to be address to make surgical simulation a reality. The challenge is for operation planning, the computational model must be patient-specific

Acknowledgements

This conference paper was made possible by NPRP grant # (5 - 457 - 2 - 181) from the Qatar National Research Fund (a member of Qatar foundation). The statements made herein are solely the responsibility of the author(s).

References

[1] Guillaume Picinbono, Herve Delingette,Nicholas Ayache. Non-linear anisotropic elasticity for real-time surgery simulation. Graphical Models 65 (2003) 305–321

[2] Victoria Valinluck Lao, Scott R. Steele. The role of simulation in colon and rectal surgery training. Seminars in Colon and Rectal Surgery, In Press, Accepted Manuscript, Available online 14 April 2015

[3] Hsiu-Hsia Lin, Lun-Jou Lo. Three-dimensional computer-assisted surgical simulation and intraoperative navigation in orthognathic surgery: A literature review. Journal of the Formosan Medical Association, 114 (4) 2015, 300-307

[4] N. Abe, S. Kuroda, M. Furutani, E. Tanaka. Data-based prediction of soft tissue changes after orthognathic surgery: clinical assessment of new simulation software. International Journal of Oral and Maxillofacial Surgery, Volume 44 (1) 2015, 90-96

[5] Robert D. Acton. The Evolving Role of Simulation in Teaching Surgery in Undergraduate Medical Education. Surgical Clinics of North America, In Press, Corrected Proof, Available online 23 May 2015.

[6] Kwong Ming Tse, Long Bin Tan, Shu Jin Lee, Siak Piang Lim, Heow Pueh Lee. Investigation of the relationship between facial injuries and traumatic brain injuries using a realistic subject-specific finite element head model. Accident Analysis & Prevention, 79, 2015, 13-32.

[7] Jennifer A. DeWit, Duane S. Cronin. Cervical spine segment finite element model for traumatic injury prediction. Journal of the Mechanical Behavior of Biomedical Materials, 10, June 2012, 138-150.
[8] Lei Cheng, Blake Hannaford. Finite Element Analysis for evaluating liver tissue damage due to mechanical compression. Journal of Biomechanics, 48 (6) 2015, 948-955.
[9] Skrinjar O Nabavi, A Duncan. Model driven brain shift compensation. Medical Image Analysis, 6(4), 2002. 361 – 373.
[10] Warfield SK, F. Talos, et. al. Real time registration of volumetric brain MRI by biomechanical simulation of deformation during image guided neurosurgery. Computing and Visualization in Science, 5 (1), 2001, 3 - 11.
[11] Ferrant M, Warrant SK, et al. Registration of 3D intraoperative MR images of the brain using a finite element biomechanical model. IEEE Transaction of medical Imaging, 20 (12), 2001, 1384 – 1397.
[12] A. Mendizabal, I Aguinaga, E. Sanchez. Characterization and modelling of brain tissue for surgical simulation. Journal of the Mechanical behavior of Biomedical Materials 45, 2015, 1 – 10.
[13] Sack Kaster, A . Samani. Measurement of the hyperelastic properties of ex vivo brain tissue slices. Journal of Biomechanics. 44 (6) 2011, 1158 - 1163
[14] C. Gao, K. Lister, J.P Desai. Constitutive modeling of liver tissue: experiment and theory. Annals of Biomedical Engineering 38 (2) 2010, 505-516.
[15] R.J. Lapeer, P.D Gasson, V. Karri. A hyperelastic finite element model of human skin for interactive real time surgical simulation. IEEE Transactions of Biomedical Engineering 58 (4), 1013 – 1022.
[16] J. Hug, R. Hutte, et. al. Virtual reality based simulation of endoscopic surgery. Presence: Teleoperat. Virtual Environment 9 (3), 2000, 310 – 333.
[17] Pals Martin, Natal Jorge, AJM Ferreira. A comparative study of sevral material models for predictions of Hyperelastic properties: application to silicon rubber and soft tissues. Strain 42 (3), 2006, 135-147.
[18] Richard Moran, Joshua H Smith, Jose J. Garcia. Fitted hyperelastic parameters for human brain tissue from reported tension, compression and shear tests. Journal of Biomechanics. 47, 2014, 3762-3766.
[19] A Yoshizawa, J. Okamoto, H Yamakawa, M.G. Fujie. Robot surgery based on the physical properties of the brain physical brain model for planning and navigation of a surgical robot. International conference on Robotics and Automation, April 2005. 904 – 911.
[20] Z. Liu, L. Bilston. On the viscoelastic character of a liver tissue. Experiment and modelling of the linear behavior. Biorheology 39 (6), 2000, 735 – 742.
[21] M. Sedef, E. Samur, C . Basdogan. Real time finite element simulation of linear viscoelastic tissue behavior based on experimental data. Computer Graphics and Applications, IEEE 26 (6) 2006, 28 – 38.
[22] L Yoo, V Gupta, et al. Viscoelastic properties of bovine orbital connective tissue and fat: constitutive models. Biomechanics Model Mechanobiology. 8 (1), 2011, 901 – 914.
[23] S. Misra, K. ramesh et al. Modeling of nonlinear elastic tissues for surgical simulation. Computational methods In Biomechanics . Biomedical engineering. 13 (6), 2010, 811-818.
[24] B Ahn, J. Kim. Efficient soft tissue characterization under large deformation in medical simulations. International Journal of Precision Engineering Manufacturing. 10 (4), 2009, 115 – 121.
[25] D. Brands, G. Peters, P. Bovendeerd. Design and numerical implementation of the brain tissue during impact. Journal of Biomechanics. 37(1), 2004, 127 – 134.
[26] M Hrapko et al. The mechanical behavior of brain tissue: large strain response and constitutive modelling. Biorheology. 43(5), 2006, 623 – 636.
[27] L.E Bilston, Z. Liu, N. Phan Thien. Large strain behavior of brain tissue in shear. Some experimental data and differential constitutive model. Biorheology 38 (4) 2001, 335 – 345.
[28] Karol Miller, Adam Wittek, Grand Joldes. Biomechanics of the brain for computer-integrated surgery. Acta of Bioengineering and Biomechanics. 12 (2), 2010, 25-37
[29] T J Horgan, M D Gilchrist.The creation of three-dimensional finite element models for simulating head impact biomechanics. International Journal of Crash. 8(4), 2003, 353–366.

Review on the Recent Developments of Photovoltaic Thermal (PV/T) and Proton Exchange Membrane Fuel Cell (PEMFC) Based Hybrid System

Afzam Zulkepli[1], Hassan Ibrahim[1], Azri Alias[1], Zafri Azran[2], Firdaus Basrawi[1]

[1]*ESFG, Faculty of Mechanical Engineering, UMP Pekan, Pahang, Malaysia*

[2]*Kuliyyah of Allied HealthSciences, IIUM, Kuantan, Pahang, Malaysia*

Abstract. Photovoltaic Thermal (PV/T) system emerged as one of the convenient type of renewable energy system acquire the ability to generate power and thermal energy in the absence of moving parts. However, the power output of PV/T is intermittent due to dependency on solar irradiation condition. Furthermore, its efficiency decreases because of cells instability at high temperature. On the other hand, fuel cell co-generation system (CGS) is another technology that can generate power and heat simultaneously. Integration of PV/T and fuel cell CGS could enhance the reliability and sustainability of both systems as well as increasing the overall system performance. Hence, this paper intended to present the parameters that affect performance of PV/T and Proton Exchange Membrane Fuel Cell (PEMFC) CGS. Moreover, recent developments on PV/T-fuel cell hybrid system are also presented. Based on literates, mass flow rate of moving fluid in PV/T was found to affect the system efficiency. For the PEMFC, when the heat is utilized, the system performance can be increased where the heat efficiency is similar to electrical efficiency which is about 50%. Recent developments of hybrid PV/T and fuel cell show that most of the studies only focus on the power generation of the system. There are less study on the both power and heat utilization which is indeed necessary in future development in term of operation strategy, optimization of size, and operation algorithm.

1 Introduction

Continuous concerns on global and local environmental problems and depletion of conventional energy source have encouraged the transition towards renewable energy resources. Research on green energy power generation system continues to gather momentum when the remaining reserve of oil and gas rapidly dwindles as the world population increases. Among the renewable energy power generation systems, Fuel cell and PV/T are promising systems as both are clean and can generate electrical and heat simultaneously. Various studies have been developed to penetrate these systems on residential application in either grid tied or standalone distributed generation. Furthermore, these two technologies also play the key role in the currently developing smart grid system that allows two-way connection between utilities and consumers[1].

Hybrid power systems combine two or more sources of renewable energy into one system which can be merged or differentiated through different configuration or architecture[2]. Electrical output of a single renewable energy source is limited and is not consistent but due to their complementarities the integration could overcome the drawbacks. The purpose of a hybrid system is to produce as much energy from renewable energy sources to ensure load demand and to reduce the dependency on only one system which has intermittent output power. The intermittent output power of photovoltaic due do the dependency on weather circumstances can be covered by merging photovoltaic system with other renewable source such as fuel cell. Fuel cell system is not dependent on solar radiation, the technology could be a backup power generator for photovoltaic during the night or when power output of photovoltaic is deficient. Fuel cell system is a promising power generation technology to provide energy for rural areas when there is no access to grid system[3]. Furthermore, using hydrogen as the reactant makes the technology environmentally clean and noiseless[4,5]. This paper presents the review of parameters that affect both PV/T and PEMFC CGS. Furthermore, recent developments on PV/T-PEMFC based hybrid system are also presented.

2 Photovoltaic Thermal (PV/T)

PV/T system consists of two main components, namely; photovoltaic panel which generates electricity from solar energy through photoelectric effect and solar thermal collector which absorbs solar radiation and transforms the energy into thermal energy[6]. The absorber plate as shown in Fig.1 underneath the solar cell has twofold functions which are to cool the solar cell and to collect

[a] Corresponding author: mfirdausb@ump.edu.my

the heat energy from solar irradiation. The idea of merging these two technologies is because the increase of solar cell temperature will decrease the efficiency of photovoltaic system proportionally. In order to resolve this problem, combining solar collector with photovoltaic will sustain the temperature of the solar cell at optimal range, thus, increasing the efficiency of photovoltaic. On the other hand, thermal energy can be used for heating and thus increase the overall system efficiency[7].

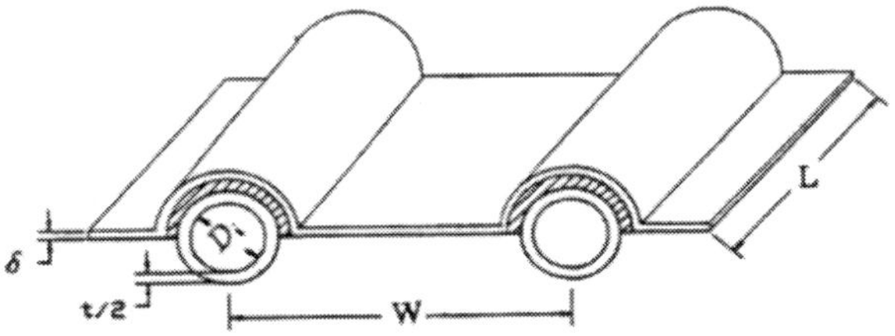

Figure 1. Absorber plate of PV/T[8]

PV/T system can be classified by the different type of coolant used, namely: air, water and refrigerant. In the previous study as shown in table 1, refrigerant coolant type has the highest efficiency which is 56%-74%. Based on physical design, it can be distinguished by flat plate type, concentrated or building integrated type.

Table 1. PV/T efficiency with different type of coolant.[9,10,11]

PV/T models	Average efficiency (%)
Air based PV/T type	24-47
water based PV/T type	33-59
refrigerant PV/T type	56-74

2.1 Controlled parameters on PV/T

2.1.1 Solar cell

The material of photovoltaic cell provides a big impact on the overall performance of PV/T. This type of solar cell has been comprehensively reviewed by Moradi et.al [12]. Multijunction type of solar cell is found to have the highest efficiency with 32% and currently the most broadly used is single crystalline silicone solar cell.

2.1.2 Packing factor

The study made by Vats et.al [13] showed that the increase of packing factor is not always proportional with the efficiency of the system. An optimal packing factor is crucial for the PV/T performance as the high value of packing factor will cause the increase of solar cell temperature and thus decrease the electrical efficiency. In contra, low value of packing factor will reduce the solar radiation absorber area and also leads to low electrical efficiency.

[a] Corresponding author: mfirdausb@ump.edu.my

2.1.3 Mass flow rate

As the efficiency of photovoltaic decreases when the temperature of solar cell increases, a good convection heat flow is vital by means of mass flow rate of the moving fluid. The moving fluid underneath the solar cell can be naturally moved [14] or by force from a pump [15].The mass flow rate is influenced by the type of material used for the circulating fluid, its velocity, and also thegeometry of the PV/T system. The mass flow rate used in previous studies range from 0.01kg/s to 0.3182kg/s as stated by Shahsavar and Ameri[16].

2.1.4 Absorber design

Ibrahim et.al [17]has conducted a study on several absorber design such as oscillatory, direct flow, modified serpentine-parallel flow, spiral flow, serpentine flow, parallel-serpentine flow and web flow. Spiral flow design showed the best result with total efficiency of 68%.

2.1.5 PV/T configuration

The configuration of PV/T also affects the performance of the system. A study on various configurations had been carried out by Tiwari and Soda [18] who investigate the glazed and unglazed PV/T. The efficiency of all the configurations were compared and the glazed without tedlar showed the best performance. Dubey and Tiwari [19] investigated the performance of solar flat-plate collector partially covered using theoretical modeling.The result showed that partially covered collector is better when the main load is thermal and for fully covered collector, it is greater when the main load is electricity. An explicit dynamic model has been introduced by Chow [20] to investigate the performance of single-glazed sheet-and-tube collector. Zondag et.al [21] compared the performance of separately photovoltaic and solar collector with combined PV-thermal collector. Combined PV and solar collector is found to produce more energy than separately installed PV and solar collector.

Various studies have been conducted previously in investigating the parameters that influencing the performance of PV/T. This result in many aspects should be considered in designing the system. Mass flow rate of PV/T was found to affect the system performance because it affects the temperature of the solar cell and also an important factor to be considered in absorbing the heat energy. PV/T system itself is however not sustainable as it depends on weather condition which needs to be integrated with other energy sources. This is necessary as it may provide opportunity in exceeding the current system performance.

3. Fuel cell

Table 2. Comparison of fuel cell with other power generation system[22].

	diesel engine	turbine generator	Photovoltaic	Wind turbine	Fuel cell
capacity range	500kW-50MW	500kW-5MW	1kW-1MW	10kW-1MW	200kW-2MW
efficiency	35%	29-42%	6-19%	25%	40-85%
capital cost ($/kW)	200-350	450-870	6600	1000	1500-3000

Table 3. Fuel cell types [23].

Fuel Cell Type	Operating Temperature (°C)	Electrical Efficiency(%)	Combined heat and power efficiency (%)
Proton Exchange Membrane	50-100	58	32
Alkaline Fuel Cell	90-100	60	20
Phosporic Acid Fuel Cell	150-200	39	43
Molten Carbonate Fuel Cell	600-700	44	36
Solid Oxide Fuel Cell	800-1000	42	42

As shown in Table 2, fuel cell has the highest efficiency when compared to other power generation systems.

Fuel cell can be distinguished by the type of electrolyte used and their operating temperature as shown in Table 3. PEMFC operates at the lowest temperature whereas Solid Oxide Fuel Cell at the highest temperature. Generally, higher output can be achieved at higher temperature. Table 3 demonstrates the electrical efficiency and combined heat and power efficiency of various types of fuel cell.

Among all types of fuel cell, PEMFC is found to be the best application for residential and transportation as it operates under low temperature, fast start up, low emission of oxides of nitrogen and sulfur and produce low noise [24]. A study on the feasibility of utilizing PEMFC in residential sector has been carried out by Oh et.al [25] where a 1-kW PEMFC was installed on an apartment house with average monthly electric consumption of 370kWh. It is showed that 20% savings can be achieved by PEMFC utilization. To commercialize PEMFC, several technical problems in PEMFC must be solved and one of them is proper thermal management. Heat generation by PEMFC is estimated similar to its electrical efficiency which is about 50%. In order to increase the overall efficiency of PEMFC, this wasted heat generation must be fully utilized. Heat generation of PEMFC is mainly caused by entropic heat of reaction, irreversibilities of the electrochemical reactions and ohmic resistances, as well as water condensation[26].Obara and Tanno[27] carried out a study on 3kW PEMFC co-generation system for residential application considering the regional characteristics in Japan. The system performance is evaluated based on exergy flow and exergy efficiency. Obara [28] conducted a study on the efficiency of PEMFC on the low load region in order to maintain the efficiency of the power generation, thus avoiding the high cost. Numerical analysis was done in determining the fuel consumption and also the carbon dioxide emission. It was found that power generation efficiency was higher at the high load factor value.

PEMFC is potentially suitable in becoming power generation system for residential with its characteristic; fast start up, low temperature, low emission, etc. Consuming hydrogen gas as its fuel makes it sustainable and not reliable on weather condition. The utilization of heat from PEMFC is an important parameter to take into account. Heat efficiency is found to be similar with power efficiency which is about 50%. All of the above features make PEMFC a great system to be integrated with PV/T system. As both systems generate electricity and heat simultaneously, study on these two output energy utilization is necessary in optimizing the system performance.

4. Different hybrid system with photovoltaic and fuel cell

4.1 Hybrid PV-FC

4.1.1 Hybrid PV-FC-Battery

In this configuration as shown in Fig.3, photovoltaic is the primary source during daylight and fuel cell act as a back-up during the night or when load exceed the output power of photovoltaic. During the peak hours, load demand is totally supported by the photovoltaic and surplus energy is sent to the battery as energy storage. Battery and fuel cell work simultaneously when output power of photovoltaic is zero. Maximum power is extracted from PV panels by using Maximum Power Point Tracker (MPPT).

[a] Corresponding author: mfirdausb@ump.edu.my

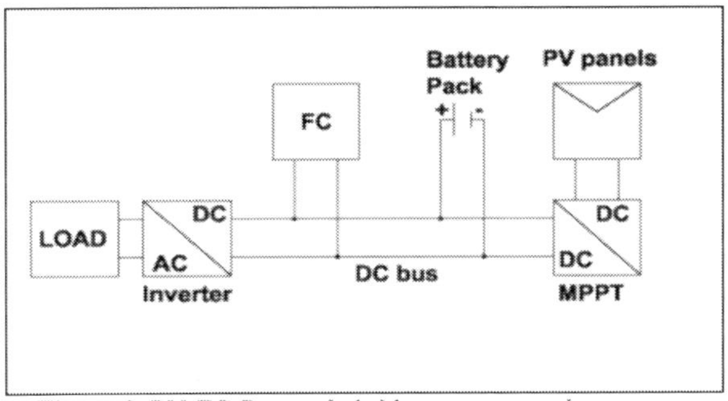

Figure 3. PV-FC-Battery hybrid power generation system layout[29].

Zhang and Huang [30] proposed a fault detection method to diagnose the fault of hybrid systems. The proposed system consists of PV panel, PEMFC, and lithium-ion battery. All the parts are connected to the DC bus. PV system is the primary source while PEMFC will meet the deficient power.

Bruni et.al[29] conducted experimental and numerical study on hybrid PV-FC-Battery to optimize the system by means of rule based control strategy. The control strategy succeeded in increasing the lifetime of components and also cutting down the cost.

Bruni et.al[31] proposed a new novel definition of the energy efficiency in system of integration PV, FC and battery. The new definition is related to the system sizing and the value influenced by the fossil energy consumption.

4.1.2 Hybrid PV-FC-Battery-Electrolyzer

This system consists generally of a photovoltaic, electrolyzer, hydrogen storage tank, fuel cell and power conditioning unit. Surplus energy from the photovoltaic will be sent to the electrolyzer for hydrogen production process. Fuel cell will be utilized when Photovoltaic energy is deficient and works as an auxiliary generator.

Gencoglu et.al [32] designed a hybrid system as shown in Figure 4 for total power load demand 6.23kW. The hybrid system output power is 7.3kW with 2.5kW from PV and 4.8kW from fuel cell.

Castaneda et.al [33] studied a stand-alone hybrid system which only use a PV as the main source. A new sizing method was proposed and three control strategies were performed for energy management.

4.1.3 Hybrid PV-FC-Electrolyzer

In this configuration, as the cost of battery is high, instead of using battery, the surplus energy is totally stored in hydrogen storage tank. Fuel cell acts as a backup during night or when PV power is deficient.

Hwang et.al [34] developed a dynamic model to simulate a stand-alone PV-FC hybrid power systemas shown in Fig.5. In this system, PV is designed to meet the load demand. During the daylight, when there is excess energy from PV, it will be used to produce hydrogen through electrolyzer. During the low radiation period, fuel cell will be the backup system by consuming hydrogen from hydrogen tank.

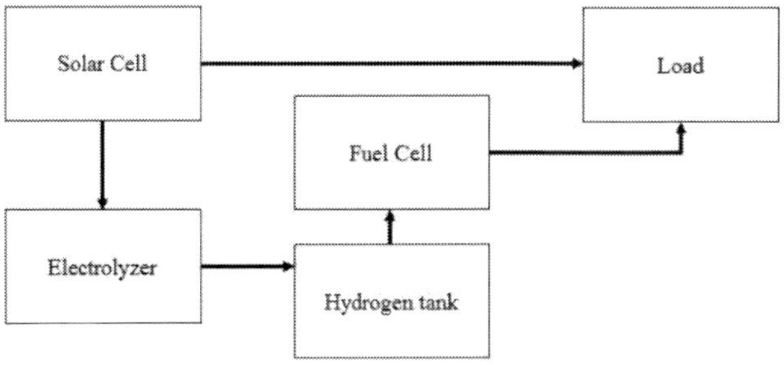

Figure 5. Schematic diagram of PVFC-Electrolyzer hybrid system

El-Shatter et.al [35] designed a hybrid system composed of PV, electrolyzer and fuel cell. A fuzzy regression model (FRM) is utilized to extract maximum power from PV. Tesfahunegn et.al [36]proposed an algorithm to suppress short term power fluctuation in PV generation.

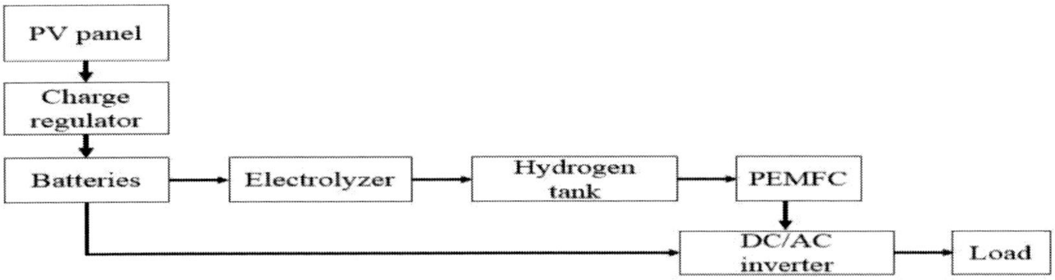

Figure 4. Hybrid PV-FC-Battery-Electrolyzer

[a] Corresponding author: mfirdausb@ump.edu.my

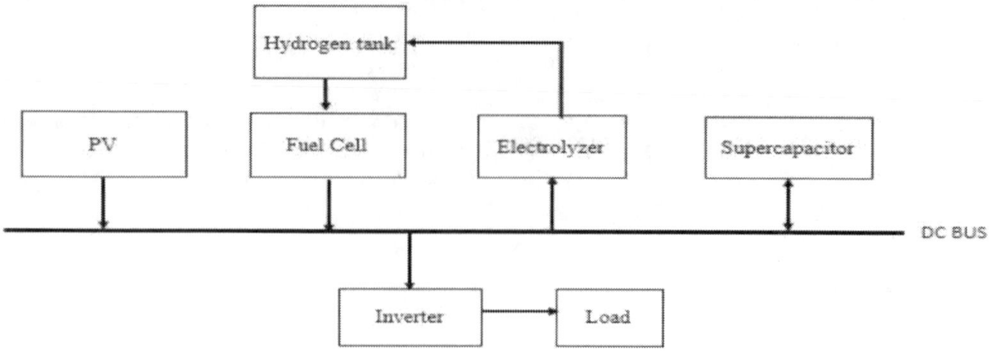

Figure 6. Block diagram of the proposed system

4.1.4 Hybrid PV-FC-Supercapacitor

Several hybrid systems utilize super capacitor for storage device due to the slow response of fuel cell. Super capacitor is greater than battery in term of power density. By using super capacitor, fuel cell will not need to support all the load demand, thus, sizing up the fuel cell is unnecessary. This will eventually prevent the gas starvation in fuel that may affect the performance and lifetime.

Uzunoglu et.al[37] proposed a dynamic modeling of hybrid PV/FC/SC as shown in Fig. 6. The system is designed for stand-alone micro grid user application. This modeling can be used for analysing each component as well as analysing the overall system performance and also for the system design.

Thounthong et al[38]designed a mathematical model of hybrid PV/FC/SC. A dynamic optimization is proposed using intelligent fuzzy logic controller. Several others proposed the same system with the energy management approach[38,39]

Various configurations have been proposed by previous studies which can be concluded that they can be classified in term of energy storage method. In the early stage a simple system which only consisting of battery is used. Battery is however rapidly degrading and not reliable in long term as well as not economic. This leads to the utilization of electrolyzer and hydrogen storage tank which is more reliable and great in storing higher capacity of energy. Besides that, supercapacitor is also used on several studies for its high power density and long lifetime characteristic.

4.1.5 Hybrid PV-FC-Wind Turbine (WT)

Generally, PV and wind turbine are the primary energy sources and FC acts as the backup. The wind turbine is proposed in the system to fully utilize the surround renewable source and to cover the intermittent power of PV.

Ahmed et.al [41]proposed a system with the combination of PV, FC and WT to increase the output power and to generate more constant power. The system is connected to grid to reduce the dependency on the grid system and to provide continuous power supply when the grid system is interrupted.

Mezzani et al[42] investigated the hybrid system consisting of photovoltaic/wind/fuel cell. A mathematical model topology was introduced with the power management of the global system control. The simulation of the proposed system has been done using Matlab/Simulink software.

Wang et al[43] carried out a study on a hybrid energy system which consists of PV, FC, WT, battery and electrolyzer. The PV and wind act as the primary power sources and the FC acts as the backup unit. The battery is used for a short-time backup.

The same configuration is also proposed with the utilization of three individual dc-dc boost converters to manage the power flow to the load[44].

Instead of using PV and FC, several studies have included wind turbine sub system in their configuration to minimize the size of PV and also generate more output power. Furthermore, combination of more sub-system provides more advantage to the system in term of reliability and sustainability aspect.

4.2 Hybrid PV/T-FC (PEMFC) Co-generation

Zafar and Dincer[45] proposed a hybrid system consisting of PV/T and fuel cell. PV/T is the primary source which converts solar energy into electricity and heat. During off peak hours, surplus energy PV is sent to the electrolyzer which then generates hydrogen. Fuel cell generates electricity, heat energy and drinkable water to the residential sector when the load demand exceeds the PV/T maximum output power. The study showed that energy and exergy of the overall system increase to 5.65% and 19.80% respectively when the heat generation of fuel cell is utilized.

[a] Corresponding author: mfirdausb@ump.edu.my

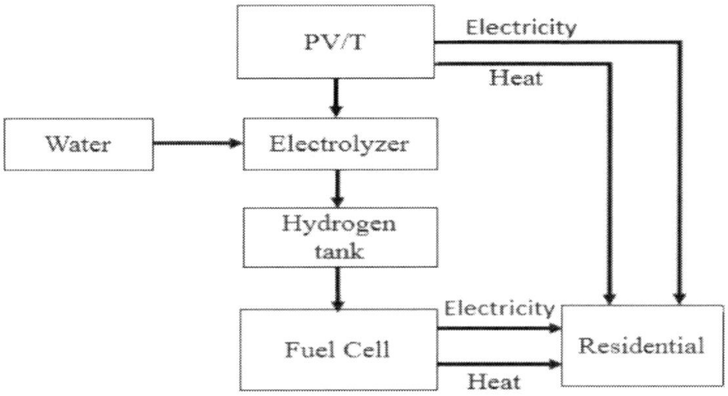

Figure 7. Schematic diagram of the combination PV/T and fuel cell co-generation system

Hosseini et.al [46] studied a combined PV and fuel cell. PV generates electricity for hydrogen production and supply load demand. High temperature gases from the fuel cell stack is flowed to a heat recovery steam generator (HRSG) to produce steam. The HRSG produce steam for the absorption chiller.

Based on previous studies, research on hybrid PV/T-PEMFC CGS is still not yet comprehensively explored. Most of the previous studies focus only on the power generation of the hybrid system where the waste heat was not utilized. There is necessary in using the waste heat as it could improve the overall system performance. This hybrid co-generation system can be further studied for several other aspects including operation strategy, optimization of size, and operation algorithm.

5 Conclusion

A review on the PV/T system, fuel cell and different type of hybrid system of PV/T and fuel cell has been done. The present study helps to understand the important parameters that affect both systems. Besides that, recent developments of the hybrid system of PV/T and fuel cell are also presented to identify the insufficient study in this field. Mass flow rate of PV/T was found to affects the system performance as it is an important parameter to decrease the solar cell temperature and also to absorb heat energy efficiently. Various configuration of PV/T-fuel based hybrid can be distinguished by their energy storage method. Mostly used was by storing the energy using electrolyzer and hydrogen tank. Previous studies mostly focus on the power generation where the waste heat was not utilized. Study on PV/T-PEMFC in simultaneous heat and power generation system are indeed necessary and significant for future development. Utilization of heat for combined heat and power (CHP) generation system significantly increases the overall system efficiency compared to single power generation system while improving both economic and environmental performance.

Acknowledgement

The author would like to acknowledge the authority of Malaysia Ministry of Education and University Malaysia Pahang upon the funding provide for the research under the grant (FGRS/2/2013/TK06/UMP/02/1) or RDU130133.

References

1. V. C. Güngör, D. Sahin, T. Kocak, S. Ergüt, C. Buccella, C. Cecati, and G. P. Hancke, "Smart grid technologies: Communication technologies and standards," *IEEE Trans. Ind. Informatics*, vol. 7, no. 4, pp. 529–539, 2011.
2. D. Rekioua, S. Bensmail, and N. Bettar, "Development of hybrid photovoltaic-fuel cell system for stand-alone application," *Int. J. Hydrogen Energy*, vol. 39, no. 3, pp. 1604–1611, 2014.
3. S. Mekhilef, R. Saidur, and a. Safari, "Comparative study of different fuel cell technologies," *Renew. Sustain. Energy Rev.*, vol. 16, no. 1, pp. 981–989, 2012.
4. a. Larrosa-Guerrero, K. Scott, I. M. Head, F. Mateo, a. Ginesta, and C. Godinez, "Effect of temperature on the performance of microbial fuel cells," *Fuel*, vol. 89, no. 12, pp. 3985–3994, 2010.
5. W. Schmittinger and A. Vahidi, "A review of the main parameters influencing long-term performance and durability of PEM fuel cells," *J. Power Sources*, vol. 180, no. 1, pp. 1–14, May 2008.
6. F. Sarhaddi, S. Farahat, H. Ajam, a. Behzadmehr, and M. Mahdavi Adeli, "An improved thermal and electrical model for a solar photovoltaic thermal (PV/T) air collector," *Appl. Energy*, vol. 87, no. 7, pp. 2328–2339, Jul. 2010.

[a] Corresponding author: mfirdausb@ump.edu.my

7. H. Zondag, "Flat-plate PV-Thermal collectors and systems: A review," *Renew. Sustain. Energy Rev.*, vol. 12, no. 4, pp. 891–959, May 2008.
8. P. G. Charalambous, G. G. Maidment, S. a. Kalogirou, and K. Yiakoumetti, "Photovoltaic thermal (PV/T) collectors: A review," *Appl. Therm. Eng.*, vol. 27, no. 2–3, pp. 275–286, Feb. 2007.
9. S. C. Solanki, S. Dubey, and A. Tiwari, "Indoor simulation and testing of photovoltaic thermal (PV/T) air collectors," *Appl. Energy*, vol. 86, no. 11, pp. 2421–2428, 2009.
10. T. H. Lin, W. C. Hung, and F. S. Sun, "PERFORMANCE EVALUATION OF SOLAR PHOTOVOLTAIC / THERMAL SYSTEMS," vol. 70, no. 5, pp. 443–448, 2001.
11. J. Ji, G. Pei, T. T. Chow, K. Liu, H. He, J. Lu, and C. Han, "Experimental study of photovoltaic solar assisted heat pump system," *Sol. Energy*, vol. 82, no. 1, pp. 43–52, 2008.
12. K. Moradi, M. Ali Ebadian, and C.-X. Lin, "A review of PV/T technologies: Effects of control parameters," *Int. J. Heat Mass Transf.*, vol. 64, pp. 483–500, Sep. 2013.
13. K. Vats, V. Tomar, and G. N. Tiwari, "Effect of packing factor on the performance of a building integrated semitransparent photovoltaic thermal (BISPVT) system with air duct," *Energy Build.*, vol. 53, pp. 159–165, Oct. 2012.
14. T. T. Chow, W. He, and J. Ji, "Hybrid photovoltaic-thermosyphon water heating system for residential application," *Sol. Energy*, vol. 80, no. 3, pp. 298–306, Mar. 2006.
15. A. Tiwari and M. S. Sodha, "Performance evaluation of hybrid PV/thermal water/air heating system: A parametric study," *Renew. Energy*, vol. 31, no. 15, pp. 2460–2474, Dec. 2006.
16. a. Shahsavar and M. Ameri, "Experimental investigation and modeling of a direct-coupled PV/T air collector," *Sol. Energy*, vol. 84, no. 11, pp. 1938–1958, 2010.
17. A. Ibrahim, M. Y. Othman, M. H. Ruslan, M. A. Alghoul, M. Yahya, A. Zaharim, and K. Sopian, "Performance of Photovoltaic Thermal Collector (PVT) With Different Absorbers Design," vol. 5, no. 3, pp. 321–330, 2009.
18. A. Tiwari and M. S. Sodha, "Parametric study of various configurations of hybrid PV/thermal air collector: Experimental validation of theoretical model," *Sol. Energy Mater. Sol. Cells*, vol. 91, no. 1, pp. 17–28, Jan. 2007.
19. S. Dubey and G. N. Tiwari, "Analysis of PV/T flat plate water collectors connected in series," *Sol. Energy*, vol. 83, no. 9, pp. 1485–1498, Sep. 2009.
20. T. T. Chow, "Performance analysis of photovoltaic-thermal collector by explicit dynamic model," *Sol. Energy*, vol. 75, no. 2, pp. 143–152, Aug. 2003.
21. H. A. Zondag, D. W. De Vries, W. G. J. Van Helden, and R. J. C. Van Zolingen, "T he yield of different combined PV-thermal collector designs," vol. 74, pp. 253–269, 2003.
22. a. Kirubakaran, S. Jain, and R. K. Nema, "A review on fuel cell technologies and power electronic interface," *Renew. Sustain. Energy Rev.*, vol. 13, no. 9, pp. 2430–2440, 2009.
23. K. Rajashekara, "Hybrid fuel-cell strategies for clean power generation," *IEEE Trans. Ind. Appl.*, vol. 41, no. 3, pp. 682–689, 2005.
24. M. Y. El-Sharkh, a. Rahman, M. S. Alam, P. C. Byrne, a. a. Sakla, and T. Thomas, "A dynamic model for a stand-alone PEM fuel cell power plant for residential applications," *J. Power Sources*, vol. 138, no. 1–2, pp. 199–204, Nov. 2004.
25. S. D. Oh, K. Y. Kim, S. B. Oh, and H. Y. Kwak, "Optimal operation of a 1-kW PEMFC-based CHP system for residential applications," *Appl. Energy*, vol. 95, pp. 93–101, 2012.
26. S. G. Kandlikar and Z. Lu, "Thermal management issues in a PEMFC stack – A brief review of current status," *Appl. Therm. Eng.*, vol. 29, no. 7, pp. 1276–1280, May 2009.
27. S. Obara and I. Tanno, "Exergy analysis of a regional-distributed PEM fuel cell system," *Int. J. Hydrogen Energy*, vol. 33, no. 9, pp. 2300–2310, 2008.
28. S. Obara, "CO2 discharge characteristic of PEM-FC/hydrogen-gas-engine hybrid cogeneration," *Int. J. Hydrogen Energy*, vol. 32, no. 7, pp. 819–831, 2007.
29. G. Bruni, S. Cordiner, M. Galeotti, V. Mulone, M. Nobile, and V. Rocco, "Control strategy influence on the efficiency of a hybrid photovoltaic-battery-fuel cell system distributed generation system for domestic applications," *Energy Procedia*, vol. 45, pp. 237–246, 2014.
30. L. Zhang and A. Q. Huang, "Model-based fault detection of hybrid fuel cell and photovoltaic direct current power sources," *J. Power Sources*, vol. 196, no. 11, pp. 5197–5204, 2011.
31. G. Bruni, S. Cordiner, and V. Mulone, "Domestic distributed power generation: Effect of sizing and energy management strategy on the environmental efficiency of a photovoltaic-battery-fuel cell system," *Energy*, vol. 77, pp. 133–143, 2014.
32. M. T. Gencoglu and Z. Ural, "Design of a PEM fuel cell system for residential application," *Int. J. Hydrogen Energy*, vol. 34, no. 12, pp. 5242–5248, 2009.
33. M. Castañeda, A. Cano, F. Jurado, H. Sánchez, and L. M. Fernández, "Sizing optimization, dynamic modeling and energy management strategies of a stand-alone PV/hydrogen/battery-based hybrid system," *Int. J. Hydrogen Energy*, vol. 38, no. 10, pp. 3830–3845, 2013.

[a] Corresponding author: mfirdausb@ump.edu.my

34. J. J. Hwang, L. K. Lai, W. Wu, and W. R. Chang, "Dynamic modeling of a photovoltaic hydrogen fuel cell hybrid system," *Int. J. Hydrogen Energy*, vol. 34, no. 23, pp. 9531–9542, 2009.
35. T. F. El-Shatter, M. N. Eskandar, and M. T. El-Hagry, "Hybrid PV/fuel cell system design and simulation," *Renew. Energy*, vol. 27, no. 3, pp. 479–485, 2002.
36. S. G. Tesfahunegn, Ø. Ulleberg, P. J. S. Vie, and T. M. Undeland, "PV fluctuation balancing using hydrogen storage - A smoothing method for integration of PV generation into the utility grid," *Energy Procedia*, vol. 12, no. 1876, pp. 1015–1022, 2011.
37. M. Uzunoglu, O. C. Onar, and M. S. Alam, "Modeling, control and simulation of a PV/FC/UC based hybrid power generation system for stand-alone applications," *Renew. Energy*, vol. 34, no. 3, pp. 509–520, 2009.
38. P. Thounthong, S. Sikkabut, a. Luksanasakul, P. Koseeyaporn, P. Sethakul, S. Pierfederici, and B. Davat, "Fuzzy logic based DC bus voltage control of a stand alone photovoltaic/fuel cell/supercapacitor power plant," *2012 11th Int. Conf. Environ. Electr. Eng. EEEIC 2012 - Conf. Proc.*, pp. 725–730, 2012.
39. P. Thounthong, V. Chunkag, P. Sethakul, S. Sikkabut, S. Pierfederici, and B. Davat, "Energy management of fuel cell/solar cell/supercapacitor hybrid power source," *J. Power Sources*, vol. 196, no. 1, pp. 313–324, 2011.
40. P. Thounthong, A. Luksanasakul, P. Koseeyaporn, and B. Davat, "Intelligent model-based control of a standalone photovoltaic/fuel cell power plant with supercapacitor energy storage," *IEEE Trans. Sustain. Energy*, vol. 4, no. 1, pp. 240–249, 2013.
41. N. a. Ahmed, a. K. Al-Othman, and M. R. Alrashidi, "Development of an efficient utility interactive combined wind/photovoltaic/fuel cell power system with MPPT and DC bus voltage regulation," *Electr. Power Syst. Res.*, vol. 81, no. 5, pp. 1096–1106, 2011.
42. N. Mezzai, D. Rekioua, T. Rekioua, a. Mohammedi, K. Idjdarane, and S. Bacha, "Modeling of hybrid photovoltaic/wind/fuel cells power system," *Int. J. Hydrogen Energy*, vol. 39, no. 27, pp. 15158–15168, 2014.
43. C. Wang, S. Member, and M. H. Nehrir, "Power Management of a Stand-Alone Wind / Photovoltaic / Fuel Cell Energy System," vol. 23, no. 3, pp. 957–967, 2008.
44. N. a. Ahmed, M. Miyatake, and a. K. Al-Othman, "Power fluctuations suppression of stand-alone hybrid generation combining solar photovoltaic/wind turbine and fuel cell systems," *Energy Convers. Manag.*, vol. 49, no. 10, pp. 2711–2719, 2008.
45. S. Zafar and I. Dincer, "Thermodynamic analysis of a combined PV/T-fuel cell system for power, heat, fresh water and hydrogen production," *Int. J. Hydrogen Energy*, vol. 39, no. 19, pp. 9962–9972, 2014.
46. M. Hosseini, I. Dincer, and M. a. Rosen, "Hybrid solar-fuel cell combined heat and power systems for residential applications: Energy and exergy analyses," *J. Power Sources*, vol. 221, pp. 372–380, 2013.

1. Luigi T.De Luca, *Propulsion physics* (EDP Sciences, Les Ulis, 2009)
2. F. De Lillo, F. Cecconi, G. Lacorata, A. Vulpiani, EPL, 84 (2008)

[a] Corresponding author: mfirdausb@ump.edu.my

MATEC Web of Conferences 74, 00020 (2016) DOI: 10.1051/matecconf/20167400020
ICMER 2015

Accelerated Life Test Design for Tractor Powertrain Front Axle

Azianti Ismail[1,a], Won Jung[2] and Qiang Liu[2]

[1]*Faculty of Mechanical Engineering, Universiti Teknologi MARA Johor, Pasir Gudang campus, 81750 Johor, Malaysia*
[2]*Department of Industrial and Management Engineering, Daegu University, 712-714 Gyeongsan, South Korea*

Abstract. Accelerated Life Test (ALT) has been applied in the manufacturing for many years due to rapid changing technologies, more complex products, speedier product development, and more demanding customer requirements. These reasons have pushed the manufacturers to acquire reliability information faster. ALT allows reducing the time needed to show the reliability of the product. The purpose of this study is to design accelerated life testing which involved determination of normal test time, acceleration factor, acceleration test time, and finally experimental setup. This case study provides the basis for ALT designs for the tractor front axle based on customer usage and field failure analysis, which were conducted to estimate the current reliability, especially on the B_{10} life during the operational stage of the product. The accelerated life test conducted has guaranteed the B_{10} life of 4,000 hours with 90% confidence level for lesser time needed for testing to show the reliability of the front axle assembly.

1 Introduction

Engineers need to understand the usage environment, whereby the final product must endure to fulfil customer satisfaction. The response in an ALT is directly related to the lifetime of the product. Typically, ALT data is right censored because the test is terminated before all units fail. LuValle pointed out that it is likely to miss failure mode during accelerated testing efforts. This might happen if the true operating environments are not replicated during the test. It should be made to understand and to utilize the correct operating environment when planning and conducting accelerated testing to ensure minimum risk due to this consequence [1].

In the ALT, failures occurred more quickly from the actual environment in less time acquired. A controlled environment for the accelerated test was developed to simulate the use rather than testing in lengthy real-time tests. Important parameters for ALT are length of test, number of samples, confidence level, shape parameter (β), and acceleration factor. Campean et al. have discussed on a generic approach to life prediction modeling for automotive components, which aims to establish a correlation between the degradation mechanism, customer usage profile and rig life testing to design ALT [2]. Meeker et al. have mentioned that with good characterization of field use conditions, it may be possible to use ALT results to predict the failure distribution in the field [3]. On the other hand, Attardi et al. have presented case study regarding the reliability analysis of some automotive components based on field failure warranty data in which can be used to design ALT [4].

As for agricultural tractors, starting from the 1970s, studies were performed to estimate the fatigue life of a powertrain starting from a load spectrum derived from experimental measurements [5] and more recently, with the goal of estimating the loads during normal use of the tractor, defining the lifetime load spectra for agricultural transmissions [6]. The agricultural tractor market is constantly evolving, due to the changes caused by globalization that oblige farmers to reduce their production costs by using more specialized equipment and agricultural tractors. As a result, agricultural tractor manufacturers have to reduce time-to-market and development costs.

Agricultural tractors are used for many kinds of farm work under various soil and field conditions. The most common tractor uses, however, are tillage and ploughing. To take into account of the workload in the design of the tractor powertrain, the operational characteristics of tractors such as annual working hours, working speed, engaged gears, and load distribution must be analyzed. It is particularly important to know the number of cycles of loads and their magnitudes under all working conditions. Furthermore, the agricultural environment adds additional factors that complicate the development of reliability improvement of agricultural vehicles; for example, the diversity of areas of operation, the presence of obstacles in fields, uneven terrains, and the varied number of tasks to be carried out. Equally, soil quality, variation of crop physical characteristics and environmental conditions such as rain, fog, and dust need to be considered [7].

Among the main components of the tractor, the powertrain manufacturing cost is one third of the total cost of the finished product. Therefore, a high percentage

[a] Corresponding author: azianti106@johor.uitm.edu.my

of cost reduction can be achieved through the improvement of the manufacturing process of the powertrain due to the re-evaluation of its reliability and durability, and through the changes in its design. The powertrain is different from the exterior component of the tractor because, even with a simple breakdown, it requires a lengthy repair period and incurs a high cost. Thus, from the outset, the powertrain has to be manufactured with the best quality. The tractor powertrain assembly module can be divided into two parts, which are the transmission assembly with the rear axle assembly, and the front axle assembly. For this case study, the example of front axle assembly is shown.

2 Methodology

The data obtained from the analysis of usage environment for this case study based on questionnaires answered by the farmers, site visits and repair data from the service centers. Usage environment such as working conditions, variety of climatic conditions, environmental conditions; crop variety, soil, and water, were considered [8]. It also covered customer usage patterns to determine the possible failure modes, geographic location, purpose, and use of the tractors to provide the exact conditions. Potential failures occurred more quickly from the actual environment in less time acquired in ALT. In Figure 1, the methodology to design the ALT is shown.

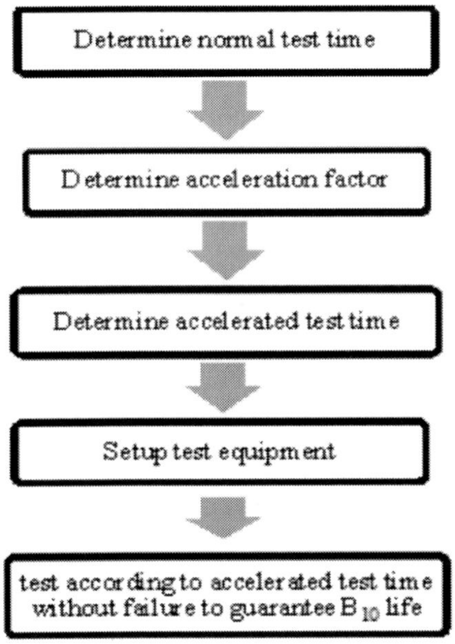

Figure 1. ALT design methodology.

Acceleration Factor (AF) can be found through test-to-fail at different stress levels or may be from available relationships, such as Linear, Eyring, Inverse Power Law, Arrhenius and others. The AF provides a quantitative estimate of the relationship between the field condition and the test condition.

2.1 Normal test time

The first step was to determine the normal test time. For this case study, the normal test time (see Eq. (1)) for t_n based on B_{10} life warranty for each part, 90% confidence level (CL=0.90, p=0.10), β parameter (determined by Minitab analysis) is summarized in Table 1. Normal test time calculation, t_n is shown Eq. (1) [9]. The failure data of powertrain is in the form of failure observations from the truncated time of 3,000 hours. Using the field failure data, the failures were investigated using Minitab. In order to estimate the suitable statistical distribution of the field failure data, Anderson-Darling goodness of fit test was 582.361 and correlation coefficient for Weibull was 0.934, in which it showed that it best suited for the distribution. β shows how the failure rate is increased in time.

$$t_n = B_i \left[\frac{\ln(1-CL)}{n \ln(1-p)} \right]^{\frac{1}{\beta}} \tag{1}$$

2.2 AF and Accelerated test time

By applying the AF, accelerated test time could be compressed and the results could be obtained faster. The calculated accelerated test time was used in conducting the ALT for the powertrain. The test was designed or setup up to provide the same type of stress in the field based on the surveys of profile usage and field visits. The data from the tests were projected over the life cycle to mimic the normal operating field conditions.

Table 1. Normal test time for for front axle assembly.

Target lifetime warranty, B_{10} life	Shape parameter, β	Normal test duration t_n
4,000 hours	2.02	14,166 hours

In this environment, there were two AFs used; inverse power law and Arrhenius relationship. Many failures mechanisms exhibited time to failure models, which fitted the inverse power law, Arrhenius or a combined inverse power law-Arrhenius model [10]. The inverse power relationship is widely used to model life as a function of pressure-like accelerating variables (e.g., stress, pressure, voltage stress). This relationship is generally considered to be an empirical model because it has no formal basis from knowledge of the physics or chemistry of the modeled failure modes [11]. Thus the AF for the front axle assembly which is involved in gear loading and speed is:

$$AF = \left(\frac{Torq_{test}}{Torq_{field}} \right)^4 \left(\frac{\mathrm{Re}\,v_{test}}{\mathrm{Re}\,v_{field}} \right)^1 \exp\left(\frac{E_a}{k} \left(\frac{1}{TempMax_{field}} - \frac{1}{TempMax_{test}} \right) \right) \tag{2}$$

$Torq_{test}$ is the accelerated test torque (kgf.m), and $Torq_{field}$ is the torque in the normal condition gathered from field data (kgf.m). Rev_{test} is the accelerated test speed (rpm), and Rev_{field} is the speed in normal condition gathered from the field data (rpm). E_a is the activation energy in eV. $TempMax_{field}$ is the normal condition temperature gathered from the field data and it was 343K or 70 degree Celsius. $TempMax_{test}$is the temperature in the ALT and it was set at 363K or 90 degree Celsius. The AF and the accelerated test time for the front axle assembly are shown in Table 2 based on Eq. (2). Through the application of AF, the accelerated test time could be compressed and results could be obtained faster. For front axle assembly, the accelerated test time is 250 hours to guarantee B_{10} life of 4,000 hours.

Table 2. AF for front axle assembly.

$Torq_{field}$ (kgf.m)	$Torq_{test}$ (kgf.m)	Rev_{field} (rpm)	Rev_{test} (rpm)	AF	Accelerated test time
191.8	350	26.1	26.8	56.6	250 hrs

2.3 Test setup of ALT

In general, it is required to run thousands of hours to perform tests until the failure of the weak links is identified. In order to speed up the reliability test duration, the test conditions on proving ground became more demanding, which were mainly realized by building more severe test conditions. Thus, the design must be market-valuable and the products with "too long" or "insufficient" design life are usually too costly and comatose in the market, so it is necessary to include customers' requirements for the design, development or test phases.

As for the test, only one sample was provided. The sample test consisted of differential bevel gear and reduction gear ratio of 22:1. The target of the test was the differential and the reduction gear of the bevel gear that should operate without failure during the accelerated test time. The engine was assembled and fixed with a chassis dynamometer each on both sides of the hubs on the front wheel axle, as shown in Figure 2. The DaTAQ Pro software and control system were used for chassis dynamometer driving torque load control. With the engine on maximum speed condition at each gear level, equivalent load was applied as chassis dynamometer's constant driving load torque at the rear wheels during the accelerated test.

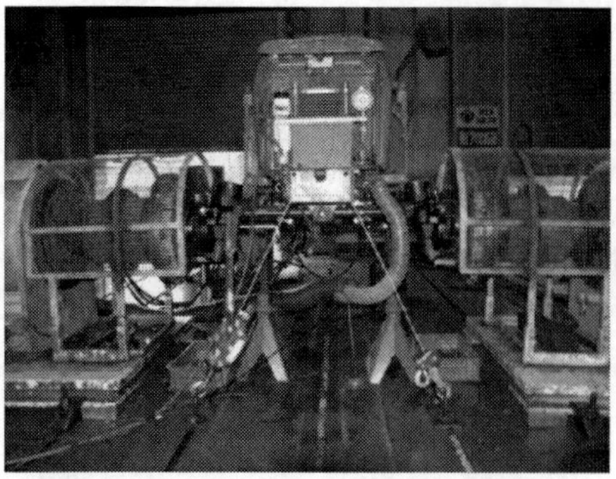

Figure 2. Test setup for front axle assembly.

Front axle assembly is shown in Figure 3 which consists of four main gears 17T, 11T, 40T and 10T. During the test, the condition for all these gears needs to be observed. Any damaged occurred to the gears is considered as failure and test need to be repeated.

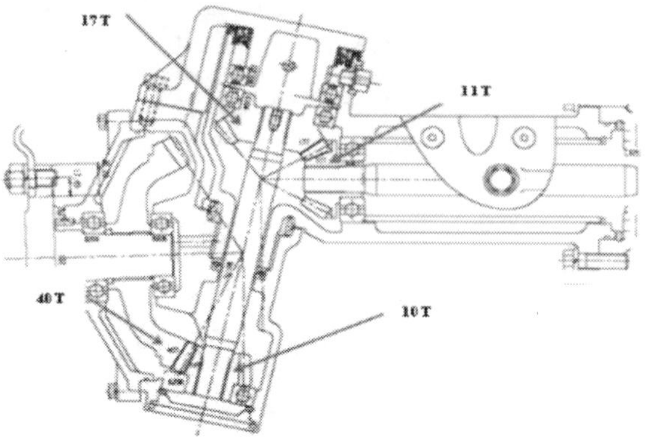

Figure 3. Schematic diagram for front axle assembly.

3 Results and discussions

As for the front axle assembly, the first two tests had to be stopped before the accelerated test time due to failure. From Table 3, the first failure for the test was due to the damaged 10T and 40T gear at 70.8 hours, as shown in Figure 4. Since the required accelerated test time is 250 hours, another test was conducted on the same sample after repaired

Table 3. Results from the first test for front axle assembly.

	Results
Test time and progress	70.8 hours & 28.3%
Failure	10T and 40T gear damage
Root cause analysis	Assumption of poor quality condition of gear meshing during applied load
Countermeasure	Review and inspect of the related case flow and gear parts

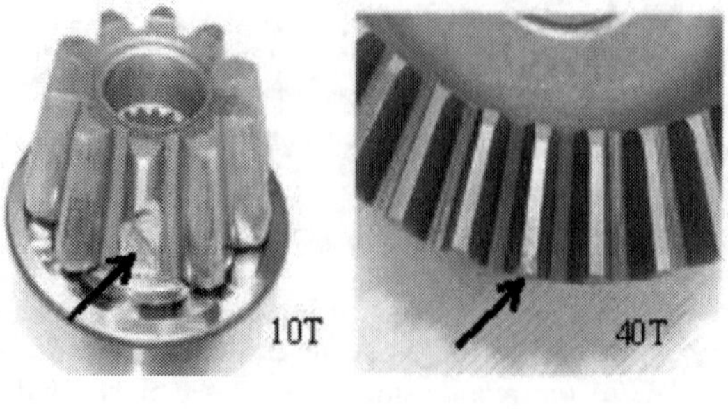

Figure 4. 10T and 40T gear tooth breakage from the first test

From Table 4, the second failure of the test was due to the damaged 10T gear at 146.1 hours, as shown in Figure 5 for second test. Since the required test time is 250 hours, another test was conducted on the same sample after repaired.

Table 4. Results from the second test for the front axle assembly.

	Results
Test time and progress	146.1 hours & 58.4%
Defect location	10T gear damage
Root cause analysis	Severe gearing coupling symptoms occurred while applying load
Countermeasure	Gearing was changed from fine-tune to shim method (40T gear bearing assembly part applied shim method). Re-checked the assembly quality characteristics table and revise.

Figure 5. 10T gear tooth breakage from the second test.

As for the third test, no failure occurred after 250 hours of the test time. Thus, the result showed that the B_{10} life of 4,000 hours was guaranteed for the front axle assembly for powertrain with 90% confidence level. On final improvement, the shim value was adjusted for 10T and 40T for backlash and abutting position level for the bevel gear inside the front axle assembly as summarized in Table 5. The front axle assembly was operated without failure once the recommended improvement was implemented. Moreover, the B_{10} life of 4,000 hours was guaranteed with 90% confidence level. As in Figure 6, no sign of gear damage was detected after the third test from the post mortem.

Table 5. Recommended improvement.

Current standard	Improvement
Checked bevel gear backlash and abutting position level inside the front axle case Regulation criteria: 0.15 to 0.35 mm	Adjusted shim value: 40T bevel gear part: 0.4t 10T bevel gear part: 1.6t

4 Conclusions

It is inconvenience to test according to the normal test time required which is 14,166 hours for the front axle assembly to guarantee B_{10} life of 4,000 hours with 90% confidence level. By designing an ALT, the time required is 250 hours for closely demonstrating the front axle assembly's life. The ALT has been conducted three times which finally emerged with failure free results.

Figure 6. Examination of the gears after the third test.

Each time the test failed, some improvements such as adjustment of shim values and level or position of the gears need to be made. None of the gears involved showed any damaged in the third test after 250 hours. Thus, it can be concluded the front axle assembly B_{10} life of 4,000 hours with 90% confidence level is guaranteed. The result obtained from this test was useful in predicting the reliability of the front axle assembly. ALT programs should be designed and conducted by teams including knowledgeable individuals about the product and its use environments, the physical/mechanical aspects of failure mode, and the statistical aspects of the design and analysis of reliability testing.

Acknowledgements

The authors would like to be obliged to Daegu University for providing laboratory facilities and financial assistance.

References

1. M.J. LuValle, Identifying Mechanisms That Highly Accelerated Tests Miss, *IEEE Trans Reliab*, **56**, 349–59 (2007).
2. I.F. Campean, A.J. Day, S. Wright, Camshaft timing belt reliability modelling, *Proc. Int. Symp. Prod. Qual. Integr. IEEE*; 377–83 (2001).
3. W.Q. Meeker, L.A. Escobar, Y. Hong, Using Accelerated Life Tests Results to Predict Product Field Reliability, *Technometrics*, **51**, 146–61 (2009).

4. L. Attardi, M. Guida, G. Pulcini, A mixed-Weibull regression model for the analysis of automotive warranty data, *Reliab Eng Syst Saf,* **87**, 265–73 (2005)..
5. K. Renius, Application of cumulative damage theory to agricultural tractor design elements. *Konstruktion*, **29**, 85–93 (1977).
6. D.C. Kim, I.H. Ryu, K.U. Kim, Analysis of tractor transmission and driving axle loads. *Trans ASAE,* **44**, 751–757 (2001).
7. Z.Q. Xi,Z.L. Zhou, Y. Li, Fuzzy Control Strategy of Powershift Transmission of Tractor, *Appl Mech Mater,* **241**, 1959–63 (2013).
8. W. Jung, A. Ismail, Method of Determining Acceleration Factors for Agricultural Tractors, *Qual. Reliab. Risk, Maintenance, Saf. Eng. IEEE,* 346–348 (2011).
9. B. Bertsche, G. Lechner, Reliability in Automotive and Mechanical Engineering. Berlin: Springer Berlin Heidelberg, (2008).
10. J.W. Evans, J.Y. Evans, editors, *Product Integrity and Reliability in Design*, London: Springer-Verlag, (2001).
11. L.A. Escobar, W.Q. Meeker, A Review of Accelerated Test Models. *Stat Sci*, **21**, 552–577 (2006).

ANALYSIS OF REFRIGERATION CYCLE PERFORMANCE WITH AN EJECTOR

Wani J. R.[1], Aklilu T. Baheta[1,a], Abraham D. Woldeyohannes[2], and Suhaimi Hassan[1]

[1]Department of Mechanical Engineering, Universiti Teknologi PETRONAS, 31750 Tronoh, Perak, Malaysia
[2]Caledonian College of Engineering, PO Box 2322, CPO Seeb 111, Sultanate of Oman

Abstract. A conventional refrigeration cycle uses expansion device between the condenser and the evaporator which has losses during the expansion process. A refrigeration cycle with ejector is a promising modification to improve the performance of conventional refrigeration cycle. The ejector is used to recover some of the available work so that the compressor suction pressure increases. To investigate the enhancement a model with R134a refrigerant was developed. To solve the set of equations and simulate the cycle performance a subroutine was written on engineering equation solver (EES) environment. At specific conditions, the refrigerant properties are obtained from EES. At the design conditions the ejector refrigeration cycle achieved 5.141 COP compared to 4.609 COP of the conventional refrigeration cycle. This means that ejector refrigeration cycle offers better COP with 10.35% improvement compared to conventional refrigeration cycle. Parametric analysis of ejector refrigeration cycle indicated that COP was influenced significantly by evaporator and condenser temperatures, entrainment ratio and diffuser efficiency.

Nomenclature

A	area (m2)
h	enthalpy (kJ/kg)
$\dot{m}$	mass flow rate (kg/s)
p	pressure (kPa)
ρ	density (m3/kg)
η	efficiency
s	entropy (kJ/kg.K)
ω	entrainment ratio

Subscripts

d	diffuser
mc	mixing chamber
n	nozzle
sn	suction nozzle
s	isentropic process exit
1,2,3....9	state points

1 Introduction

In a conventional vapor compression refrigeration cycle, the pressure drop from the condenser to the evaporator is done by throttling. Throttling process is always associated with loss and it is one of the thermodynamic losses. Throttling is an isenthalpic process and during this expansion process the resulted kinetic energy gain is lost by friction into the refrigerant in the form of heat. Hence, the isenthalpic expansion process forms large amount of vapor at the end of the expansion compared to isentropic expansion. Consequently, a reduction of refrigeration load in the cycle is observed. The use of two-phase ejector has become a promising cycle modification recently [1]. The potential kinetic energy is recovered by the ejector during the expansion process. The ejector also reduces the compressor work by increasing the compressor inlet pressure higher than that of the evaporator pressure which leads to a coefficient of performance (COP) improvement. There are two types of ejectors according to the position of the nozzle. These are constant pressure and constant area mixing ejectors. It was found that the performance of constant area mixing ejector is lower than the constant pressure mixing ejector [2].

There have been many attempts to study the use of ejector on performance enhancement of refrigeration cycle with different assumptions and ejector design. For instance, the COP of a refrigeration cycle with an ejector expansion device

[a] Corresponding author: aklilu.baheta@petronas.com.my

using R12 refrigerant was investigated by Kornhauser [3]. To represent the ejector process constant mixing pressure model was used and found 21% COP improvement over the conventional refrigeration cycle (CRC) under standard conditions. With R134a refrigerant, Disawas and Wongwises [4] experimentally investigated the performance of refrigeration cycle with ejector where there was no upstream expansion valve of the evaporator so that the evaporator was flooded with the refrigerant. At low heat sink temperature the test showed COP improvement over CRC. Using R134a and constant area ejector, Bilir and Ersoy [5] showed that COP increases as the difference between condenser and evaporator temperatures increases. Chen et al. [6] proposed an ejector model to predict the ejector performance at critical and sub-critical operation modes.

The performance of a refrigeration cycle also depends on the type of working fluid. As suggested by Takeuchi et al.[7], R134a gives better performance compared to R404A, R407 and R410 using ejector as an expansion device in a refrigeration cycle.

Though there have been many attempts, detailed study and analysis is still needed on ejector refrigeration cycle performance improvement compared to conventional refrigeration cycle. Thus, a refrigeration cycle with constant pressure-mixing zone ejector is analyzed and simulated using a model. Different operating conditions are considered and the cycle performance with R134a refrigerant is compared to the conventional R134a refrigeration cycle. In addition, parametric analysis is done to select the influential parameter/s.

2 Ejector refrigeration cycle modeling

Figure 1 (a) shows schematic representation of the ejector refrigeration cycle (ERC). It consists of evaporator, ejector, separator, throttling device, compressor, and condenser constituting two loops. The refrigeration loop and the power loop supply cooling effect and the driving force, respectively. First the primary (motive) fluid is accelerated in a convergent–divergent nozzle until its velocity reaches supersonic. The primary fluid entrains the secondary evaporated fluid from the evaporator, and the two fluids mix in the mixing chamber. In the diffuser, at the expense of the kinetic energy of the mixed fluid its pressure is increased to the separator pressure as shown in Figure 1 (b).

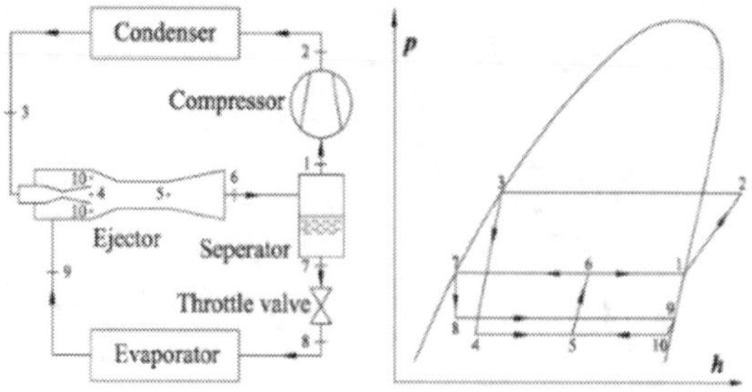

(a) (b)

Figure 1. (a) Ejector refrigeration system; (b) *p-h* diagram

Assumptions

- pressure drop in the evaporator and condenser are assumed to be negligible,
- expansion efficiencies of the motive and suction streams are constant,
- refrigerant kinetic energies at the ejector inlet and exit are negligible, and
- shock wave loss effect is considered in the ejector diffuser efficiency.

To represent the process of the cycle, each component has been modeled using first principle.

2.1 Nozzle Model and Analysis

The primary nozzle exit enthalpy is determined from its isentropic relationship and given as

$$h_4 = h_3 - \eta_n (h_3 - h_{4s}) \quad (1)$$

Velocity of the motive fluid and cross-sectional area at the nozzle exit are given by,

$$V_4 = \sqrt{2(h_3 - h_4)} \quad (2)$$

$$A_4 = \frac{1}{(1+\omega)\rho_4 V_4} \quad (3)$$

The entrainment ratio, $\omega = \dot{m}_{10}/\dot{m}_3$ where $\dot{m}_{10}$ and $\dot{m}_3$ are entrained and the primary fluid mass flow rates, respectively.

The suction chamber enthalpy at the nozzle exit is determined from its isentropic relationship and given as

$$h_{10} = h_9 - \eta_{sn}(h_9 - h_{10s}) \quad (4)$$

Velocity of the secondary fluid and the cross

sectional area at the suction nozzle exit are,

$$V_{10} = \sqrt{2(h_9 - h_{10})} \quad (5)$$

$$A_{10} = \frac{\omega}{(1+\omega)\rho_{10}V_{10}} \quad (6)$$

For a unit total mass flow rate in the ejector diffuser, the primary and the entrained mass flow rates in terms of the entrainment ratio of the ejector are given as

$$\dot{m}_4 = \omega/(1+\omega) \text{ and } \dot{m}_{10} = 1/(1+\omega)$$

(7)

2.2 Mixing chamber analysis

At the ejector mixing chamber, the mixing process is done at a constant pressure. Applying the momentum and energy conservations on the mixing chamber would result relations for its exit velocity and enthalpy.

$$V_5 = \sqrt{\eta_{mc}}\left(\frac{1}{(1+\omega)}V_4 + \frac{\omega}{(1+\omega)}V_{10}\right) \quad (8)$$

$$h_5 = \frac{1}{(1+\omega)}\left(h_4 + \frac{V_4^2}{2}\right) + \frac{\omega}{(1+\omega)}\left(h_{10} + \frac{V_{10}^2}{2}\right) - \frac{V_5^2}{2} \quad (9)$$

2.3 Diffuser analysis

Similarly applying the conservation of energy and isentropic relation gives the following equations for the diffuser.

$$h_6 = h_5 + \frac{V_5^2}{2} \quad (10)$$

$$h_{6s} = h_5 + \eta_d(h_6 - h_5) \quad (11)$$

where η_d is the diffuser efficiency and h_{6s} represents the enthalpy obtained at state 6 under isentropic process through the diffuser.
The pressure and the quality of the refrigerant at the ejector exit are determined using Eqs. (12) and (13).

$$p_6 = p(h_{6s}, s_5) \quad (12)$$

$$x_6 = \frac{1}{1+\omega} \quad (13)$$

2.4 Compressor model analysis

For a given isentropic compression process, the actual enthalpy at the exit of the compressor becomes

$$h_2 = h_1 + \frac{h_{2s} - h_1}{\eta_{comp}} \quad (14)$$

where η_{comp} refers to the isentropic efficiency of the compression process which is given by Eq. (15) [8].

$$\eta_{comp} = 0.874 - 0.0135(p_2 / p_1) \quad (15)$$

The rate of compressor work input is expressed by

$$\dot{W}_{comp} = \dot{m}_p(h_2 - h_1) \quad (16)$$

The refrigeration capacity, $\dot{Q}_{evap}$ is calculated as

$$\dot{Q}_{evap} = \dot{m}_s(h_9 - h_8) \quad (17)$$

The coefficient of performance of the ejector refrigeration cycle is defined as

$$COP = \dot{Q}_{evap}/\dot{W}_{comp} \quad (18)$$

2.5 Simulation input

In order to simulate the performance of the ejector refrigeration cycle a subroutine was written in EES environment. This software has most of the refrigerant properties including R134a. Furthermore, for simulation purpose typical data given in Table 1 is used as a design conditions which was taken from [9].

Table 1. Input design conditions.

Parameters	Temperature (°C)	Pressure (bar)
Compressor exit	80	7.7
Condenser exit	30	7.7
Evaporator	-5	1.64
Parameters	**Specifications**	
Ejector area ratio	14	
Ejector diffuser efficiency	0.85	
Nozzle efficiency	0.90	
Mixing chamber efficiency	0.85	
Entrainment ratio	0.53	
Refrigerant	R 134a	
Primary mass flow rate	0.005773 kg/s	
Secondary mass flow rate	0.01089 kg/s	

3 Results and discussion

Maintaining other parameters, COP of ERC was investigated for various evaporator temperatures ranging from -15°C to 0°C. As shown in Figure 2 COP improves as the evaporator temperature increases. This is because as the evaporator temperature increases, the pressure difference between the nozzle exit and the evaporator (p_9-p_4) increases. This resulted an increases in the secondary flow rate and hence, COP and cooling capacity of the system also increased. However, the temperature of the evaporator is determined by the space to be cooled. The CRC was simulated for the same condenser and evaporator conditions. As shown in the same Figure 2, for a given evaporator temperature the COP of ERC is higher than the COP of CRC. At the design conditions the ERC achieved 5.141 COP compared to 4.609 COP of the CRC which is 10.35% improvement compared to CRC.

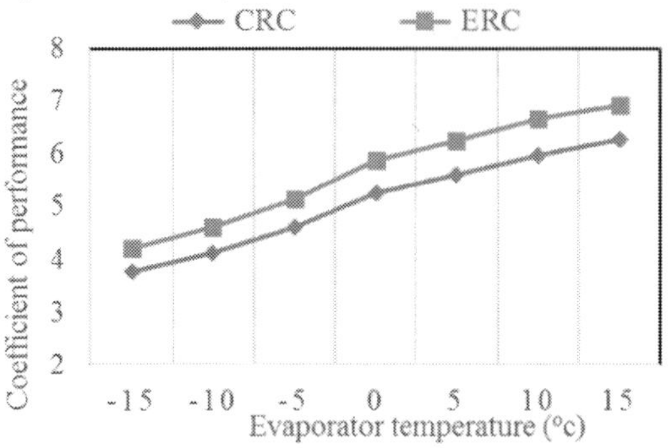

Figure 2. Variation of COP with respect to evaporator temperature

Similarly maintaining other parameters constant, the condenser exit temperature was varied from 25 to 50°C to study its effect on COP of the cycle. As shown in Figure 3 COP decreased with increasing condenser temperature. This is because as the condenser exit temperature increases, the enthalpy of refrigerant at the inlet to the evaporator increases. Meanwhile, the evaporator exit enthalpy remains constant and hence causing low refrigeration effect and low COP. On the other hand, to reject the heat to the surroundings the temperature of the condenser should not be below the surroundings temperature that will limit the minimum temperature. Furthermore, the COP of ERC is higher than COP of CRC at a given condenser exit temperature.

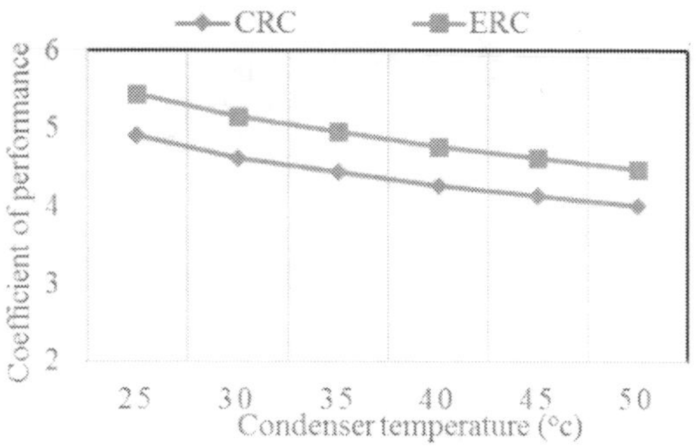

Figure 3. Variation of COP with respect to condenser temperature

Figure 4 shows COP with respect to entrainment ratio. It indicates that COP increases with increase in entrainment ratio. This is because as the entrainment ratio increases, more refrigerant is entrained by the primary fluid from the evaporator. Consequently, the mass flow rate in the evaporator increases which in turn increases the cooling capacity and the coefficient of performance.

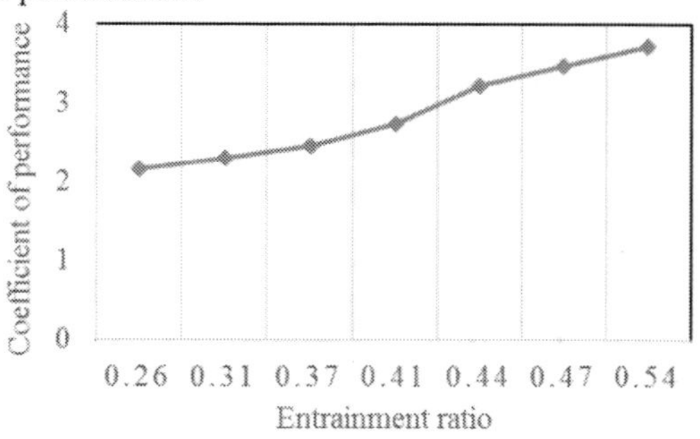

Figure 4. Variation of COP with respect to entrainment ratio

The effect of diffuser efficiency on the COP was investigated. Figure 5 indicates the higher the diffuser efficiency, the smaller the work of compression. Since better efficiency has boosted the compressor inlet pressure, less compression work is needed by the compressor for a given overall compression ratio. As a result, the COP increases as the diffuser efficiency increases as shown in Figure 5.

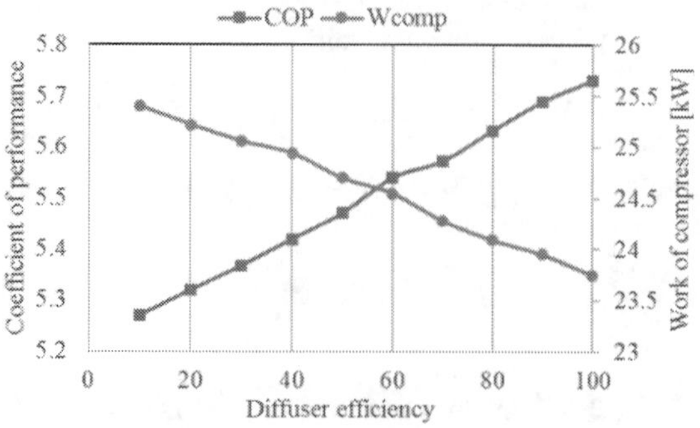

Figure 5. Variation of COP and compressor work with respect to diffuser efficiency

4 Conclusions

Refrigeration cycle with ejector was modeled. In order to simulate the performance of the cycle a subroutine was developed in EES environment. It was simulated for a given design conditions based on the data taken from literature. At the design conditions the ejector refrigeration cycle achieved 5.141 COP compared to 4.609 COP of the conventional refrigeration cycle. Hence, ejector refrigeration cycle offers better COP with 10.35% improvement compared to conventional refrigeration cycle. From the parametric analysis of ERC, it was found that COP is influenced by evaporator and condenser temperatures, entrainment ratio and diffuser efficiency. Based on these outcomes, it is hoped that a better understanding of operation of ejector refrigeration cycle can be achieved. With the identification of the parameters that affect the COP significantly, future design of ejector refrigeration cycle can be improved.

Acknowledgements

This work was supported by Ministry of Higher Education, FRGS grant number 0153AB-I70. The authors thank Universiti Teknologi PETRONAS for providing supports to conduct the project.

References

[1] Jahar Sarkar, Ejector enhanced vapor compression refrigeration and heat pump systems: A review, Renewable and Sustainable Energy Reviews, Vol. **16**, pp. 6647–6659 (2012).

[2] Riffat SB, Jiang L, Gan G., Recent development in ejector technology—a review, International Journal Ambient Energy, Vol. **26**, pp 13–26 (2005).

[3] Kornhauser AA., The use of an ejector as a refrigerant expander. Proceedings of USNC/IIR-Purdue refrigeration conference, USA, (1990).

[4] Disawas S, Wongwises S., Experimental investigation on the performance of the refrigeration cycle using a two-phase ejector as an expansion device, International Journal Refrigeration, Vol. **27**, pp. 587–94 (2004).

[5] Bilir N, Ersoy HK, Performance improvement of the vapour compression refrigeration cycle by a two-phase constant area ejector, International Journal Energy Research, Vol. **33**, pp. 469–80 (2009).

[6] Chen, W., Liu, M., Chong, D., Yan, J., Little, A.B., Bartosiewicz, Y. A, 1D model to predict ejector performance at critical and sub-critical operational regimes, International Journal of Refrigeration, Vol 36 **(6)**, pp. 1750-1761 (2013).

[7] Takeuchi H, Kume Y, Oshitani H, Ogata G. Ejector cycle system, U.S. Patent 6, 438, 993 B2 (2002).

[8] Huashan Li, Fei Cao, Xianbiao Bu, Lingbao Wang, and Xianlong Wang, Performance characteristics of R123yf and R134a ejector-expansion refrigeration cycle, Applied energy, Vol **121**, pp. 96-103 (2014).

[9] K. Ganesh Babu and K. Ravi Kumar, Usage of a two-phase ejector as expansion device to improve the performance of R-134a used refrigeration system (College of Eng, Chevella, Ranga Reddy-515002, Praesh, India 1977).

Optimization and Economic Analysis of Grid-Photovoltaic Electric Boat Charging Station in Kuala Terengganu

N.A. S. Salleh[2], W. M. W. Muda[1, 2] and Roslan Umar[3]

[1] *Renewable Energy Research Group*
[2] *School of Ocean Engineering, University Malaysia Terengganu, Malaysia*
[3] *East Coast Environment Research Institute, University Sultan Zainal Abidin, Malaysia.*

Abstract. This study evaluates the feasibility of developing grid-photovoltaic electric boat charging station in Kuala Terengganu using simulation-based method. The main focus is on reducing the dependency on subsidy spent by the government in fisheries sector and encouraging green technology in commercial sector. All data such as solar radiation, amount of subsidy received by fishermen, and fishing activities were collected for the selected area. Economic analyses of the proposed system are discussed based on payback period and net present cost (NPC). The comparison between the proposed system and the grid-only system is done based on the production and consumption of electricity per year, the NPC and emission of pollutant. The system also generates high income from selling energy to the grid with tariff rates RM 1.49/kWh. It is concluded that the proposed system is feasible to be developed in the selected area with the payback period and the NPC are 8.2 years and RM 759,098, respectively. The results also show that the performance of the proposed system is better than grid-only system in all the interested parameters.

1 Introduction

The escalating demand of energy and the uses of conventional technology in producing electrical power normally resulted in the pollution that affected everyone. It is because of the emission content which can harm the consumers. Thus, many studies have been proposed regarding renewable energy (RE) sources to replace and reduce the consumption of conventional energy [1-3]. RE is the energy obtained from natural environment like solar, wind, biomass and tidal energy. RE does not only benefit the environment but also for the economy.

Malaysia has a strong potential to exploit RE technologies as the fact that Malaysia is endowed with abundance of natural resources [3]. The potential for solar power generation is very high due to its tropical climate with approximately 4000-5000 Wh/m2 of daily average solar radiation [4, 5]. To encourage the use of RE, Malaysia plays their role by lifting it up as the fifth fuel [6]. In Tenth Malaysia Plan [7], Malaysia target to increase the use of RE sources to 985 MW by 2015, which is 5.5% to Malaysia's total electricity generation mix. In addition, the Feed-in tariff (FiT) has been introduced to promote the use of RE among consumers [7]. Malaysia also has assigned the Ministry of Energy, Green Technology and Water Malaysia (KeTTHA) that was established in 2009 [8] to focus on the development of green energy.

Since Malaysia is mostly surrounded by seas and ocean, fisheries are one of the contributing sectors in building up the nation's economy. There are almost 90,000 fishermen in Malaysia and 47% of them involve in offshore fisheries [9]. The fishing activities have been affected by the increase in fuel price. Since most of them are poor [10], the local government has given the fuel subsidies that worth RM 2,133 million in 2007 to almost 33,000 fishermen [9]. The use of electric boat (EB) may solve the abovementioned problem. Using the EB, fishermen do not rely on the fuel, thus the government can reallocate the fuel subsidies to other useful sectors such as agriculture, health and education. In addition, the use of EB may reduce air, marine and noise pollution.

In efforts to promote green technology in fisheries and transportation sectors, EB is one of convincing electric vehicles (EVs) that have potential to reduce pollution and dependency on fuel. Currently, EV charging station systems are connected directly to the grid. It is called Vehicle-to-Grid (V2G) and Grid-to-Vehicle (G2V) systems [11]. However, the uncontrolled charging demand from the EV could damage the grid [12]. Thus, the development of RE charging station may reduce the damage.

In this study, an EB charging station with hybrid grid-PV system is proposed. The main focus is on the charging station itself. Similar to other countries that have been using EVs [13], only the charging station infrastructure is provided by the government for the EV's user. The cost of EV is assumed to be incurred by the

[2] Corresponding author: gsk2357@pps.umt.edu.my

user. Thus, the cost of EB is not considered in this study. It is also well-known that the starting cost of developing RE technology is very expensive [14], so it is important to study the economics analysis in order to know the potential of the proposed system. HOMER software is used to find the optimal size of each component in the system. In addition, the payback period is computed. The income generated from the proposed system by selling excess electricity using FiT rates is then compared with the subsidy spent by government. Furthermore, the performances of PV-grid and grid only system are also compared in terms of net present cost (NPC), electricity production and consumption per year, and emission of pollutant. The proposed system is targeted to be used for offshore fishermen only.

2 Charging Station System Configuration

The proposed system configuration is as shown in Figure 1. The electricity generated by PV is used to charge the EB battery which is load in this system. The excess electricity generated by PV is stored in batteries. Any other excess electricity is sold to the grid by following FiT scheme. Grid also is used as backup where it supplies electricity to the load when there is insufficient energy from PV and batteries. A converter is used in this configuration to convert AC to DC and vice versa as the EB battery used in this system is a DC load.

A selected area for this study is Kuala Terengganu Utara (KTU) with latitude and longitude are 5.31 and 103.12, respectively [5]. HOMER software requires some input parameters to calculate the optimization results for a specific configuration of system. The input parameters are daily solar radiation (kWh/m2/d) and load profile (kW). In addition, the specification for each component used such as PV, batteries, converter and grid are also needed.

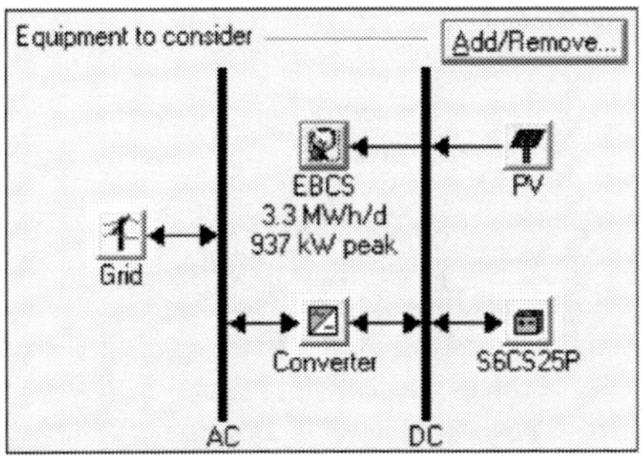

Figure 1. System configuration of grid-PV system in HOMER

2.1 Load Profile/ Load Demand

To determine the load profile, 16 out of 242 fishermen in KTU were selected for interview sessions to know about their fisheries activities. Another interview session with Lembaga Kemajuan Ikan Malaysia (LKIM) Chendering [15] was conducted to validate the data obtained from interview with fishermen. Besides, the information on the total amount of fuel subsidy and government incentive was also acquired from them. All the collected data are summarized in Table 1.

Table 1. Summarization of data obtained from interview sessions

Subject	Data
No. of fishermen	242
Fuel subsidy (Petrol)	RM 1.65/litre
Engine power (horsepower, hp)	15, 30, 40, 60
Operating hours (for offshore fishermen only)	Approximately 6.30 a.m. – 2.00 p.m. (except for Nov-Feb, no fishing activities)
Distance from land	≈ 20 miles = 32 km
Duration taken to fishing area	≈ 1 hour
Total subsidy for KTU in 2014 (RM/yr)	RM 1,228,325.70

From Table 1, engine power with 60hp is considered in determining the load profile of the proposed system. To replace 60hp engine, the EB needs 45kW electric motor. Inspired by [16], a 20kWh Li-ion battery is chosen in this study to be used to supply electricity to electric motor. Since 1 unit EB battery needs approximately one hour to fully charge, and the charging station is used after operating hours which is 2 p.m. till 12 p.m. only, the total EB battery can be charged in one unit charging station is 10 units/day. In order to provide the facilities to all fishermen in KTU, 25 unit charging stations are needed. Assuming that all 25 units are used to charge 20kWh battery at one time, so the maximum amount of load in a day for this system is 500kW for 10 hours.

Figure 2 shows the load profile of the case study area. There is no charging activity during the day until 2 p.m. because the fishermen are out for their fishing activities. The load profile starts to be at peak after 2 p.m. as the fishermen are home to recharge their battery boat at the charging station. During Northeast Monsoon from November to February, the load profile becomes null because the fishermen not be able to go out for fishing. Thus, no charging activities occur for four months. In order to avoid confusion between the EB battery and batteries used to store energy in the system, the EB battery is referred to load for the rest of this paper.

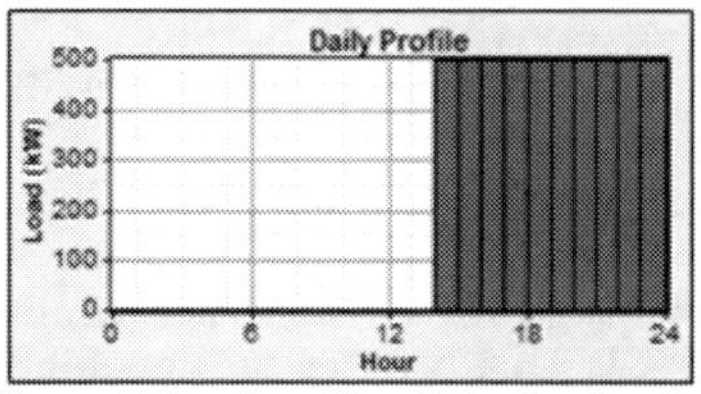

Figure 2. Load profile of March - October

2.2 Solar Resources

Solar radiation data was collected from UNISZA weather station [17] and the solar radiation for selected area is shown in Figure 3.

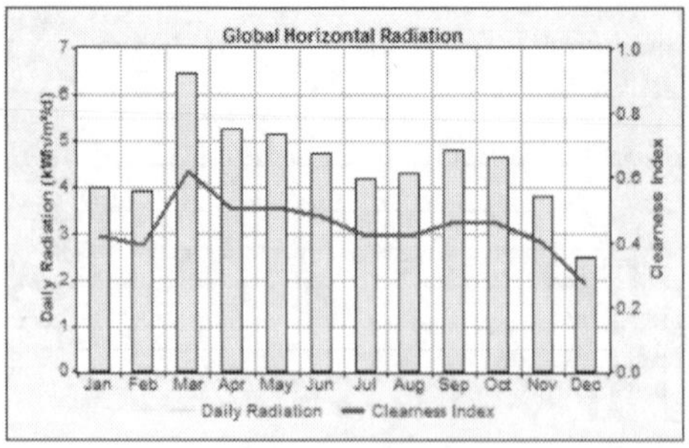

Figure 3. The annual solar radiation and the clearness index of Kuala Terengganu, Terengganu.

According to Figure 3, the clearness index of solar radiation data is calculated by HOMER with the following daily radiation data entered. The range of solar radiation is from 2.493 kWh/m2/day to 6.447 kWh/m2/day and the average of solar radiation is 4.472 kWh/m2/day. From the graph plotted, it is noticeable that solar irradiance from March to October is high meanwhile; it is low from November to February due to the Northeast Monsoon season.

2.3 Grid Utility and Feed-in-Tariff

There are two types of grid utility system in Malaysia which are, a single phase system with 240 V/50 Hz AC and 3-phase system with 415 V/50 Hz AC [18]. In case if there is deficit energy from the renewable energy source system, grid acts as backup supplier of energy. Tenaga Nasional Berhad (TNB) [19] is responsible in managing grid utility in Malaysia. Meanwhile, Sustainable Energy Development Authority (SEDA) [20] of Malaysia is a statutory body to administer and manage the implementation of the FiT mechanism which is mandated under the Renewable Energy Act 2011 (Act 725).

This project uses the Tariff C1; Medium Voltage General Commercial Tariff [19] with the commercial rate and maximum demand RM 0.365/kWh and RM 30.3/kW/month, respectively. The FiT rate is fixed at RM 1.49/kWh [21].

2.4 Photovoltaic (PV) system

The PV specification for this system is obtained from [21, 22] which consider the PV price at RM 2.55/W. Considering mounting and installation cost of 3 kW PV is RM 20,000, the initial cost of PV is RM 27,650. The replacement cost is assumed to be the same. Since the peak of load demand is 937 kW, and the load requires 3.3 MWh/day, the sizing range of PV is considered from 0 – 3400 kW, where 0 kW is used to see the possibility of configuration without PV panel. As Terengganu itself has a monsoon season with heavy rainfall, the cleaning cost can be negligible because it does not require regular cleaning process. The derating factor of this PV module is 80% with a lifetime of 25 years [21].

2.5 Inverter

4 units of 250 kW Sunny Central 250U inverter are used in this system to satisfy the total output of 1 MW load profile. The configuration of the inverter are based on [23, 24] which cost RM 448,920 with operating and maintenance approximately 0.3% from its cost. The inverter has a 20 years lifetime with output power factor 0.97.

2.6 Battery storage

The model Surrete 6CS25P [25] battery are chosen as storage element in this system. The arrangement of battery is important to ensure the energy capacity produced is high to support the electricity gained from the PV and grid. Thus, the sizing range of the battery is considered from 0 to 10 units as the storage of energy.

3 System Analysis

A few analyses can be carried out in HOMER from a grid-PV charging station system. In term of economics feasibility, NPC and payback period can be analysed. Moreover, this study compares between the cost needed to spend on 25 years of the charging station and the subsidy spent by the government on fuel for the fishermen.

3.1. Net present cost (NPC)

The NPC considers the cost throughout the system lifetime including the initial costs, replacement costs, operating and maintenance costs and grid rate costs. Generally, total NPC is calculated using the following equation (1) to equation (3) of [21];

$$NPC = \frac{C_{ann,tot}}{CRF(i,N)} \quad (1)$$

$$CRF(i,N) = \frac{i(1+i)^N}{(1+i)^N - 1} \quad (2)$$

$$i = \frac{i' - f}{1+f} \quad (3)$$

From the equation presented above, all cost of capital, replacement, annual operating and maintenance are included in the total annualized cost, $C_{ann,\ tot}$. N is the number of years. Meanwhile, CRF is the capital recovery factor which the annual interest rate, *i*, can be calculated using equation (3) where *i'* is the nominal interest rate and *f* is the annual inflation rate.

Until March 2015, Department of Statistics Malaysia reported the inflation rate in Malaysia is 0.90 % and the nominal interest rate is 3.25 % [26, 27]. As a result, the annual real interest rate is 2.33%.

 DOI: 10.1051/matecconf/20167400022

3.2 Payback period

The Payback Period is the length of time it takes for a project to generate enough cash flow to pay back the initial investment. Payback period is the ratio of the extra first cost, ΔP to the annual savings, *S* as written in equation (4) [28]:

$$Payback\ period = \frac{Extra\ first\ cost, \Delta P\ (\$)}{Annual\ saving, S\ (\$/yr)} \tag{4}$$

4 Results and analysis

The outcome shows the grid-PV system and the grid-only system based on 25 years project lifetime with annual interest rate of 2.33%. The simulation results were generated by HOMER with configuration of 3000 kW grid-PV modules and 1000 kW converter, representing the most feasible system. The optimization results are shown in Figure 4.

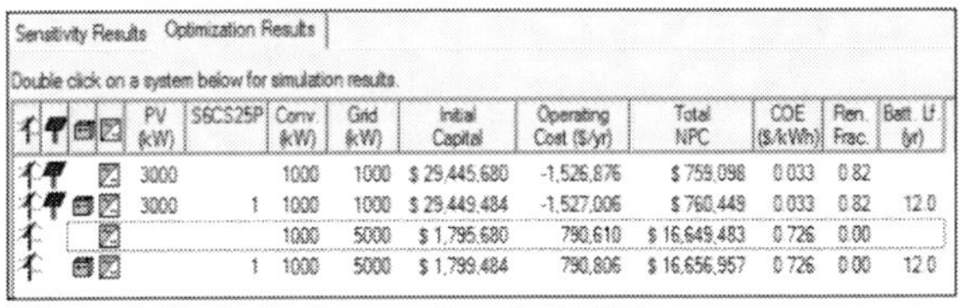
Sensitivity Results | Optimization Results
Double click on a system below for simulation results.

PV (kW)	S6CS25P	Conv. (kW)	Grid (kW)	Initial Capital	Operating Cost ($/yr)	Total NPC	COE ($/kWh)	Ren. Frac.	Batt. Lf. (yr)
3000		1000	1000	$ 29,445,680	-1,526,876	$ 759,098	0.033	0.82	
3000	1	1000	1000	$ 29,449,484	-1,527,006	$ 760,449	0.033	0.82	12.0
		1000	5000	$ 1,795,680	790,610	$ 16,649,483	0.726	0.00	
	1	1000	5000	$ 1,799,484	790,806	$ 16,656,957	0.726	0.00	12.0

Figure 4. Optimization results of grid-PV system

Grid-PV system is the most optimal system with total NPC is the lowest at RM 759,098 with cost of electricity (COE) at RM 0.033/kWh. The negative value of operating cost is obtained because of the power generated from the system is sold to the grid is much higher than the power purchased from the grid to the load. This can be seen clearly as shown in the cash flow summary of grid-PV system in Figure 5. Grid-PV system with battery has a higher NPC than grid-PV system due to the cost of battery. Note that, the system only require 1 unit battery although the sizing range is set up to 10 units because selling extra electricity to the grid is more profitable than store it in the battery. For grid- only system, it might not invest much in capital cost but shows the highest total NPC at RM 16,649,483, and COE at RM 0.726/kWh. Note also that, the COE for grid only system is higher than the tariff given by TNB because the system requires converter that contributes to the increasing of COE.

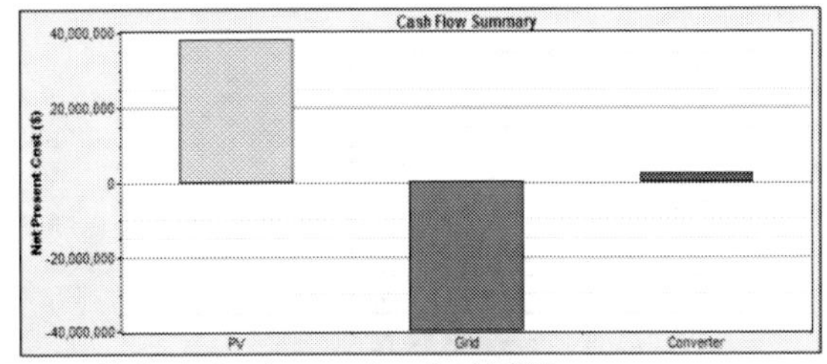

Figure 5. Optimization results of grid-PV system

The production and consumption of electricity for grid-PV and grid-only systems are shown in the following Figure 6 and Figure 7, respectively. For grid-PV system, the figures show that PV module produced 3,914,504 kWh/year of electricity with 82% from electricity produced by the system and another 18% of electricity was purchased from the grid with 849,924 kWh/year. Loads used up to 1,220,886 kWh/year of electricity with approximately 34% of electricity production while 66% of excess electricity was sold to the grid at 2,419,648 kWh/year with rates RM 1.49/kWh. Electricity produced by the grid-only system is 1,436,381 kWh/year and electricity consumption by the load is 1,220,923 kWh/year with 215,458 kWh/year electricity losses. No energy was sold to the grid because the amount of energy purchased from the grid was used up by the loads.

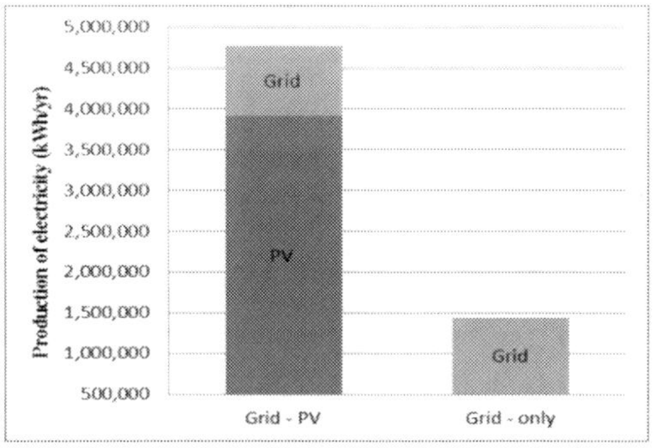

Figure 6. Production of electricity

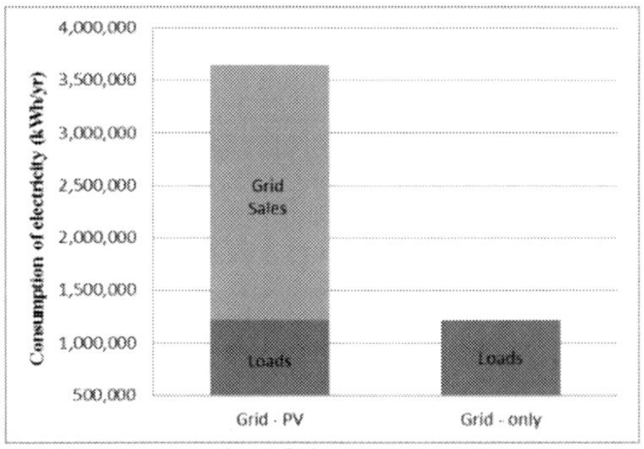

Figure 7. Consumption of electricity

This system has no exception in generating pollutants. In fact, greenhouse gases are emitted due to the energy from grid as the energy is generated with fossil fuel [29]. Because of this system still needs to purchase energy from the grid as supporting energy; this contributes to the emission of greenhouse gasses. However, with grid-PV system, there was a reduction of emission compared to the grid-only system to 373,034 kg/year that is 40% in reduction of emission. The results are shown in Table 2.

Table 2. The pollutant emission of systems.

Pollutant	Emissions of systems (kg/yr.)	
	Grid-only	**Grid-PV**
Carbon dioxide	907,793	537,152
Carbon monoxide	0	0
Unburned hydrocarbon	0	0
Particulate matter	0	0
Sulphur dioxide	3,936	2,329
Nitrogen oxide	1,925	1,139
Total emission	**913,654**	**540,620**

4.1 Grid-PV system

Figure 8 shows the production of electricity from PV in yellow colour. The electricity from the grid is still need to be purchased to supply electricity to the load especially at night. However, during Northeast Monsoon, the electricity produced by PV are fully sold to the grid because no fishing activities during this season. Thus, the amount of power sold to the grid is higher than the grid purchased for the loads. The consumption of electricity by the loads is 1,220,886 kWh/year which was approximately 34% from the production of electricity. Therefore, the 66% of excess electricity were sold to the grid at 2,419,648 kWh/year with rates RM 1.49/kWh.

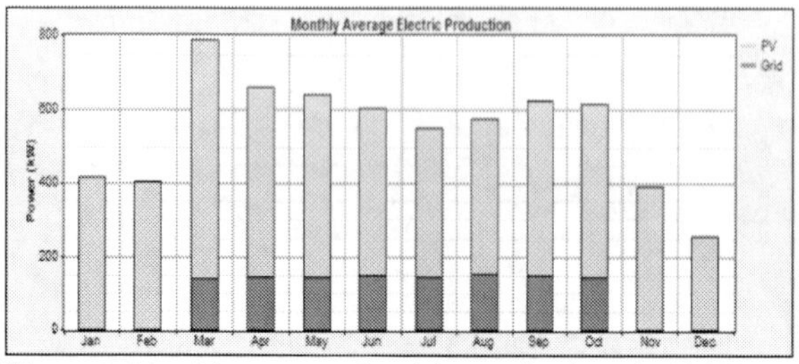

Figure 8. Monthly average electricity production of grid-PV system

4.2 Payback period

Payback period of the system can be determined from equation (4). With rates RM 1.49/kWh and cost of energy sold to the grid-PV system, the annual saving of the system is RM 3,605,275.52/year. Thus, the payback period is 8.2 years

4.3 Malaysia Subsidy Scheme

Based on [15], the amount of fuel subsidy received by the fishermen at Kuala Terengganu Utara area in 2014 was RM 1,228,325.70. Total income from selling the electricity of the proposed system is higher than the amount of subsidy spent by the government which was RM 3,605,275.52/year. Thus, this initiative is one of ways to reduce the government spending on fuel subsidy in fisheries sector and to raise fund in other sector such as education and medical sector.

5 Conclusions

In order to reduce fuel consumption and subsidies from the government, the idea of grid- PV charging stations system is produced. Besides being able to reduce the cost of subsidies, it can also reduce fuel consumption and thus reduce pollution by using cleaner energy from renewable energy sources, namely solar energy. It is feasible to develop EB charging station in Kuala Terengganu. Moreover, grid-PV system is much better than grid-only system in term of cost and emission as mentioned before. Even though the installation cost of PV is expensive, but it only takes 8.2 years to gain the payback period of the system. In addition, the revenue of the system is higher than the subsidy spending by the government on fuel. All in all, this system is beneficial in aspects of cost, emission and good returns in the future.

Acknowledgement

The financial support of this work by the Fundamental Research Grant Scheme (FRGS) and Universiti Malaysia Terengganu is gratefully acknowledged.

References

1. S. Mekhilef, A. Safari, W. E. S. Mustaffa, R. Saidur, R. Omar, M. A. A. Younis, Solar energy in Malaysia: Current state and prospects. *Renewable and Sustainable Energy Reviews* **16**, 386-396 (2012)
2. B. Parida, S. Iniyan, R. Goic, A review of solar photovoltaic technologies. *Renewable and Sustainable Energy Reviews* **15**, 1625-1636 (2011)
3. R. Ali, I. Daut, S. Taib, A review on existing and future energy sources for electrical power generation in Malaysia. *Renewable and Sustainable Energy Reviews* **16**, 4047-4055 (2012)
4. S. Mekhilef, M. Barimani, A. Safari, Z. Salam, Malaysia's renewable energy policies and programs with green aspects. *Renewable and Sustainable Energy Reviews* **40**, 497-504 (2014)
5. W. M. W. Muda, C. A. Fong, Simulation-based method to evaluate pv-wind hybrid renewable energy system in Terengganu. *Energy and Sustainability V* **186**, 91-102 (2015)
6. Pusat Tenaga Malaysia (PTM), Energy Sector Embracing Climate Change. Paper presented at *National Conference on Climate Change Preparedness Towards Policy Changes*, (2007)
7. *Tenth Malaysia plan 2011-2015* (2010)
8. Kementerian Tenaga Teknologi Hijau dan Air, (KeTTHA). Available: http://www.kettha.gov.my/portal/index.php [Accessed 16th May 2015]
9. A. M. Yusof, Z. Mahbar, in *Proceeding of PERKEM III*, 268-279 (2008)
10. N. H. Sa'at, Social mobility among the coastal community: A case study in kuala terengganu. *Kajian Malaysia* **29**, 95–123 (2011)
11. G. Fabbri, G. Tarquini, L. Pasquali, L. Anniballi, S. Odoardi, S. Teodori, E. Santini, in *Industrial Electronics (ISIE), 2014 IEEE 23rd International Symposium*,1677-1682 (2014)
12. Vítor Monteiro, J. C. Ferreira, J. L. Afonso, paper presented at the 1st International Conference on Smart Grids and Battery IT Systems, (2012)
13. L. Liu, F. Kong, X. Liu, Y. Peng, Q. Wang, A review on electric vehicles interacting with renewable energy in smart grid. *Renewable and Sustainable Energy Reviews* **51**, 648-661 (2015)
14. T. Khatib, K. Sopian, H. A. Kazem, Actual performance and characteristic of a grid connected photovoltaic power system in the tropics: A short term evaluation. *Energy Conversion and Management* **71**, 115-119 (2013)

15. Lembaga Kemajuan Ikan Malaysia, (LKIM). Available: http://www.lkim.gov.my/ [Accessed 16th May 2015]
16. S. Minami, K. Koizumi, T. Hanada, N. Matsuda, K. Ishizu, J. Nishi, T. Fujiwara, in *Electric Vehicle Symposium and Exhibition (EVS27), 2013 World*, 1-4 (2013)
17. East Coast Environmental Research Institute (ESERI), Universiti Sultan Zainal Abidin (UniSZA). Available: http://www.unisza.edu.my/eseri/ [Accessed 10 February 2015]
18. M. A. A. F. Yousif, E. Alfegi, N. Asim And K. Sopian, paper presented at the 7th International Conference on Renewable Energy Sources (RES '13), Kuala Lumpur, Malaysia, (2013)
19. Tenaga Nasional Berhad (TNB). Available: https://www.tnb.com.my/ [Accessed 30 March 2015]
20. Sustainable Energy Development Authority (SEDA). Available: http://www.seda.gov.my/#tab-A [Accessed 30 March 2015]
21. R. N. S. R. Mukhtaruddin, H. A. Rahman, M. Y. Hassan, in *Clean Electrical Power (ICCEP), 2013 International Conference,* 577-583 (2013)
22. Sharp Electronics Corporation. Sharp-ND204QCJ datasheet. Available: www.sharpusa.com/solar [Accessed 21st May 2015]
23. W. Y. Teoh, S. Y. Khu, C. W. Tan, I. H. Hii, K. W. Cheu, Technoeconomic and carbon emission analysis for a grid-connected photovoltaic system in malacca. *ISRN Renewable Energy* **2012**, 8 (2012)
24. SMA America Inc. Sunny Sentral 250U/500U Datasheet. Available: https://www.civicsolar.com [Accessed 21st May 2015]
25. M. S. Ngan, C. W. Tan, Assessment of economic viability for PV/wind/diesel hybrid energy system in southern Peninsular Malaysia. *Renewable and Sustainable Energy Reviews* **16**, 634-647 (2012)
26. Trading Economics. Available: http://www.tradingeconomics.com/ [Accessed 6th May 2015]
27. Malaysia Energy Information Hub (MEIH). Available: http://meih.st.gov.my/statistics [Accessed 6th May 2015]
28. M. M. Gilbert, in *Renewable and efficient electric power systems*. (John Wiley and Sons, New York, 2004), chap. Chapter 5: Economics of Distributed Resources, 240-254 (2014)
29. B. Norton, P. C. Eames, S. N. G. Lo, Full-energy-chain analysis of greenhouse gas emissions for solar thermal electric power generation systems. *Renewable Energy* **15**, 131-136 (1998)

A review on model updating of joint structure for dynamic analysis purpose

S.N. Zahari[1], A.A.R. Zakaria[1] M.S.M. Sani[1,a] and I. Zaman[2]

[1]*Advanced Structural Integrity and Vibration Research (ASIVR), Faculty of Mechanical Engineering, Universiti Malaysia Pahang, 26600 Pekan, Pahang, Malaysia*
[2]*Department of Engineering Mechanics, Faculty of Mechanical and Manufacturing Engineering, Universiti Tun Hussein Onn Malaysia 86400 Parit Raja, Batu Pahat, Johor, Malaysia*

Abstract. Structural joints provide connection between structural element (beam, plate etc.) in order to construct a whole assembled structure. There are many types of structural joints such as bolted joint, riveted joints and welded joints. The joints structures significantly contribute to structural stiffness and dynamic behaviour of structures hence the main objectives of this paper are to review on method of model updating on joints structure and to discuss the guidelines to perform model updating for dynamic analysis purpose. This review paper firstly will outline some of the existing finite element modelling works of joints structure. Experimental modal analysis is the next step to obtain modal parameters (natural frequency & mode shape) to validate and improve the discrepancy between results obtained from experimental and the simulation counterparts. Hence model updating will be carried out to minimize the differences between the two results. There are two methods of model updating; direct method and iterative method. Sensitivity analysis employed using SOL200 in NASTRAN by selecting the suitable updating parameters to avoid ill-conditioning problem. It is best to consider both geometrical and material properties in the updating procedure rather than choosing only a number of geometrical properties alone. Iterative method was chosen as the best model updating procedure because the physical meaning of updated parameters are guaranteed although this method required computational effort compare to direct method.

1 Introduction

Joints are essential parts of complex structure and play a vital role in the assembled structure's behaviors (e.g. flexibility and damping). There are various kinds of joints used in the engineering structure, such as welded joints, bolted joints and riveted joints. These connecting elements contribute significantly to dynamic behaviors of a complete structure. Assuming rigid connections between substructures in modelling without considering the joints effects may lead to different characteristics compared to actual physical structure.

Dynamic properties of a joint structure can be investigated via two methods which by numerical and experimental analysis. The most commonly used numerical method is finite element (FE) method which is used to simulate the behavior of real systems and several models have been presented in this literature. The purpose of numerical modelling can be classified into three main categories which are analysis, prediction and design. Normally, the numerical model for FE analysis of the associated actual engineering structure is constructed on the basis of highly idealized engineering design that may not fully represent all the physical and geometrical aspects of the actually built structure analysis. Therefore, FE predictions are often called into question when they are in conflicts with experimental result. Based on previous studies [1, 2], the discrepancy between numerical and experimental results may exceed 10% or sometime even 40% error. Incorrect modelling of boundary condition, modelling of joints and difficulties modelling with damping that led to existing of difference between numerical and test result.

This has led to the development of model updating technique also known as model calibration or in simplified terms, parameter estimation or identification. The main purpose of model updating is to reduce the errors between numerical and test results. Model updating often implemented by analyzing the degree to which a finite element model represent a single set of experimental data [3, 4]. There are two methods of model updating which are direct method and sensitivity method. In model updating, the selection of updating parameter is the most important task. If numerical predictions are insensitive to a chosen parameter, then updating will result in a change to the parameter of uncertain value, because the difference between predictions and results has been reconciled by changes to other (more sensitive) parameters that might be less in need of updating. The result, in that case will be an updated model which

[a] Corresponding author: mshahrir@ump.edu.my

replicates the measurements but lacks physical meaning. The updated parameters should be justified physically and the quality of the final model should be assessed within the operating range. There are three aspects that lead to a credibility for updated structural; robustness to uncertainties, fidelity to data and confidence in predictions [5, 6].

This review paper first outlines on FE modelling for different kinds of joints structure from previous studies with attention given to spot welded modelling joints due to its continuous relevance to this review paper on model updating. The identification of updating parameters and stochastic model updating will be emphasized in the second section. The following section will provide a brief explanation on the method of model updating using sensitivity approach.

2 FE modelling of joints structure

Many engineering structures were assembled from components by using variety of connections such as bolted, riveted, fastener and welded joints. For a simple structure, modelling and numerical predictions have been developed for decades. However prediction of dynamic characteristic for structure with joints is not an easy task due to its complexities in modelling the joints. Joints add damping to the structural system and decrease overall stiffness thus changing the overall dynamics characteristic of the system. Due to this effect, modelling of the joint has become more and more significant.

Bolted joints are widely used in joining the component especially in modern aircraft application due to its characteristics; easily disassembled, maintained and inspected. Many researches on joints have been conducted by scientists and engineers since early 1970 [7-9]. On top of that, they have tried to understand the characteristics of joints and simulate their findings into analytical modelling. Gaul and Nitsche [7] concentrated on describing nonlinear transfer behaviour of frictional interface using constitutive and phenomenological models. Constitutive models are based on interface physics by describing the friction phenomena in local manner while phenomenological model depends on experimental observation. The paper covered on Jenkin-element models (also known as an elasto-slip model) that consisting of linear springs and Coulomb friction element were used to represent the friction interface of the bolted joints. It is shown that this model has a potential to simulate friction force in a bolted joint very well. Iranzad [10] also utilized a constitutive model with a thin layer of elasto-plastic material to model nonlinear behaviour of bolted joints. The thin layer elastic material properties represent the joint linear behaviour at low vibration levels. Yoo et al. [8] proposed cone frusta method for jointed part and spring elements to represent the contact effects occurring in the interface area in simulating the dynamic behaviour of bolted joints structure with a large interface structure. Another bolted joint model presented by Wang et al. [9] introduced the strain rate dependant elastic modulus into the spring mass model. On the other hand, Rutman et al. [11] identified the modelling techniques of the bolted joint used to connect with different types of components in NASTRAN such as spring elements (CELAS and CBUSH), connector elements (CBAR and CBEAM) and rigid connection element (RBE2). The combination of element used to idealise the bending and the shear of the fastener shank. It also bring elastic bearing stiffness of the plate and fasteners at the contact surface. These proposed models are capable to predict the strain rate dependant stiffness and strength of composite bolted joint under static and dynamic loading.

Compared to bolted and riveted joints, welded joints are permanent and reliable evaluation of the behaviour of the welds is always a concern. There are many local effects such as geometrical irregularities, residual stresses, materials inhomogeneity and defects during welding process that are not to be considered by FE modelling lead to difficulties to model the welded joints. In real structure, there are thousands of spot welds and modelling the detailed model will be a challenge. A very detailed model produces a detailed and smooth stress field, but it may provide less accurate prediction of stiffness for real spot welds and their effect on the rest of the structure. However for vibration analysis purposes, only simple models that represent the stiffness characteristic needed to predict their influence on the rest of the structure. Several previous research [12-15] focused on modelling of spot weld been discussed by [16] for stress analysis and stiffness simulation. For latest vibration analysis and model updating purposes, the spot weld modelling being carried out by some of these researchers [17-19].

Abu Husain et al. [17] developed FE model of spot welds using CWELD element for dynamic prediction. CWELD element represented by two nodes special shear flexible beam type element with 12 DOFs (six for each node) and all nodes are connected to their corresponding patches with constraints from the Kirchhoff shell theory [20]. This type of model can be used to represent laser spot weld with good accuracy by selecting the right updating parameters. When this CWELD element modelling to be used in dynamic analysis of structure of similar constructions, it is recommended to put the value for Young modulus of the patch three times the value for Young modulus of weld dynamic analysis of structure. Another spot weld model was proposed by Kuratani [18]. The model known as Area Contact Model 2 (ACM2) using multi point constraint (MPC) that designed in ANSYS software that having a same features as in NASTRAN. The ACM2 model consisted of a single solid element connecting the upper and lower shell element with RBE3 (rigid connection element). The RBE3 element is an interpolation element and automatically generates internal MPC equations in finite element analysis process. These models can be employed to investigate the effects of mesh size on modal properties (natural frequency and mode shape). ACM2 model is relatively sensitive to mesh size in patch area. When the size of mesh in patch area increases, the natural frequency also increases. This leads to the increasing stiffness at the centre of the patch due to the patch area size increase. The higher and lower mesh size will cause the loss of stiffness. Therefore, patch area must be meshed with care. From this study, it concludes that the proper shell element size in the patch area is dependent on the solid element size determined from the diameter of a weld nugget. The recommended range of the ratio of element sizes of the shell to the solid elements is between 1.0 and 1.5. On the other hand, Alvarez [19] also proposed new model of spot weld that imposes a surface to surface connection between two structures using simple spring element and multipoint constraint connection (MPC) therefore the coincident meshes are not required. Area of the spot weld is consistent to spot weld dimension whereas rotational stiffness is proportional to the spot weld radius. This model develops a connection model by adding an equivalent rotational stiffness to the system using an array of translational spring instead of adding stiffness to rotational degree of freedom (DOF). This type of model built to avoid the problem of sensitivity to element size that leads to poor convergence. When the proposed spot model had been used, the average sensitivity of element size is reduced almost five times from the actual value.

3 Experimental modal analysis

Experimental modal analysis or modal testing has grown steadily in popularity for the past several decades. Modal testing is defined as the study of dynamics characteristic of a mechanical structure. The setup and the instrumentation used influence the experimental result in modal testing [21]; therefore to obtain the accurate result, modal testing should be carried out in a free-free boundary condition. EMA is used to extract modal parameters (natural frequency, mode shape, damping ratio, modal vectors and modal scaling). The structure will be excited by certain inputs (hammer, shaker) and sensing mechanism will be used to measure the input force in order to produce a set of frequency response function (FRF's) that contain inherent dynamic properties of a structure. EMA plays an important part in design and analysis of structures and conducted to validate the results from simulation models before they can be used for further detailed analysis. Finite element model updating will be performed to reduce the discrepancy between modal testing and numerical results.

4 Identification of updating parameters

The choice of parameters is a critical part in model updating and to produce well-conditioned updating problem, it is necessary to select those updating parameters which will be most effective in producing a genuine improvement in modelling the structure [3]. Parameters that have been selected will be adjusted to minimize a penalty function based on residual between an experimental set and the corresponding numerical method. Normally the parameter that sensitive to the data should be taken into consideration as updating parameters but this is cannot be the main reason for selecting updating parameters. Updating parameters should be chosen with the aim of correcting recognized uncertainties in the model and the data should be sensitive to them.

There are two main groups of uncertainties in analytical modelling; physical uncertainties and numerical uncertainties. Boundary and initial condition, material properties, geometry and load are classified as physical uncertainties meanwhile for conceptual and mathematical modelling, discretization error, numerical solution and human mistakes known as numerical uncertainties [4]. Once the uncertainties are being considered, a deterministic problem will then change to non-deterministic problem (stochastic). It is very useful and highly recommended to explore numerical predictions on behavior of structure with uncertainties and many previous researches studied on stochastic model updating approach using two methods [5-15]; Monte Carlo simulation and perturbation method. For this paper, it focuses on spot welded structure and as simplified in the first section, the joints structure could be considered less accurately modelled due to the uncertainties in the structural parameters such as modulus elasticity and diameter of the weld, mass density, boundary condition etc. Therefore, the parameterization of the inaccurate parts of the welded structure is important.

The material parameters, thickness and cross sectional dimensional tends to be most powerful updating parameters because they often apply throughout a FE mesh affecting a large number of elements therefore a small change in these parameters will affect the natural frequency very considerably [16]. There is a study by Abu Husain et al [12] that shows that selecting some of material properties as the updating parameters provide better convergence than those updated by using only the thickness parameter. From these findings, it stated that in numerical predictions it is best to consider both geometrical and material properties in the updating procedure rather than choose only a number of geometrical properties alone.

Then, Abu Husain et al. also presented in the next research [13] on welded structure as the main uncertainties employed a perturbation method that being used by Haddad Khodaparast et al. [11] to investigate variability that exist between a set of nominally identical structures. The perturbation methods are used for estimation of means and covariance of updating parameters and two approaches of parameter weighting matrix assignment are explained; one approach using three parameters from the welds and the other used eight parameters which are five from the component itself and three are from the weld. The latter approach is in a very good agreement with experimental data and excellent correlation between predicted and measured covariance of the output is achieved compare to the first approach that only demonstrates good correlation between the predicted mean natural frequencies and their measured data, but poor correlation is obtained between the predicted and measured covariance of the outputs [13].

5 Model Updating Procedure

The FE method is a well-known technique that is used to analyze the behavior of a structure subjected to a variety of loads. The computation of undamped natural frequency and mode shapes is usually conducted using normal mode analysis (SOL103) in NASTRAN [17]. The result from analysis is then compared with experimental counterparts for validation purposes. The correlation between the results from these two methods for natural frequencies obtained directly but for mode shapes, there should be further analysis to validate mode shapes. Modal assurance criterion (MAC) values range from 0 and 1 can be used to predict the correlation and to pair the mode shapes vectors from experimental and simulation result. The comparisons between the numerical modal properties (i.e., the natural frequencies and mode shapes) and their measured counterparts will normally reveal some discrepancies between the two approaches, thus the FE model updating method [3] should be performed to minimise the errors. Model updating can be categorized into two different classes: direct (non-iterative) method and sensitivity (iterative) method. Direct method is known as representational method because their ability in replicating measured data [3]. But the measurement and numerical data will show some discrepancies due to existing error such as noise and model inadequacies. Using direct method, if the updated model exactly reproduces inaccurate measurement, any subsequent analysis will be inaccurate. Although these methods are computationally cheaper and reproduce the measured model data exactly, they violate structural connectivity and updated structural matrices are difficult to interpret.

Iterative method also known as sensitivity method provides wide choices of updating parameters, structural connectivity can be easily maintained and corrections suggested in the selected parameters can be physically interpreted. Iterative methods are based on minimizing an objective function that is generally a nonlinear function of selected updating parameters. Iterative method used either eigendata or frequency response function (FRF) to construct objective function. For analytical model updating, Collins [18] focused on the eigendata sensitivity meanwhile Lin and Ewins [19] used FRF data. Modak et al. [20] and his next research [21] neglecting the damping, only focusing on comparison of response function method (RFM) and inverse eigensensitivity method with an objective to study the accuracy with which they predicted the corrections required in an FE model. One of sensitivity method is the Design Sensitivity and Optimization Code (SOL 200) in Nastran [22] is used for updating, and an objective function is based on residuals between measurement data (natural frequency, mode shapes, FRF, etc.) and their predictions are set for minimisation procedure. The objective functions based on eigenvalues [3] is defined as Eq. (1):

$$J = \sum_{i=1}^{n} W_i \left(\frac{\lambda_i^{fe}}{\lambda_i^{exp}} - 1 \right)^2 \qquad (1)$$

Where:

J is the objective function,
W_i is the weighting coefficient,
λ_i^{fe} is the ith predicted eigenvalue from finite element analysis,
λ_i^{exp} is the ith experimental eigenvalue,
n is the machining time in minutes.

The procedure will be repeated until the convergence has been accomplished where the difference between objective function (J) value from consecutively iteration become smaller. SOL 200 in NASTRAN is employed for model updating through procedures [23] illustrate in Figure 1. The optimisation algorithm uses partial derivatives of a function to assist in a numerical search for optimisation procedure. The optimisation algorithms in NASTRAN generally belongs to the gradient based methods [24].

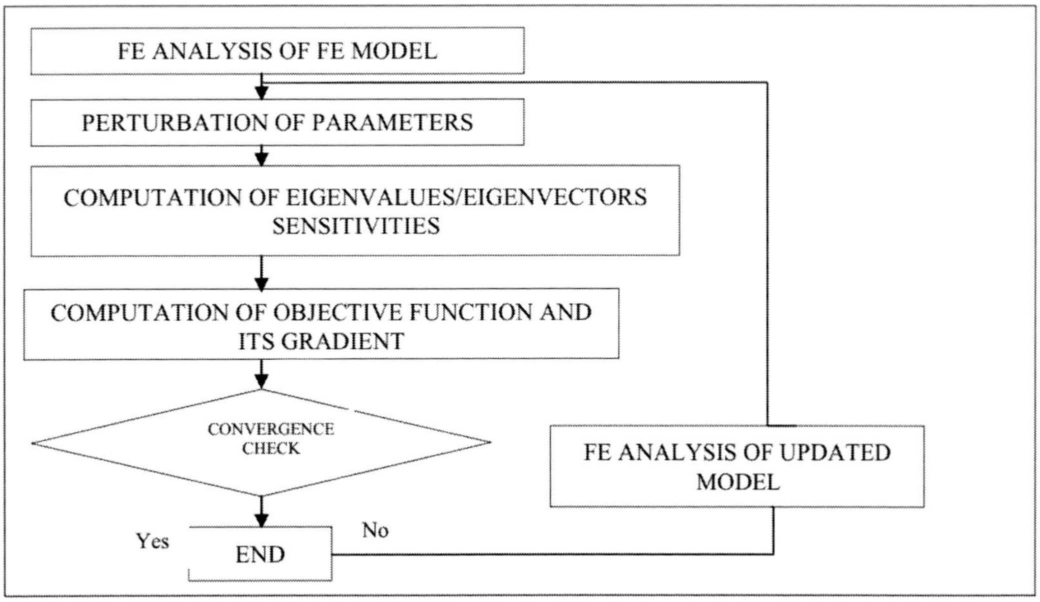

Figure 1. Procedure of model updating using SOL200 in NASTRAN

6 Conclusion

In this paper, a simple review of model updating on the joint structure has been presented. Currently, FE modelling for joints regarded as an important part in model updating. Without accurate modelling, model updating cannot be carried out significantly. The selection of updating parameters also plays a crucial part in performing model updating. It is best to consider both geometrical and material properties in the updating procedure rather than choosing only a number of geometrical properties alone. SOL200 in NASTRAN software recommended as the one of iterative methods for model updating procedure by minimizing objective function in order to decrease the difference between results obtained through predicted model parameters and experimental counterparts. This is due to their iterative nature that provides guarantee for physical meaning of updated parameters even though this method required computational effort.

Acknowledgements

The author would like to greatly acknowledge the support by focus group of Advanced Structural Integrity of Vibration Research (ASIVR) and Universiti Malaysia for providing all the equipment used for this project and special thank for Knowledge Transfer Programme (KTP) and Ministry Of Education for financial assistance support.

References

1. Brownjohn, J.M. and P.-Q. Xia, *Dynamic assessment of curved cable-stayed bridge by model updating.* Journal of Structural Engineering, (2000). **126**(2): p. 252-260.
2. Živanović, S., A. Pavic, and P. Reynolds, *Finite element modelling and updating of a lively footbridge: The complete process.* Journal of Sound and Vibration, (2007). **301**(1–2): p. 126-145.
3. Friswell, M. and J.E. Mottershead, *Finite element model updating in structural dynamics*. Vol. 38. (1995): Springer Science & Business Media.
4. Dascotte, E. *The use of FE model updating and probabilistic analysis for dealing with uncertainty in structural dynamics simulation*. in *Japan Modal Analysis Conference, Sept.* (2003).
5. Schuëller, G.I., et al., *Uncertainty Analysis of a Large-Scale Satellite Finite Element Model.* Journal of Spacecraft and Rockets, (2009). **46**(1): p. 191-202.
6. Mares, C., J. Mottershead, and M. Friswell, *Stochastic model updating: part 1—theory and simulated example.* Mechanical Systems and Signal Processing, (2006). **20**(7): p. 1674-1695.
7. Mottershead, J., et al., *Stochastic model updating: part 2—application to a set of physical structures.* Mechanical Systems and Signal Processing, (2006). **20**(8): p. 2171-2185.
8. Fang, S.-E., W.-X. Ren, and R. Perera, *A stochastic model updating method for parameter variability quantification based on response surface models and Monte Carlo simulation.* Mechanical Systems and Signal Processing, (2012). **33**(0): p. 83-96.
9. Rui, Q., H. Ouyang, and H.Y. Wang, *An efficient statistically equivalent reduced method on stochastic model updating.* Applied Mathematical Modelling, (2013). **37**(8): p. 6079-6096.
10. Bao, N. and C. Wang, *A Monte Carlo simulation based inverse propagation method for stochastic model updating.* Mechanical Systems and Signal Processing, (2015).
11. Khodaparast, H.H., J.E. Mottershead, and M.I. Friswell, *Perturbation methods for the estimation of parameter variability in stochastic model updating.* Mechanical Systems and Signal Processing, (2008). **22**(8): p. 1751-1773.
12. Husain, N.A., H.H. Khodaparast, and H. Ouyang. *Parameter selections for stochastic uncertainty in dynamic models of simple and complicated structures*. in *Proceedings of the 10th International Conference on Recent Advances in Structural Dynamics, University of Southampton, Southampton*. (2010).
13. Husain, N.A., et al. *Application of the Perturbation Method With Parameter Weighting Matrix Assignments for Estimating Variability in a Set of Nominally Identical Welded Structures*. in *ASME 2010 10th Biennial Conference on Engineering Systems Design and Analysis*. (2010). American Society of Mechanical Engineers.
14. Govers, Y. and M. Link, *Stochastic model updating—Covariance matrix adjustment from uncertain experimental modal data.* Mechanical Systems and Signal Processing, (2010). **24**(3): p. 696-706.
15. Chen, H.-P. and T.S. Maung, *Regularised finite element model updating using measured incomplete modal data.* Journal of Sound and Vibration, (2014). **333**(21): p. 5566-5582.
16. Mottershead, J.E., M. Link, and M.I. Friswell, *The sensitivity method in finite element model updating: a tutorial.* Mechanical Systems and Signal Processing, (2011). **25**(7): p. 2275-2296.
17. Sitton, G., *MSC/NASTRAN basic dynamic analysis user's guide.* Macneal-Schwendler Co, (1997).
18. Collins, J.D., et al., *Statistical Identification of Structures.* AIAA Journal, (1974). **12**(2): p. 185-190.
19. Lin, R.M. and D.J. Ewins, *Analytical model improvement using frequency response functions.* Mechanical Systems and Signal Processing, (1994). **8**(4): p. 437-458.
20. Modak, S.V., T.K. Kundra, and B.C. Nakra, *Comparative study of model updating methods using simulated experimental data.* Computers & Structures, (2002). **80**(5–6): p. 437-447.
21. Modak, S.V., T.K. Kundra, and B.C. Nakra, *Model updating using constrained optimization.* Mechanics Research Communications, (2000). **27**(5): p. 543-551.

[a] Corresponding author: mshahrir@ump.edu.my

22. Husain, N.A., et al., *Finite-element modelling and updating of laser spot weld joints in a top-hat structure for dynamic analysis.* Proceedings of the Institution of Mechanical Engineers, Part C: Journal of Mechanical Engineering Science, (2010). **224**(4): p. 851-861.
23. Moore, G.J., *MSC/NASTRAN design sensitivity and optimization: user's guide, version 68.* (1994): MacNeal-Schwendler Corporation.
24. Mares, C., M. Friswell, and J. Mottershead, *Model updating using robust estimation.* Mechanical Systems and Signal Processing, (2002). **16**(1): p. 169-183.

Parametric study of multiple configurations of pico hydrokinetic turbines using CFD

Yogaraja Balakrishnan[1] and At-Tasneem Mohd Amin[2,a]

[1]*Faculty of Mechanical Engineering, Universiti Malaysia Pahang, 26600 Pekan, Pahang, Malaysia*
[2]*Fellow Researcher, Centre for Earth Resources Research & Management (CERRM), Universiti Malaysia Pahang, Lebuhraya Tun Razak, 26300 Gambang, Kuantan, Pahang, Malaysia*

Abstract. This paper aims to study the river flow characteristics over pico hydrokinetic turbines with variation of arrangement using computational fluid dynamics (CFD) software. This study is required to obtain the optimum spacing and angle between the turbines which leads to higher turbine effective utilisation and performance in terms of power generated. In this study, a river model is created in CFD software to simulate the water flow over the turbines as they are placed in a river to obtain the water flow characteristic. Different types of array arrangements are simulated in the river model. Multiple turbines are used to accumulate more power. The turbine model which consists of eight turbines is arranged in series with different spacing, ranging from a size of diameter (1D) to four times diameter (4D) of turbine is simulated to identify the optimum spacing between the turbines. Then, the simulation is continued using a sufficient spacing of 0.5D with angles ranging of 10° to 60° from datum of original position to minimise the disruption of the aquatic environment. The velocity profiles of each turbine are obtained and analysed. The 4D spacing and 40° angle displayed higher average velocities compared to other arrangements. Thus, from this study, the 4D and 40° are deduced as the optimum spacing and angle, respectively.

1 Introduction

The term pico hydropower is a small-scale hydropower with power generation of less than 5 kW. It has made significant contributions to remote and off-grid settlements [1]. Installing and operating a run of river application, such as a pico hydrokinetic system is fully environment friendly as it does not alter the environment's natural course. This technology is favourable because of its constant supply of electrical energy and cost effectiveness in terms of low investment and maintenance costs [2]. The turbines can be installed in any flow with a velocity greater than 0.5 m/s [3]. Kinetic energy from the flowing water can be converted into mechanical energy by rotating the turbine blade which then generates electricity.

Most computational researches on array arrangements are conducted for tidal turbines, which is comparable to run of river turbines [4]. Computational fluid dynamics (CFD) models are used to study and analyse wake effects. The simulations of the turbines are done using the boundary conditions as employed by [4, 5] to obtain the wake characteristics and comparing it later with experimental data.

A wake is the region of disturbed flow or recirculating flow immediately behind a stationary or moving solid body, caused by the fluid flowing around the body. The flow in this trail is slowed down and quite turbulent when compared with the flow arriving in front of the turbine. In incompressible fluids such as water, it results in a wave. Similar to all wave forms, it then spreads outward from the source until the energy it possesses is overcome or lost, normally by friction or dispersion [6].

According to A.S Bahaj et. al. [7], near wake may take the form of vortices shed from the tip of the blades or from the support structure and will bound the slower moving flow from the stream. A discontinuity in the stream velocity profile is created by these vortices. Usually the near wake exists from 0D - 3D/4D, in which time the surrounding turbulence of the free flowing stream breaks down the bounding vortices.

Based on the study conducted by Wu et. al. [8], the formation and recovery of turbine wakes are known to be affected by the incoming flow characteristics, the operational states of a turbine and the turbine-generated turbulence characteristics. Both numerical and experimental studies of single turbine wakes in turbulent boundary layer flows has shown that the turbulence intensity level in the incoming flow can have a great effect on the wake recovery rate. A.S Bahaj et. al. [7] mentioned that the faster the free stream flows, the greater the convection of the wake. This means that the wake is carried further downstream in the same period of time. Therefore, the speed of the free stream is a key parameter in analysing the wake velocity.

[a] Corresponding author: tasneem@ump.edu.my

Study by Churchfield et. al. [9] reveals that the way in which the turbulent inflow is simulated greatly affects the wake propagation and power production of the array flow. From the study, it shows that the low speed wakes of the first row of turbines have an effect upon the second row. Hence, the wakes of the second row of turbines exhibit low stream velocities. The second row of turbine wakes also appears to contain more turbulent motions and meander significantly.

The work of Bai et. al. [5] presents the relative effect of increasing the inter row spacing and row spacing between TEC turbines. The results display that the array with larger width spacing shows the fluid entering the central TEC to be less influenced by the wakes from the upstream TEC row. Thus, this variation in performance highlights that there is an opportunity to modify spacing to fully optimise array performance.

The objectives of this study are to create a river model using CFD software and to simulate the water flow over multiple hydrokinetic turbines over different array arrangements to obtain the water flow characteristics. The aim is to obtain the optimum spacing and angle between the turbines which leads to higher turbine effective utilisation and performance in terms of power generated.

2 Mathematical formulations

Governing equations

In this study, the flow field is assumed to be incompressible, steady and non-isothermal flow. Therefore, the governing equations for the continuity, momentum and energy can be expressed as [10].

Continuity equation for incompressible flow:

$$\frac{\partial u}{\partial x}+\frac{\partial v}{\partial y}+\frac{\partial w}{\partial z}=0 \tag{1}$$

$$\vec{\nabla}\cdot\vec{V}=0 \tag{2}$$

where u, v and w are velocity components in the x, y and z directions, respectively, and $\vec{\nabla}$ is divergence operator.

Momentum equation:

$$\rho\frac{D\vec{V}}{Dt}=\rho\left[\frac{\partial\vec{V}}{\partial t}+\left(\vec{V}\cdot\vec{\nabla}\right)\vec{V}\right]$$
$$=-\vec{\nabla}p+\rho\vec{g}+\mu\nabla^2\vec{V} \tag{3}$$

where ρ is density of the fluid (kg/m^3), V is velocity vector of the fluid (m/s), t is time (seconds), g is gravitational acceleration (m/s^2) and μ in the fluid viscosity (kg/m.s).

Conservation of energy equation:

$$\rho\frac{DE}{Dt}=\frac{\partial(u\sigma_{xx})}{\partial x}+\frac{\partial(v\sigma_{yy})}{\partial y}+\frac{\partial(w\sigma_{zz})}{\partial z}$$
$$+\frac{\partial(u\tau_{yx})}{\partial y}+\frac{\partial(v\tau_{zx})}{\partial z}+\frac{\partial(v\tau_{xy})}{\partial x}+\frac{\partial(v\tau_{zy})}{\partial z}$$
$$+\frac{\partial(w\tau_{xz})}{\partial x}+\frac{\partial(w\tau_{yz})}{\partial y}-\frac{\partial q_x}{\partial x}-\frac{\partial q_y}{\partial y}-\frac{\partial q_z}{\partial z} \tag{4}$$

where σ is stress and τ is shear stress.

Kinetic energy (KE) is the energy of motion. The kinetic energy of an object is the energy it possesses because of its motion. Kinetic energy equation is given as:

$$KE=\frac{1}{2}\,m\,v^2 \tag{5}$$

where KE represents kinetic energy (J), m is mass (kg) and v is the velocity (m/s).

The kinetic energy can be converted into mechanical energy. Kinetic energy from the flowing water can be converted into kinetic energy of rotating shaft by rotating the turbine blade. The formula below shows that velocity is directly proportional to the angular velocity:

$$v=r\,\omega \tag{6}$$

where r is the radius (m) and ω is the angular velocity (rad/s).

The kinetic energy of a rotating object is analogous to linear kinetic energy and can be expressed in terms of the moment of inertia and angular velocity. For a given fixed axis of rotation, the rotational kinetic energy can be expressed in the form:

$$KE=\frac{1}{2}\,I\,\omega^2 \tag{7}$$

where I is the moment of inertia or rotational inertia (kg.m^2).

If a constant torque rotates a body through an angle θ, the work done is:

$$W=T\,\theta \tag{8}$$

where W is the work done (J), T is the torque (J/rad) and θ is the angle (rad).

The power of the rotating turbine blade is given by:

$$P=T\,\omega \tag{9}$$

where P is power in watts (W).

3 Numerical analysis

3.1 Turbine and blade design

The dimensions of the turbine blade are shown in Figure 1. All values are in millimetres (mm).

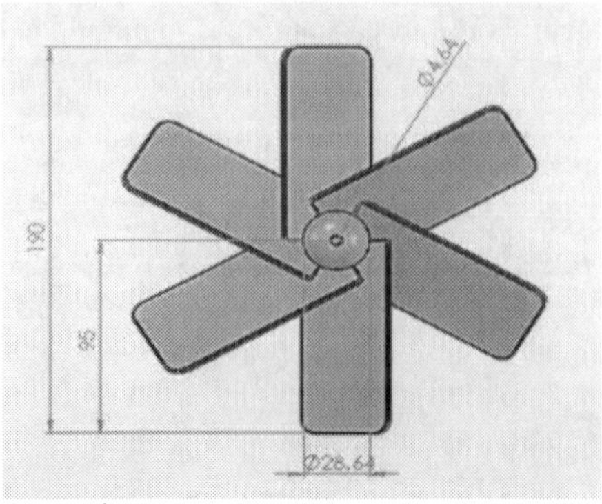

a) Front view

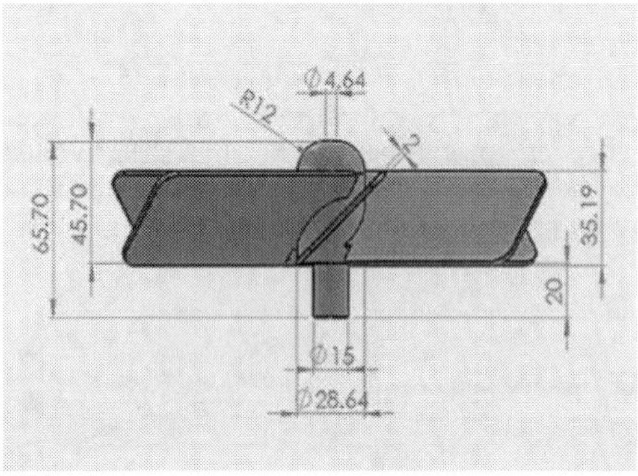

b) Side view

Figure 1. Dimensions of the blade

The blade is assembled to the turbine body as shown in Figure 2. For this study, eight identical turbines were used. The turbines were fitted through a shaft as shown in Figure 3.

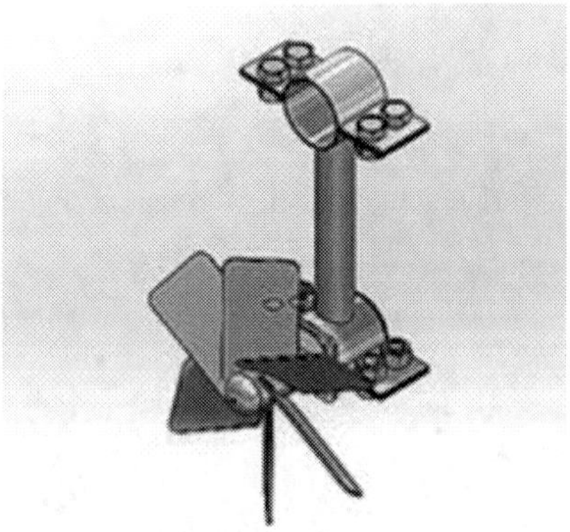

Figure 2. Turbine body

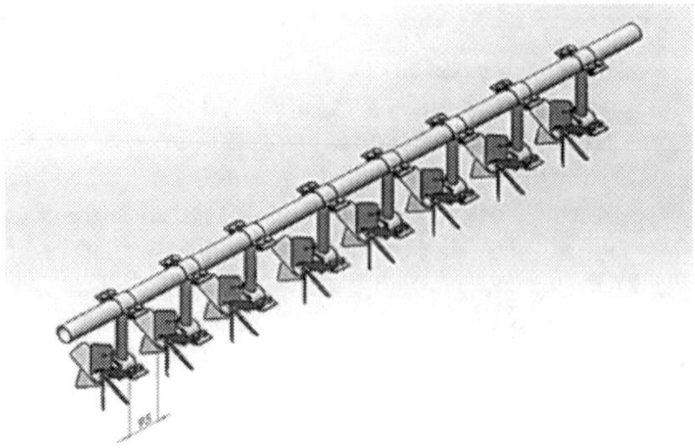

Figure 3. Full turbine arrangement

3.2 Computational Domain

The CFD software used is ANSYS 15.0 and the flow is studied using the Fluid Flow (CFX) analysis system. A river domain was set up using a box as an enclosure. The turbines were placed within the enclosure with the desired array arrangements of spacing and angle. Eight turbines are used for the simulation of different spacing while four turbines are used for the simulation of different angles.

3.3 Meshing

Fine mesh size is used on the model for every array arrangements. The number of element counts differs according to the spacing and angle between the turbines.

3.4 Boundary Conditions

For this study, the river flow was set as a steady flow fluid. As it is an open-channel flow, the fluid is composed of air (at the top) and water. However, the presence of air was neglected. The inlet velocity is 2.0 m/s with equal upstream and downstream heights. Hence, the gradient of the river is zero. The upstream height is the height of water level of the river at the inlet and the downstream height is the height of the river water level at the outlet. The boundary conditions used in the simulation are summarised in Table 1.

Table 1. Boundary conditions of river model

Boundaries	Parameter
Inlet	Velocity = 2.0 m/s
Outlet	Constant static pressure
Top surface	Zero pressure
Sidewall	No slip wall
Bottom	No slip wall

3.5 Model configuration

There are ten different array arrangements used for this study. The simulations were conducted on eight turbines which are arranged in series, as shown in Figure 4 and tested with different spacing, x. Four tests were conducted, in which x is varied from a size of diameter (1D) to four times diameter (4D) of turbine, with an increment of 1D consecutively. D represents the diameter of the turbine blade, where D equals to 190 mm.

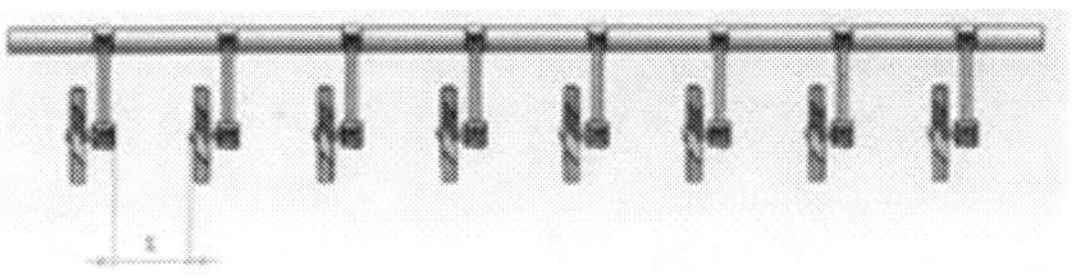

Figure 4. Turbine arranged in series

The simulation is continued by using four turbines with 0.5D spacing and tested with different angles, θ. Angles, as in shown in Figure 5 are introduced with the aim to minimise the aquatic environment. Six tests were conducted, where θ is varied with angles of 10° to 60° from datum of original position, with an increment of 10° consecutively.

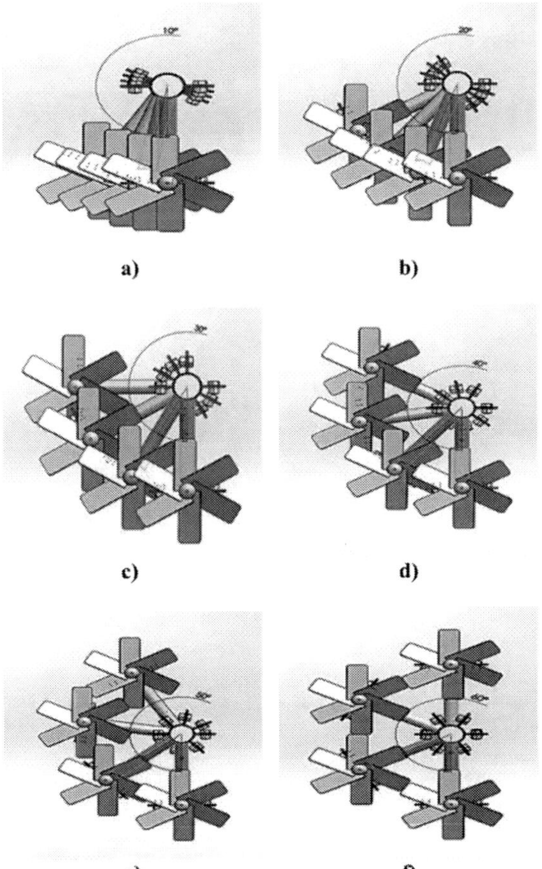

a) b) c) d) e) f)

Figure 5. Turbines arranged under the influence of angles; a) 10°, b) 20°, c) 30°, d) 40°, e) 50° and f) 60°

4 Results and discussion

The CFD simulations and analysis were conducted to study the river water flow through the multiple hydrokinetic turbines of different spacing and angles. A comparison data is conducted to present the differences of velocity of each turbine in every arrangement. These simulations were carried out with a constant mean value of river velocity of 2.0 m/s throughout ten different array arrangements with consists of four different spacing (1D, 2D, 3D and 4D) and six different angles (10°, 20°, 30°, 40°, 50° and 60°).

For the velocity plots, velocity values at the entrance of the turbine were extracted from the line drawn across the mouth of the turbines. The values for all array arrangements were plotted for analysis. The higher the value of velocity at the turbine mouth, the higher the performance of the turbine. This is due to the higher kinetic energy in the flow. The average velocity for each array arrangement is calculated to determine the best arrangement to generate maximum power.

4.1 Spacing

Figure 6 displays the changes in velocities of the river water as it flows throughout the turbines. It shows that the velocity of the flow decreases as it passes through the turbines. Figure 6a) shows a large drop in velocity occurs as the water passes through the turbines of 1D spacing and the velocity remains low at the mouth of each turbine. However, as the spacing increases, the water velocity starts to recover slowly as it flows over a distance. Figure 6b) shows a slight increase in velocity as well as in Figure 6c), which exhibits a higher increase in velocity. Based on the four tests, the 4D spacing arrangement in Figure 6d) shows the highest velocity at the mouth of each discrete turbine.

 DOI: 10.1051/matecconf/20167400024

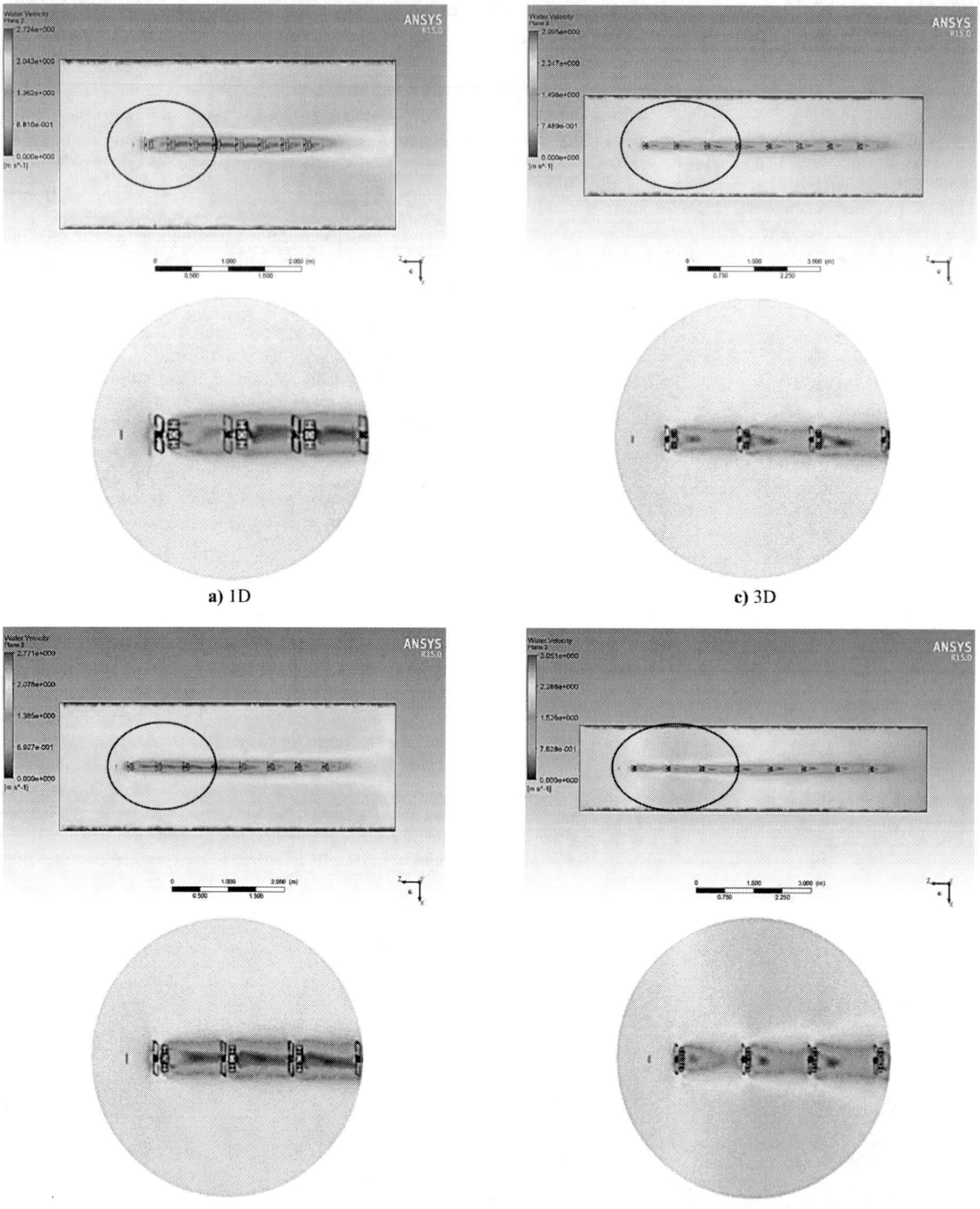

Figure 6. Velocity planes from simulation results of turbines with different spacing

4.2 Angle

Figure 7 displays the arrangement of turbines with different angles and its velocity profiles. Figure 7a) shows a large drop in velocity occurs as the water passes through the turbines with 10° angle and the velocity at the mouth of each discrete turbine remains low. However, as the angle increases, the velocity at the mouth of each discrete turbine also increases. Based on the six tests, it can be concluded that when the angle increases, the position of the turbine moves away from the low velocity region that forms behind each turbine. Hence, Figure 7f) with the 60° angle arrangement exhibits the highest velocity.

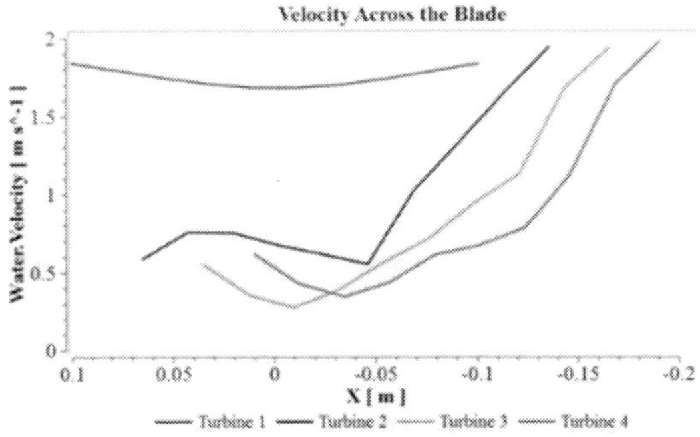

a) 10°

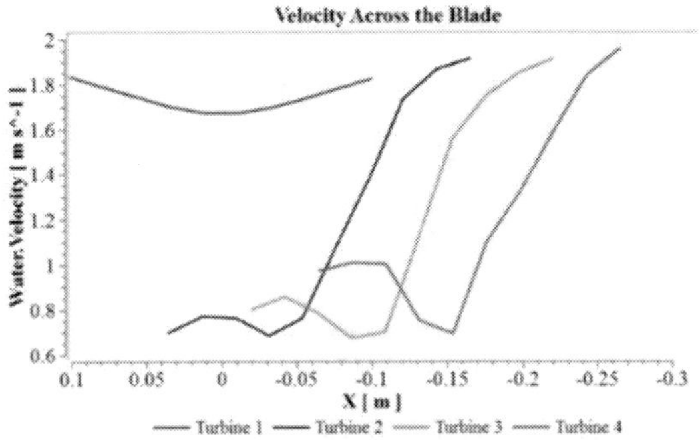

b) 20°

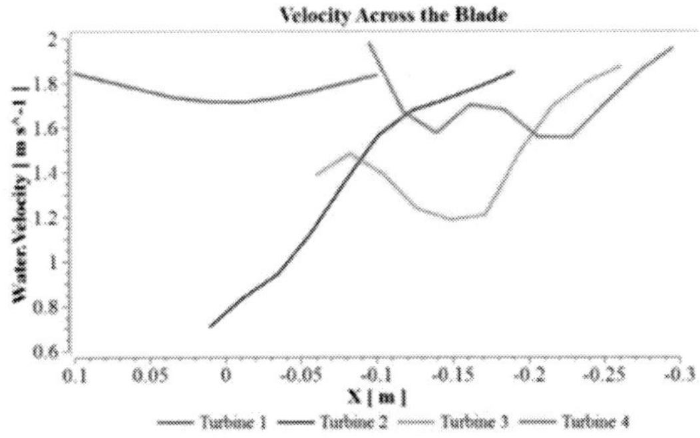

c) 30°

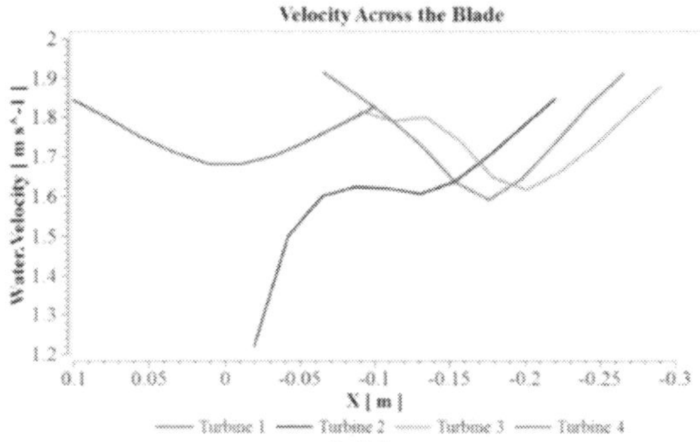

d) 40°

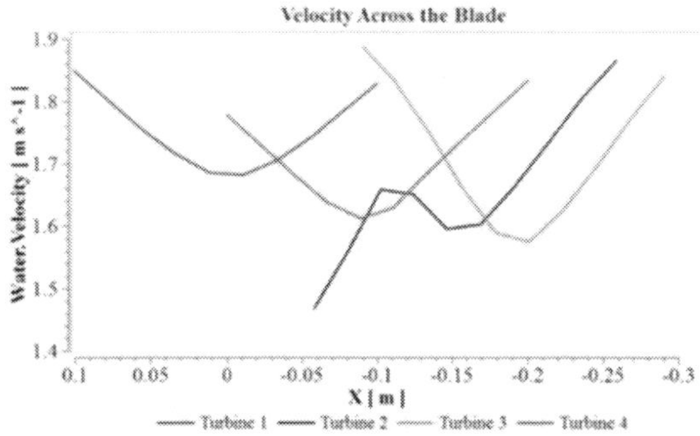

e) 50°

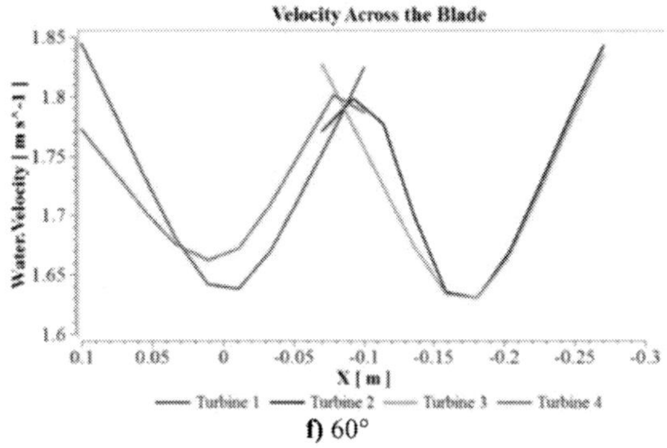

f) 60°

Figure 7. Velocity plots from simulation results of turbines with different angles

4.3 Comparison between arrangements

Based on the tests conducted, the average velocity of each spacing and angle arrangement is obtained from the velocity plot and the results are tabulated in Table 2 and Table 3 respectively.

Table 2. Average velocity respective to their spacing

Spacing (mm)	Average velocity (m/s)
190 (1D)	1.022
380 (2D)	1.156
570 (3D)	1.292
760 (4D)	1.458

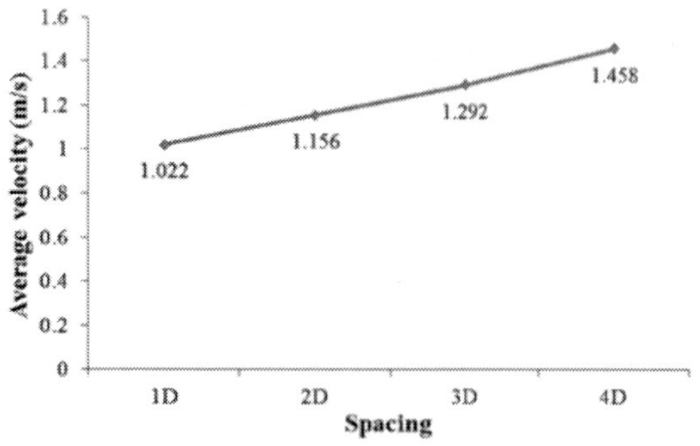

Figure 8. Graph of average velocity versus spacing

Figure 8 shows the effect of spacing onto the average velocity at the mouth of each discrete turbine. In the series arrangement of the turbines, the average velocity increases as the spacing increases. Based on the graph, the 760 mm (4D) spacing arrangement shows the highest average velocity at the mouth of each discrete turbine. Hence, the larger the spacing between the turbines, the higher the velocity obtained. Therefore, larger spacing is needed to generate maximum power.

Table 3. Average velocity respective to their angle

Angle (°)	Average velocity (m/s)
10	1.318
20	1.417
30	1.588
40	1.699
50	1.722
60	1.736

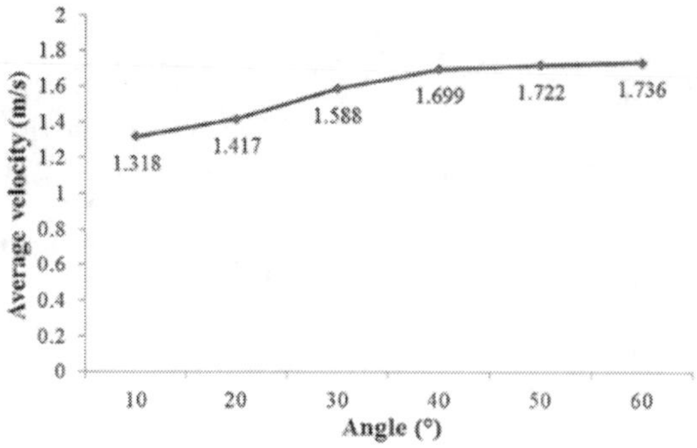

Figure 9. Graph of average velocity versus angle

Figure 9 shows the effect of angle onto the average velocity at the mouth of each discrete turbine. The average velocity increases as the angle increases. Based on the graph, the 60° angle arrangement shows the highest average velocity at the mouth of each discrete turbine. However, it can be noticed from the graph that the angle of 40° is able to achieve higher velocity compared to smaller angles and the velocity at angle greater than 40° does not increase much and remains about the same. Therefore, angle greater than 40° is needed to generate maximum power.

5 Conclusions

In this study, the effect of velocity flow over different array arrangements could be studied and analysed. Based on the simulation results obtained, it is proven that for a series arrangement of turbines, the larger the spacing between the turbines, the higher the velocity obtained. High velocity is necessary to generate maximum power. Small spacing does not show signs of good performance due to its low average velocity throughout the turbines. Based on the tests conducted on the different spacing, the results show that the spacing of 4D exhibits the highest average velocity. Therefore, it is concluded as the optimum spacing between the turbines.

However, the application of hydrokinetic turbines in rivers might be restricted to the aquatic environment available at a particular area. Therefore, when taking that into consideration, angle spacing should be considered in the turbine arrangement to minimise the area involved for the installation of the turbines. From the tests conducted on different angles, 60° angle shows the highest velocity obtained. However, it is found that the angle of 40° is sufficient enough to be considered as the optimum angle where maximum power can be generated as it is capable in achieving higher velocity compared to smaller angles. Therefore, angles greater than 40° are applicable.

All discussions and results obtained are based on software simulation only. Due to the limitations of the study, the effect of the rotation of the turbine blades to the water flow is not considered. Hence, the results of this study should not be applied thoroughly without considering other possible factors when dealing with fluid flow.

For future application of hydrokinetic turbines, spacing and angle should both be accounted in order to achieve maximum power generation and to overcome environmental limitations as well as minimising the disruption of aquatic environment.

Acknowledgements

The authors would like to thank Centre for Earth Resources Research & Management (CERRM), Universiti Malaysia Pahang for providing financial assistance under project no. RDU140345.

References

1. S. Kamaruzzaman and A.R. Juhari, Pico Hydro: Clean Power From Small Streams. *3rd WSEAS Int. Conf. on Renewable Energy Source,* 1-6 (2009)
2. R. Kapoor, Pico Power: A Boon for Rural Electrification, *Advance in Electronic and Electric Engineering, 3,* 865-872 (2013)
3. S. Kari, Small-scale Water Current Turbines for River Applications, 4 (2010)
4. M.A. At-tasneem, et al., Numerical Simulation of Multiple Arrays Arrangement of Micro Hydro Power Turbines. *World Academy of Science, Engineering and Technology International Journal of Mechanical, Aerospace, Industrial and Mechatronics Engineering*, **8**, 966-972 (2014)
5. L. Bai, R.R.G. Spence and G. Dudziak, Investigation of the Influence of Array Arrangement and Spacing on Tidal Energy Converter (TEC) Performance using a 3-Dimensional CFD Model. *8th European Wave and Tidal Energy Conference, Uppsala, Sweden,* 654-660 (2009)
6. Ximo, C.O.M, Wake, *Ject Press* (2012)
7. A.S. Bahaj, et al., Characterising the wake of horizontal axis marine current turbines. *7th European Wave and Tidal Energy Conference, Porto, Portugal,* (2007)
8. Y.T. Wu and P. Fernando, Atmospheric Turbulence Effects on Wind-Turbine Wakes: An LES Study, 1-23 (2012)
9. M.J. Churchfield, Y. Li, and P.J. Moriarty, A Large-Eddy Simulation Study of Wake Propagation and Power Production in an Array of Tidal-Current Turbines, 1-10 (2011)
10. Y.A. Cengel and J.M. Cimbala, *Fluid Mechanics: Fundamentals and Application, New York: McGraw-Hil* (2013)

H-minima transform for segmentation of structured surface

Nor Hidayawati Fauzi ismail[2, 3], Talib Ria Mohd Zaini[1, 2, a] Mohd Jaafar[3] and Nuraini Che Pin[1, 2]

[1]*Institute of Graduate Studies, Universiti Teknologi MARA, Malaysia*
[2]*Fakulti Kejuruteraan Mekanikal, Universiti Teknologi MARA (Pulau Pinang), Jalan Permatang Pauh, 13500 Permatang Pauh, Pulau Pinang, Malaysia.*
[3]*Automative Research and Testing Center, Universiti Teknologi MARA (Pulau Pinang), Malaysia.*

Abstract. The watershed segmentation method has been used in surface metrology to determine the feature parameter and the segment of areal surfaces. However, during segmentation, over-segmentation always happens, which may overshadow the significant features. Thus, a new method is required to reduce this over-segmentation, as well as to retain the necessary information for further improvement in the signal processing data. Hence, this paper proposed a new method to overcome over-segmentation for segmentation of structured surface. Over-segmentation happens when the surface area is over-segmented into a large number of insignificant, tiny, and shallow hills and dales features rather than a few significant dales/hills features. H-minima transform was proposed to overcome this issue. From the results, the data pertaining to simulated and measured surface topographies, as well as height threshold value of H-minima transform had depended on the irregularities of the surface to obtain the required features. As a result, it had been concluded that H-minima transform is possible to reduce over-segmentation.

1 Introduction

Structured surface is purposely designed to meet a specific functional requirement [1]. It contains a topography of generally high aspect-ratio and deterministic feature pattern [2]. This surface is critically important because its influence on the functionality of that surface, such as in tribological application for improving lubrication, reduction of friction and wear [3], as well as in manufactured items like micro and nanometer scale transistor, micro-electro mechanical systems (MEMS), nano-electro mechanical systems (NEMS), and optics.

Thus, in order to improve the functional performance of this structured surface, it is a must to characterize and analyze the surface. Recently, several researchers have proposed characterizing structured surfaces based on geometrical properties of individual features on the surface [4]. Hence, for surface characterization, it is necessary to identify the relevant features by segmenting the surface into regions of interest. Surface features consist of areal features (hills and dales), line features (course and ridge lines) and point features (peak, pit and saddle points). The stable extraction of significant surface features can be carried via Watershed Segmentation method [2]. According to Maxwell, a landscape surface is divided into regions of hills and dales. Hill is an area from which maximum uphill paths lead to one particular peak, while dale is an area from which maximum downhill paths lead to one particular pit. The boundaries between hills are course lines (watercourses), while the boundaries between dales are ridge lines (watershed lines). Besides, the ridge and the course lines are maximum uphill and downhill paths that originate from saddle points and end at peaks and pits.

Meanwhile, segmentation is used to determine regions of the scale-limited surface that defines the scale-limited features [5]. In this study, the features involved included hills and dales. In order to segment this surface, both features cannot exist at the same time during the segmentation is carried out. However, the existence of a large number of insignificant or irrelevant features during segmentation often occurred over-segments the surface, as shown in figure 1.

Over-segmentation may affect the extraction of the significant relevant features from the structured surface. It could distort the required structured surface data during the surface data analysis and characterization. Hence, a method is required to reduce this over-segmentation. The most common method used to reduce over-segmentation in surface metrology is Wolf Pruning [6]. This method determines the significance of regions based on height and depth. The insignificant regions are merged together until all remaining regions become significant. Even

[a] Corresponding author: dayahzaini@gmail.com

though the Pruning method is usually used to overcome over-segmentation, this method is not fully explored [7]. Then, in this context, the objective of this study was to propose a new mechanism to reduce over-segmentation on structured surface in the segmentation method. In this paper, an alternative method is introduced to reduce over-segmentation by applying H-minima transform [8][9][10][11].

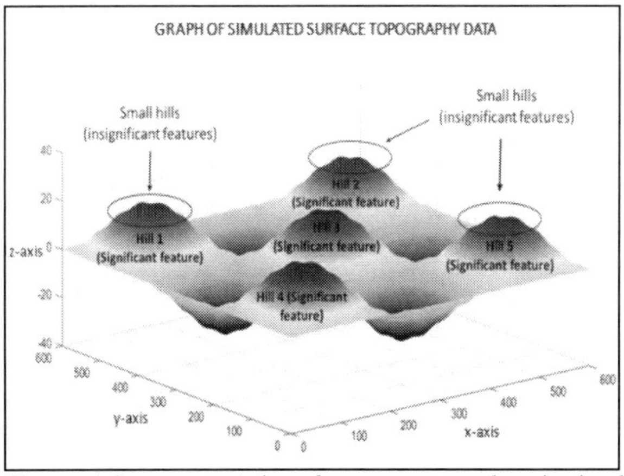

Figure 1. Over-segmentation that occurs on the simulated structured surface.

1.1. The Proposed Segmentation Procedure for Structured Surface

A few steps are involved in determining the significant features. Figure 2 shows proposed segmentation procedure for structured surface.

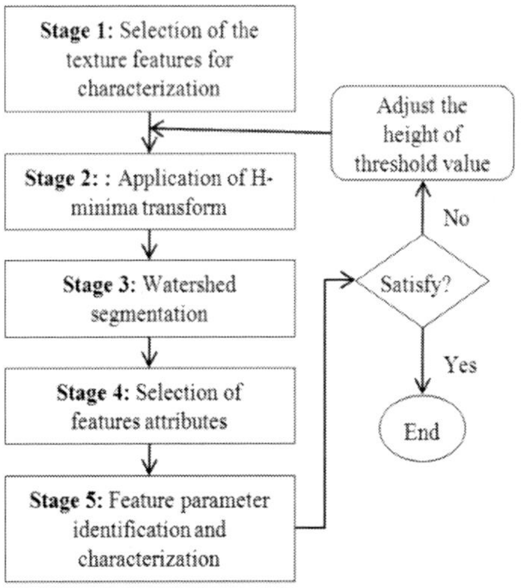

Figure 2. The proposed segmentation procedure.

From figure 2, stage 1 involves the selection of the types of structured data. Then, application of H-minima transform is introduced in stage 2. This method should be applied before the watershed segmentation to eliminate insignificant features on the surface. The existence of small and tiny dales within the large dales may overshadow relevant surface features wanted. Once the significant features are determined, watershed segmentation will segment the surface into region of interest. Meanwhile, stages 4 and 5 are the future works for feature parameter characterization to determine the functional performance of the structured surface. The process is repeated until the featured parameters of the required structured surface are satisfied.

1.2. H-Minima Transform Method

This paper proposed the H-minima transform to reduce over-segmentation. This method is applied before the watershed segmentation is carried out. H-minima transform is the method proposed to replace the Wolf pruning[6]. H-minima transform suppresses all regional minima in I, whose depth is less than or equivalent to H. I is the areal surface topography dataset, while H is the height threshold value (non-negative scalar) [12]. Besides, regional minima are connected components of pixels with the same dataset value, t, whose external boundary pixels all have a value greater than t [13].

To eliminate all regional minima, except the significant minima, H-minima transforms can be applied to specify a height threshold value by using the following equation:

$$H = N*Z \tag{1}$$

Where: H = height threshold value
Z = maximum height from mean plane
N = percentage from maximum height (%)

From the equation, it is observed that height threshold value is dependent on the percentage from maximum height. Value of N is depending on the irregularities of the measured surface topography data respectively (~5%-20%). For this method, H-minima transform only affects the regional minima; as none of the other pixel values is changed. The significant minima remain, although their heights are increased. Moreover, the size of the significant minima tends to increase and the number of regional minima decreases. Figure 3 shows the overview of H-minima transform function applied on the structured surface.

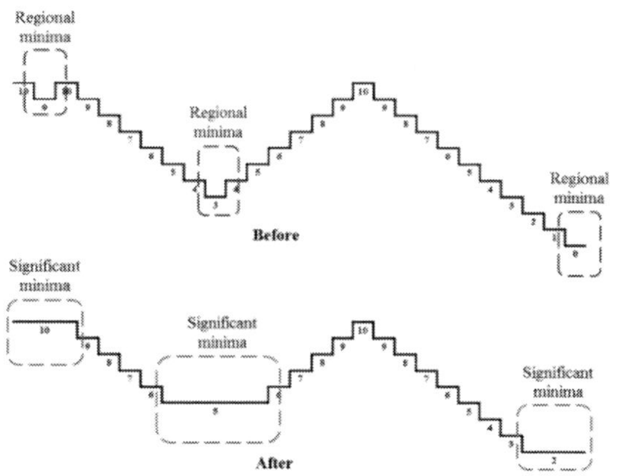

Figure 3. Overview of H-minima transform algorithm for 1-dimensional diagram.

Before application of H-minima transform, numbers 9, 3, and 0 (inside the dotted red box) were the regional minima. For number 9, its depth was 1 (from 9 to 10), while for number 3, its depth was 7 (from 3 to 10), and for 0, its depth was 10 (from 1 to 10). For example, by taking 20% of the maximum height, the height threshold value becomes 2 (also called as depth). After the application of the H-minima, the result showed that H-minima transform removed all the regional minima, which were insignificant with depth smaller or equal to 2, and increased the height of the regional minima by 2. As per figure display, before and after application of H-minima transform, the regional minima (9, 3, and 0) are merged with the adjacent minima to become significant minima. Hence, all regional minima that had been less significant were transformed by 'flattening out' any insignificant regional minima into the required significant minima. Thus, H-minima transform is defined as the reconstruction by erosion of f, and increased by a height, H as in the equation 2 [14][8]:

$$Hmin_H(f) = R_f(f+H) \qquad (2)$$

Where: $Hmin_H(f)$ = H-minima transform
$R_f(f+H)$ = reconstruction by erosion of f increases by height threshold H

2 COMPUTATIONAL SETUP

2.1 Computational procedure

Structured surface from simulated and measured data were taken for the computational setup in order to demonstrate the function of H-minima transform on the surface. The types of structured surface involved were areal surface features, which consisted of hill and dales. The simulated structured data were generated by combining the sine and cost values, whereas the measured structured data were obtained from the electroplated diamond tool specimen measured with Confocal Laser Scanning Microscope (CLSM). Furthermore, the topography data had been stated in pixel size with (data interval) dx and dy. The grid number of the topography data was #120, while the average grain size of the topography data was about 150 μm. The diamond grain distribution and the orientation were in random arrangement. Both surfaces were analyzed and evaluated by Matlab software[12].

Apart from that, both structured data focused on hills segmentation alone. The percentage of maximum height was taken as the control parameter. Both surfaces were divided into two simulations. First, both structured surfaces must undergo watershed segmentation only, while the other employed the simulation watershed segmentation plus H-minima transform methods. The reason was to demonstrate the effect of H-minima transform on the structured surface. H-minima transform specified the height threshold as per stated before. Moreover, this research started with the simulated surface topography data. As for simulated surface topography data, the number of percentage taken began with 1% and it was added up to 0.5% till 5.5%. The simulation was run for ten times in order to obtain the most stable number of segments for the structured surface. Nonetheless, the simulation was stopped when the significant features matched within the region segment.

Then, for measured surface topography data, the procedure applied was the same. Before the simulation begins, the grains on the real data are manually counted (figure 4). The reason is to validate the results before and after simulation. However, due to the large number of the dataset, the number of percentage that had been taken began with 5% and was added up to 5% till 55%. The simulation was run for 15 times in order to obtain the most stable number of segments for the structured surface. The results of the application H-minima transform for both simulation is discussed in Results and Discussion section.

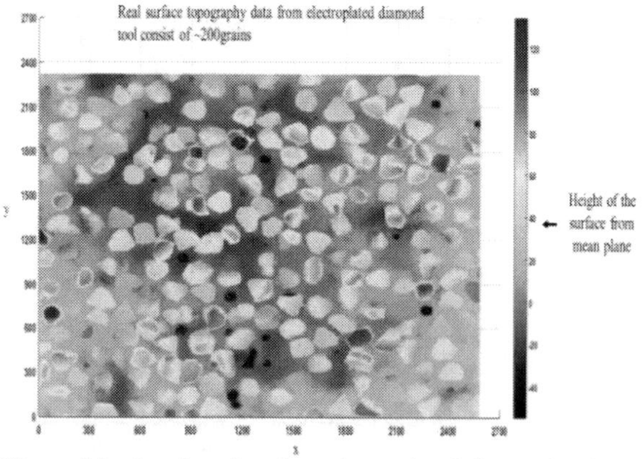

Figure 4. Real surface data from electroplated diamond tool consist of ~200 grains.

2.2. Data Structure

The dataset that contained areal surface topography information of height values (z values) arranged along the rows and the columns of a regular x, y grid with constant Δx and Δy was plotted on a three dimensional Cartesian coordinate system with an origin at 0, as portrayed in figure 5. The colors indicate the height of the surface features in the surface topographical map.

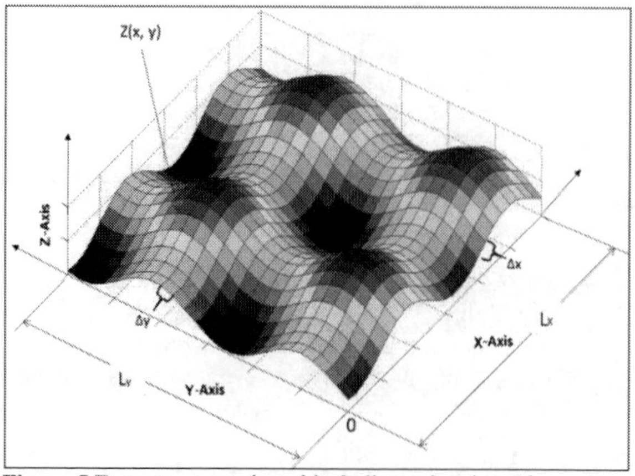

Figure 5. Data structure plotted in 3-dimensional graph.

Z (x, y) = surface height matrix consists of x and y pixel values
Δx = constant x data interval
Δy = constant y data interval
Lx = length of x values
Ly = length of y values
0 = origin value on Cartesian coordinate system

3 RESULTS AND DISCUSSIONS

In this section, results of the H-minima transform applied on the simulated and the measured surface topography data are demonstrated. Graph number of segments versus percentage of N from maximum height for simulated and measured data had been plotted.

Figure 6 to figure 9 shows the structured surfaces without and with application of the H-minima transform for both simulated and structured surfaces. Figure 6 represents the results obtained without application of H-minima transform, whereby the simulated structured surface had been over-segmented, as well as the mixing of insignificant and significant features. The line segments on the surface also are not well segmented on the simulated surface. Total hills which are over segmented are 37 hills. Figure 7 shows the results retrieved with application of H-minima transform, where significant features were determined on the simulated structured surface at 2.5% from maximum height. Number of hills segments after application of H-minima transform is 5 hills. Each hill represents grains as shown in the previous figure (figure 1).

Figure 8 depicts results for without application of H-minima transform, where measured structured surface were over-segmented, significant and insignificant features were mixed-up together, and significant features could not be identified. Total hills over segmented are 30651 hills. Figure 9 shows with application of H-minima transform, significant features were determined on the measured structured surface at 10% from maximum height of the measured structured surface. There are 191 significant features are detected within the segmented regions. Each region contains one significant feature which has its own attributes towards those surfaces. In this research, it was found that at 10% from maximum height, the significant features detected are nearly similar with the manual calculation which is ~200 grains. For this types of surface, 10% may suit and possible to determine the relevant features.

Otherwise, from each figures (figure 6-figure 9), the color bar represents the height of the measured surface (z value) from a mean plane. The unit for z value is in μm. While for x and y, both unit are in pixels. Thus, both values need to be converted to the μm in order for feature parameter characterization. The line between the colors of the region is the line segments, which separate one feature to the adjacent feature. Each feature was evaluated and quantified in order to determine the feature parameter for future works.

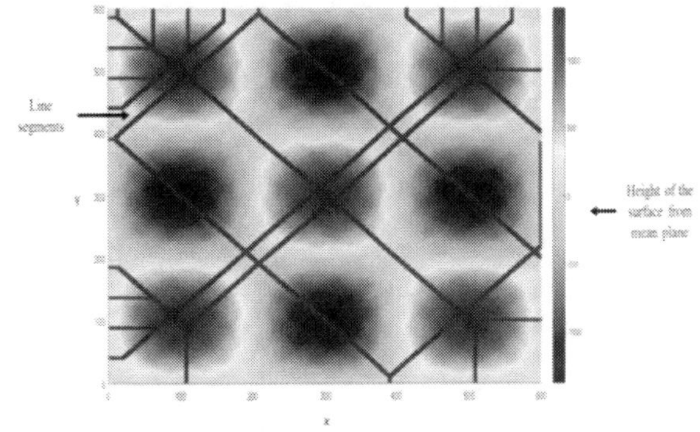

Figure 6. Over-segmentation occurred on the simulated surface topography data. Total segment is 37 hills (without H-minima transform algorithm).

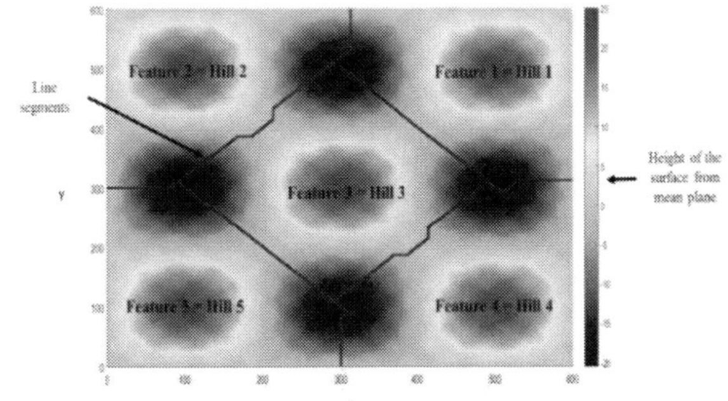

Figure 7. Significant features were determined on simulated structured surface at 2.5% H-minima transform, number of hills =5.

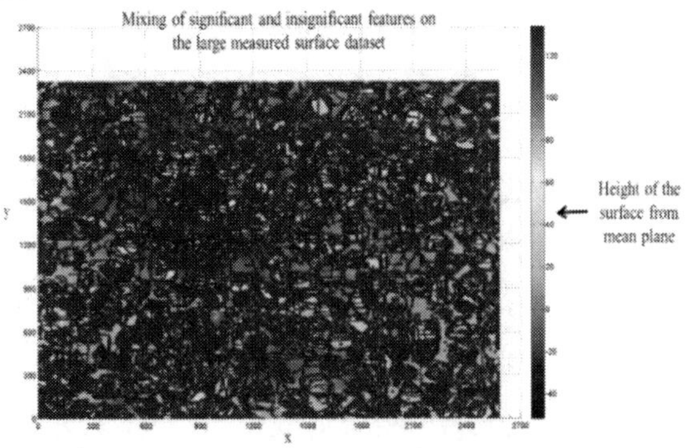

Figure 8. Over-segmentation occurred on the measured surface topography data. Total segment is 30651 hills (without H-minima transform algorithm).

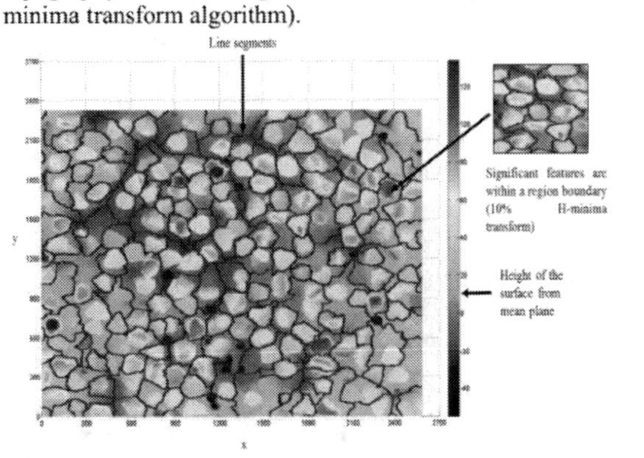

Figure 9. Significant features were determined on measured structured surface at 10% H-minima transform, number of hills =191.

The data collected from the experiment on simulated structured surface had been plotted in the graph number of segments versus percentage of N from maximum height as shown in figure 10. The number of segments had been varied when the percentages were changed. The results showed that at 2.5% to 5.5% the number of segments had become stable. The significant features are obtained at 2.5% H-minima transform. Figure 10 shows the data collected from the experiment on measured structured surface, which were plotted in the graph number of segments versus percentage of N from maximum height.

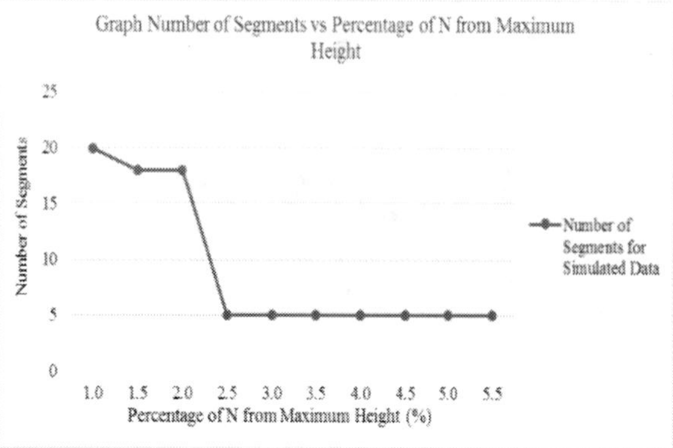

Figure 10. Graph number of segments versus percentage of N from maximum height for simulated data.

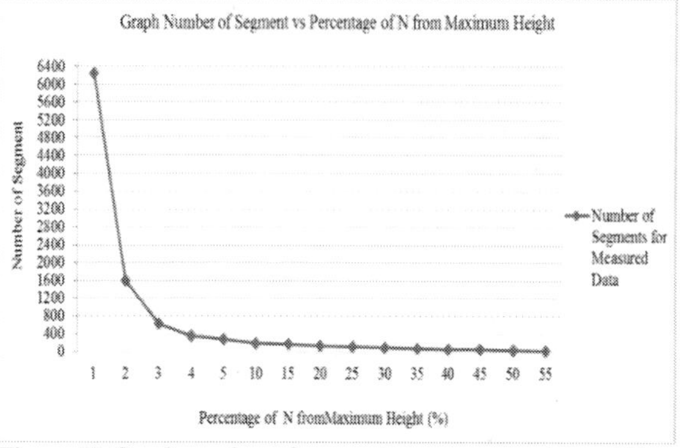

Figure 11. Graph number of segments versus percentage of N from maximum height for measured data.

Graph in figure 11 shows the number of segments is decrease as the percentage of H-minima transform is increases respectively. The number of segment is decreased sharply at 1% to 5% and become stable at 10% towards the end of the experiment. 10% is the most suitable percentage to determine significant features because the number of hills segmented is nearly similar with the manual calculation which is ~200 grains.

3.1. Limitations

The H-minima transform applied parallel with Wolf Pruning method depended on height threshold value. The height threshold value of the H-minima transform was determined by the percentage of the maximum height. Nevertheless, the percentage might not always lead to the desired segmentation result, as per discussed before. Thus, several combinations of areal data pre-processing and H-minima transform solutions are needed in order to obtain the best percentage for determination of significant features. Additionally, trial and error method is still needed for the user side to apply this application until they satisfied. Otherwise, prior knowledge about the surface also important for the surface texture characterization. However, the solution proposed in this study could help in minimizing time consumption for the segmentation method to solve the over-segmentation issue.

4 CONCLUSION

In this paper, a new method for segmentation had been proposed. The proposed method successfully segmented the surface feature into the features of interest by applying H-minima transform. The height threshold value of H-minima transform is depended on percentage from the maximum height. The results showed that H-minima transform could be used to overcome over-segmentation during watershed segmentation process on the structured surface.

ACKNOWLEDGEMENT

The authors gratefully acknowledge Universiti Teknologi MARA and the Malaysian Ministry of Higher Education for the financial support under the Fundamental Research Grant Scheme (Grant no.: FRGS/1/2014/TK01/UITM/02/6).

REFERENCES

1. E. M. Division and U. Kingdom, "Investigation of micro manufacturing process performance via innovative surface chracterisation methods," pp. 3–7, (2014).
2. X. Jiang, P. J. Scott, D. J. Whitehouse, and L. Blunt, "Paradigm shifts in surface metrology. Part II. The current shift," *Proc. R. Soc. A Math. Phys. Eng. Sci.*, vol. **463**, no. 2085, pp. 2071–2099, Sep. (2007).
3. A. A. G. Bruzzone and H. L. Costa, "Functional Characterization of Structured Surfaces for Tribological Applications," *Procedia CIRP*, vol. **12**, pp. 456–461, Jan. (2013).
4. G. Macaulay, N. Senin, C. L. Giusca, and R. K. Leach, "Marker based watersheds for segmentation of structured surfaces."
5. P. J. Scott, "Feature parameters," *Wear*, vol. **266**, no. 5–6, pp. 548–551, Mar. (2009).
6. G. W. Wolf, "A FORTRAN subroutine for cartographic generalization," *Computers & Geosciences*, vol. **17**, no. 10. pp. 1359–1381, (1991).
7. N. Senin, L. a Blunt, R. K. Leach, and S. Pini, "Morphologic segmentation algorithms for extracting individual surface features from areal

8. surface topography maps," *Surf. Topogr. Metrol. Prop.*, vol. **1**, no. 1, p. 015005, Oct. (2013).
9. P. Soille, *Morphological Image Analysis: Principles and Applications*. (1999).
10. C. Jung and C. Kim, "Segmenting clustered nuclei using H-minima transform-based marker extraction and contour parameterization," *IEEE Trans. Biomed. Eng.*, vol. **57**, no. 10 PART 2, pp. 2600–2604, (2010).
11. H. Masmoudi, S. M. Hewitt, N. Petrick, K. J. Myers, and M. a Gavrielides, "Automated quantitative assessment of HER-2/neu immunohistochemical expression in breast cancer.," *IEEE Trans. Med. Imaging*, vol. 28, no. 6, pp. 916–25, 2009.
12. F. Raimondo, M. A. Gavrielides, G. Karayannopoulou, K. Lyroudia, I. Pitas, S. M. Ieee, and I. Kostopoulos, "Automated evaluation of Her-2 / neu status in breast tissue from fluorescent in situ hybridization images," vol. **14**, no. 9, pp. 1288–1299, (2004).
13. [12] C. P. Documentation and N. Agreement, "Installation Guide R 2014 b," (2014).
14. A. Bleau and L. J. Leon, "Watershed-Based Segmentation and Region Merging," *Comput. Vis. Image Underst.*, vol. **77**, pp. 317–370, 2000.
15. Q. Wu, F. A. Merchant, K. R. Castleman, and C. Kenneth R., *Microscope Image Processing*. (2008).

[a] Corresponding author: dayahzaini@gmail.com

Overview of Steel Slag Application and Utilization

J.W. Lim[1,2], L.H. Chew[2], Thomas S.Y. Choong[2], C. Tezara[1] and M.H. Yazdi[1,3]

[1]Department of Mechanical Engineering, INTI International University, Jalan BBN 12/1, Bandar Baru Nilai, 71800 Nilai, Negeri Sembilan,Malaysia

[2]Department of Chemical and Environmental Engineering, Universiti Putra Malaysia, 43400 Selangor, Malaysia

[3]Department of Mechanical Engineering, Islamic Azad University, Mashhad Branch, Mashhad, Iran

Abstract. Significant quantities of steel slag are generated as waste material or byproduct every day from steel industries. Slag is produced from different types of furnaces with different operating conditions. Slag contains Ferrous Oxide, Calcium Oxide, Silica etc. Physical and chemical properties of slag are affected by different methods of slag solidification such as air cooled, steam, and injection of additives. Several material characterization methods, such as X-ray Diffraction (XRD), Scanned Electron Microscopy (SEM) and Inductive Coupled Plasma (ICP-OES) are used to determine elemental composition in the steel slag. Therefore, slags can become one of the promising materials in various applications such as in transportation industry, construction, cement production, waste water and water treatment. The various applications of steel slag indicate that it can be reused and utilized rather than being disposed to the landfill. This paper presents a review of its applications and utilization

1. Introduction

Steel are mainly produced from integrated steel mills which consist of Basic Oxygen Furnace (BOF) or Blast Furnace (BF). Meanwhile steel from scrap based materials are produced in Electrical Arc Furnace (EAF) operated mills. Steel slags have variety of compositions and in various forms after cooling. Most steel slags form hard rock-like materials upon cooling or solidification. However some slags can be reused to make dusty powders. There are different roles of steel slag in steelmaking process; as a sink for collecting impurities during steel refining process, to prevent diffusion N2and H2 gases from atmosphere to the molten steel, protecting steel against re-oxidation, as the barricade heat transfer from molten steel to the surrounding and as emulsifier for hot metal to promote carbon oxidation. Slags are porous, large surface area with suitable hydraulic conductivity [1] (Figure 1).The different application and utilization of steel slag are briefly described in the following section. The main by-products produced during iron and crude steel production are slags (90%), dusts and sledges that reported for three types of furnaces.

Figure 1: Slag as a by-product from steel making furnace [2]

2. Applications of Steel slag

According to the Malaysia Environmental Quality (Scheduled Waste) Regulation (2005), Steel slag is classified as non-hazardous waste which can be disposed off to appropriate landfills. The disposal of steel slag requires large landfill area which is rather unfavourable in economical terms. Therefore, several research and development have been conducted to establish the potential use of slag in different applications. However, it might have the risk that different components of slag might elute, especially on the heavy metal. Therefore, it is necessary to examine this metallurgical waste material in appropriate ways. Typical applications of steel slags are sealing aggregate (skid resistant), asphalt aggregate, base, sub-base, construction fills, subsoil drains, grit blasting and waste water treatment. The various applications for steel slag in 2010 (22.3 million tonnes) is illustrated in Figure 2

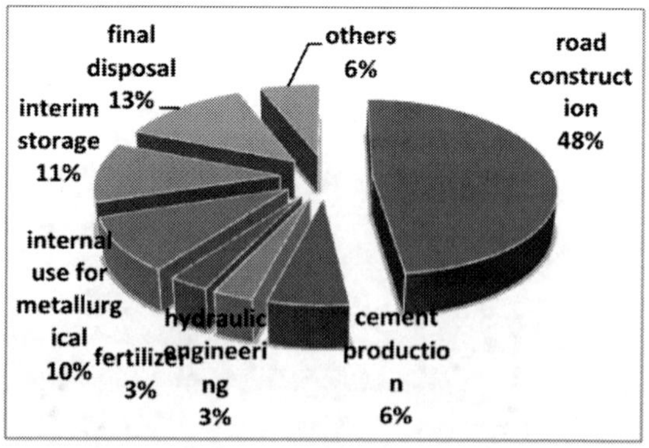

Figure 2: Statistical data of various applications of steel slag (2010) [3]

2.1 Cement Productions

In cement and concrete industries, slag can be used either as an aggregate or binder in stabilized base courses. In order to conserve natural resources and reduce environmental impact, slags can be used as an aggregate. To diminish the need for cement which is expensive in cost, slag is replaced and used as a binder. Because of these advantages, many researchers have examined adding slag content to cement and concrete [4]. Addition of slag to cement presents some important technical advantages over ordinary Portland cements. These benefits can be enumerated as development of mechanical strengths, low solubility of the hydrates and porosity, lower heat of hydration, excellent durability and stronger aggregate–matrix interface. However, there are some disadvantages such as high shrinkage, formation of micro-cracks and rapid setting. Therefore, the combination of various type of slags (activated, chemical and physical modified) to cement are investigated. For instance, it can boost the hydration ability of slag cement concretes, improving the mechanical performance of cement, enhancing the strengths of cement, autogenous deformation on the cracking, microstructure and durability, resistance.

2.2 Road Construction

Industrial by-products such as steel slag can be utilized instead of natural aggregates in order to prevent depletion of resources. Million tons of sand, till and crushed rocks were extracted from mines due to roads construction. It is well known that technical, economical and many environmental benefits are obtained when steel slag has been considered as alternative construction materials [5]. Due to prominent properties of steel slag, it is an ideal aggregate for asphalt surface course materials and road surface treatments. It has been proven that the use of steel slag in road construction mixtures has more advantage compared to conventional asphalt. The investigations of the asphalt mix and road surface have shown very good results in terms of stability, stiffness and durability. The slag asphalt, as expected, provided good friction values, noise reduction levels, enhances resistance to cracking at low temperatures with excellent performance in roughness and good resistance against water permeability[6].

2.3 Water and Wastewater Treatment

As slag has good sportive characteristics and is low cost, it is widely used in wastewater and water treatment, and may be the alternative of using granular activated carbon. Various research such as adsorptions of dye, heavy metals and organics were carried out by using steel slag. However, the uptake capacity of slag is dependent on the pH solution. The hydration of slag composition in the aqueous solutions provides a high pH. The possible hydration reactions can be occurred with different compositions of slag as followings:

$$CaO + H_2O \rightarrow Ca(OH)_2 \quad (1)$$

$$MgO + H_2O \rightarrow Mg(OH)_2 \quad (2)$$

$$2Ca_2SiO_4 + 4H_2O \rightarrow 3CaO.2SiO_2.3H_2O + Ca(OH)_2 \quad (3)$$

$$2Ca_3SiO_5 + 6H_2O \rightarrow 3CaO.2SiO_2.3H_2O + 3Ca(OH)_2 \quad (4)$$

$$CaO.7Al_2O_3 + 12H_2O \rightarrow CaO.7Al_2O_3.6H_2O + 6Al_2O_3.H_2O \quad (5)$$

With the high pH condition, the slag surface is negatively charged and adsorption metal ions especially cations are preferred. However, heavy metal absorption by using slags can be conducted with a condition of either high temperature or low pH under certain conditions. Yamashita et al. investigated the possible mechanisms of removing dissolved heavy metals from aqueous waste liquids via the converter furnace slag [7]. They suggested that the removal can be attributed to one or more of following effects: adsorption, co-precipitation, hydroxide precipitation as hydroxide, sulphide and ion exchange.

Wastewater from the steel mill contains a high concentration of heavy metals due to the accumulation of filtered particles from the filtration process of wastewater. A direct release of the backwash water into the environment may cause serious effects to both land and aquatic lives. Based on the toxilogical studies, metals are toxic and non-biodegradable and may continue to exist in these water bodies. Furthermore, the heavy metals also have the tendency to accumulate in the food chain. Hence, a strict environmental regulation has been established in an effort to mitigate the heavy metals contamination of the discharge of industrial effluent. Absorption of lead ions by two different types steel slag was carried out in column-type contact process [8]

2.4 Phosporus Removal

Phosphorus (P) from agricultural wastewater is one of the major pollutants in natural water that cause the algae growth and eutrophication of lakes. Therefore, Phosphorus is one of the major nutrients that is need to be removed from domestic wastewater before being discharged into water bodies. The application of an appropriate technology to remove the maximum capacity of nutrient is imperative. In order to remove phosphorus from wastewater, various methods or approaches have been attempted, including biological, chemical process (precipitation, metal salt addition) and physical (electro-dialysis, reverse osmosis). Among all the methods, the chemical precipitation is expensive and increasing sludge volume by up to 40%. The biological phosphorus removal needs a lot more volume or space (anaerobic unit) than the other processes. Various research on the potential use of industrial by-products to remove P from wastewater was studied in the late 1980s.

The P removal is along with the development of constructed wetlands (CW). This method is a low cost technology for pollutions treatment, which slag performs an important role in absorbing impurities especially phosphorus. By comparing most of industrial by products, steel slag is a low-cost and abundant material, which its combination with small secondary treatment systems (such as constructed wetlands) is preferred in compared to the others methods [9,10]. The slag contains various metal oxides such as iron oxides and alumina that may be effective in P reduction from domestic, agricultural effluents and municipal [11]. Steel Slag is a light weight porous medium with numerous sites for sorption. Steel slags have different physic-chemical property due to various feedstock ores, fluxes and manufacturing process. These differences caused a board range of P sorption capacity by using slag, ranging from 76.4 to 8390 mg/kg. Many experiments have been carried out in laboratory to evaluate the sorption capacity of slag. These studies have found slag to be a promising substrate for P-removal [12].

3. Steel Slag Characterization

3.1 Scanned Electron Microscopy (SEM)

SEM is a device that utilizes focused beam of high energy electrons to analyze solid material by obtaining topographical, morphological and compositional information of the tested sample. A scanning electron microscope (model S-3400N, HITACHI, Japan) was used to obtain external morphology, crystalline structure and orientation of materials making up the sample at an accelerating voltage of 20.0 kV, beam current of 50 mA and working distance of 9.0 mm – 10.0 mm using backscattered electrons (BSE) objective with aperture of 3. Two images were captured at a low magnification (300×) and high magnification (1000×) respectively.

The micrographic appearance of the unreacted EAFS shows distinct mineral phases, presented in Figure 3. The light areas contain a higher proportion of Fe, Mg, Mn and Al while the dark areas contain a higher proportion of CA and Si (Bankole et al., 2014). From Figure 3, it can be seen that the light areas are larger than the dark areas. This indicates a higher composition of Fe, Mg, Mn and Al compared to the composition of Ca and Si

Figure 3: SEM micrograph of unreacted EAFS at magnification of (a) 300x and (b) 1000x

3.2 X-Ray Diffraction (XRD)

X-Ray diffractometer (model XRD-600, Shimadzu, Japan) was used for phase identification of unreacted EAFS with Cu Kα radiation at 40 kV and 40 mA and scanned between $2\theta = 20° - 80°$ and scanning speed of 0.1°/min. The X-Ray diffraction pattern observation was generated by using powder diffraction method. The XRD pattern was then analyzed using X'Pert High Score Plus software to obtain the mineralogical and chemical composition of the sample slag. Each crystalline solid has its unique characteristic XRD pattern which may be used for its identification. X'Pert High Score Plus software has the database of these XRD patterns so that they could be used to match with the peaks appear in the XRD pattern of unreacted EAFS. Once all the peaks were identified for the chemicals present in the sample slag, the intensity under the peaks was automatically used by the software to determine their composition.

The XRD analysis of unreacted EAFS is shown in Figure 4. The diffraction pattern is heterogeneous, consisting of a mixture of crystalline phases. The analysis identifies calcium oxide (CaO), silicon oxide (SiO2), hematite (Fe2O3), manganese (III) oxide (Mn2O3), magnesium oxide (MgO), aluminium oxide (Al2O3), phosphorus oxide (P2O5), metallic iron (Fe), srebrodolskite (Ca2Fe2O5), dicalcium silicate (Ca2SiO4) and pyroxene (CaAl2SiO6).

The mineralogical composition of steel slags was studied by several researchers. The XRD analysis of steel slags shows a complex structure. Many overlapping peaks appear in XRD patterns indicates the presence of crystalline phases in the steel slag [14]. The more crystalline structured compound has its XRD peaks more easily to be identified [15]. The common mineral phases presents in steel slag include two- and three-component compositions of the type CaO-SiO2, CaO-Fe2O3, CaO-MgO-SiO2, MgO-FeO-SiO2 and CaO-Al2O3-SiO2 [14,16], and the most highly represented minerals in slag are dicalcium and tricalcium silicates, while different aluminates and silicates are likely to appear as well [17].

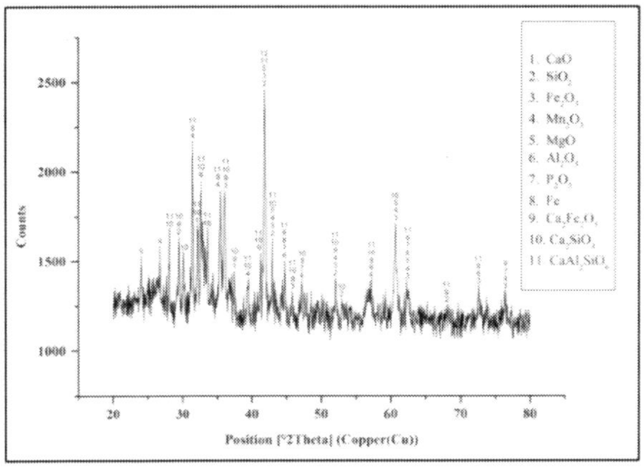

Figure 4: XRD pattern of unreacted EAFS

3.3 Inductively Coupled Plasma Optical Emission Spectrometry

Inductively Coupled Plasma Optical Emission Spectrometry (ICP-OES) is an elemental analytical technique used to determine the composition of a wide variety of materials in liquid form, with excellent sensitivity. In this study, leachates from the experiments manipulating the initial concentration of HCl and EAFS size were diluted and sent for elemental analysis using Inductively Coupled Plasma Optical Emission Spectrometer (model Optima 7300DV, Perkin-Elmer, USA) to determine the concentration of the elements Fe and Ca. Calibration curves for Fe and Ca of known concentration 0 ppm – 9 ppm and 0 ppm – 90 ppm, respectively was prepared for the determination of concentration of Fe and Ca in the samples.

4. Conclusion

In this review, widespread usage of slags as a by-product of steel making industries in various applications have been discussed. On the other hand, XRD and SEM are used to determine the mineral compostion of steel slag. Based on the literature studies, it is pretty obvious that slags can be the promising replacements for conventional way of application of adsorbent and binder in various fields such as adsorption, construction, road and waste water treatment. The ease of use, low cost, and sources availability are among the few advantages on reusing and utizing the steel slag

Acknowledgement

The authors would like to gratefully acknowledge Ministry of Higher Education (MOHE), and Universiti Putra Malaysia (UPM) for the financial support of this work.

References

[1] T.C. Alex , A.M. Kalinkin, S.K. Nath, B.I. Gurevich , E.V. Kalinkina , V.V. Tyukavkina , S. Kumar, *Utilization of zinc slag through geopolymerization: Influence of milling atmosphere*, International Journal of Mineral Processing **123** (2013)102–107.

[2] http://geology.about.com/bio/Andrew-Alden 453.htm. Accessed on 14 January 2015

[3] EUROSLAG. Retrieved from http://www.euroslag.com/products/statistics/2010/. 4 March 2015

[4] M. Frias, M. I. Sanchez de Rojas, C. Rodriguez, *The influence of SiMn slag on chemical resistance of blended cement pastes*, Construction and Building Materials , **23**(2009)1472–1475.

[5] P. Suer, J.E. Lindqvist, M. Arm, P. Frogner-Kockum, *Reproducing ten years of road ageing accelerated carbonation and leaching of EAF steel slag*, Science of The total Environment, **407**(2009)5110–5118.

[6] M. Ameri, S. Hesami, H. Goli, *Laboratory evaluation of warm mix asphalt mixtures containing electric arc furnace (EAF) steel slag*, Construction and Building Materials, **49**(2013) 611-617.

[7] K. Yamashita, T. Ikenata, K. Tate, K. Nakahara, *Method of Removing Dissolved Heavy Metals from Aqueous Waste Liquids*, US Patent, 4377483, (1983).

[8] M.R. Nilforoushan, S. Otroj, *Absorption of Lead Ions by Various Types of Steel Slag*, Iranian Journal of Chemistry & Chemical Engineering **27**(2008)69-75.

[9] A. Drizo, C. Forget, R.P. Chapuis, Y. Comeau, *Phosphorus removal by electric arcfurnace steel slag and serpentinite*. Water Research, **40**(2006)1547-1554.

[10] F. Chazarenc, J. Brisson, Y. Comeau, *Slag columns for upgrading phosphorus removal from constructed wetland effluents*. Water Science Technology **56**(2007)109-115.

[11] C. Vohla, M. Koiv, H.J. Bavor, F. Chazarenc, U. Mander, *Filter materials for phosphorus removal from wastewater in treatment wetlands-A review*. Ecological Engineering, **37**(2011)70-89.

[12] C. Pratt, A. Shilton, *Active slag filters-simple and sustainable phosphorus removal from wastewater using steel industry byproduct*. Water Science and Technology, **62** (2010)1713-1718.

[13] Bankole, L., Rezan, S. and Sharif, N. *Assessment of hexavalent chromium release in Malaysian electric arc furnace steel slag for fertilizer usage*. *IOP Conf. Ser.: Earth Environ. Sci.*, ***19***, 012004. doi:10.1088/1755-1315/19/1/012004 (2014)

[14] Yildirim, I. andPrezzi, M. *Chemical, mineralogical, and morphological properties of steel slag*. Advances In Civil Engineering, 2011, 1-13. doi:10.1155/2011/463638 (2011)

[15] Nwaubani, S. andMuntasser, T. *Hydration characteristics of cement pastes incorporating electric*

arc-furnace slag. Journal Of Civil Engineering And Construction Technology, 3(11), 291-301. doi:10.5897/JCECT12.043 (2012)

[16] Rađenović, A., Malina, J. andSofilić, T. *Characterization of ladle furnace slag from carbon steel production as a potential adsorbent.* Advances In Materials Science And Engineering, 2013, 1-6. doi:10.1155/2013/198240 (2013)

[17] Selih, J., Ducman, V., Mladenovič, a., Sever, S.A., Pavšič, P..Makarovič, M. and Legat, A. *The use of waste materials in building and civil engineering in Slovenian.* Mater.Techno. Ljubljana **38**: 79-86. (2004)

MATEC Web of Conferences 74, 00027 (2016) DOI: 10.1051/matecconf/20167400027
ICMER 2015

Energy efficiency estimation of a steam powered LNG tanker using normal operating data

Rajendra Prasad Sinha[1]

[1]Department of advanced Marine Engineeringt, Malaysian Maritime Academy

Kuala Sg Baru Post code 78300, Masjid Tanah, Melaka,Malaysia

rajendra@alam.edu.my

ph 06-3882200, fax 063876700

Abstract. A ship's energy efficiency performance is generally estimated by conducting special sea trials of few hours under very controlled environmental conditions of calm sea, standard draft and optimum trim. This indicator is then used as the benchmark for future reference of the ship's Energy Efficiency Performance (EEP). In practice, however, for greater part of operating life the ship operates in conditions which are far removed from original sea trial conditions and therefore comparing energy performance with benchmark performance indicator is not truly valid. In such situations a higher fuel consumption reading from the ship fuel meter may not be a true indicator of poor machinery performance or dirty underwater hull. Most likely, the reasons for higher fuel consumption may lie in factors other than the condition of hull and machinery, such as head wind, current, low load operations or incorrect trim [1]. Thus a better and more accurate approach to determine energy efficiency of the ship attributable only to main machinery and underwater hull condition will be to filter out the influence of all spurious and non-standard operating conditions from the ship's fuel consumption [2]. The author in this paper identifies parameters of a suitable filter to be used on the daily report data of a typical LNG tanker of 33000 kW shaft power to remove effects of spurious and non-standard ship operations on its fuel consumption. The filtered daily report data has been then used to estimate actual fuel efficiency of the ship and compared with the sea trials benchmark performance. Results obtained using data filter show closer agreement with the benchmark EEP than obtained from the monthly mini trials . The data filtering method proposed in this paper has the advantage of using the actual operational data of the ship and thus saving cost of conducting special sea trials to estimate ship EEP. The agreement between estimated results and special sea trials EEP is expected to improve further with use of better and more accurate filtering techniques.

Key words: Ship Energy Efficiency, Specific Fuel Rate, LNG, Shaft Power, Hull fouling

1 Introduction

It is common practice to conduct specially designed controlled sea trials to estimate a benchmark ship EEP to be used for future comparison during ship's actual operation. This benchmark index, although more accurate, is less related to the real life situations in which the ship actually sails because the operating environment is very different [1]. For a meaningful comparison with benchmark index an effective EEP of the ship should be extracted from the actual operating parameters which will then reflect truly on the condition of its machinery and hull fouling [2] which is the main purpose of performance monitoring.

The fuel efficiency of a ship is determined by measuring the quantity of fuel consumed to produce the required amount of propulsion and auxiliary power to move the ship at a specified speed. Lesser the quantity of fuel used the ship is considered more efficient. This efficiency has been measured in many different units but the one most universally accepted by the International shipping community is the Specific Fuel Rate (SFR). The SFR is expressed in grammes of fuel consumed for every kWh of power produced by the main propulsion machinery plus electrical generators [3], expressed as:

$$Vessel\ SFR = \frac{Fuel\ Consumed\ at\ sea\ per\ Day\ in\ grammes}{Total\,(SHP + T/G\ power)\ kWh} \left[\frac{gm}{kWh}\right]$$

Typical sample plots of SFR estimates from daily report of January and February 2014 for the ship under investigation are shown in figure 1. These plots show very inconsistent variations in the SFR of the ship. It is observed, on a particular day the ship performs well but then next day the SFR rises very high followed by again a sudden drop indicating improvement in fuel consumption.

Corresponding author: rajendra@alam.edu.my

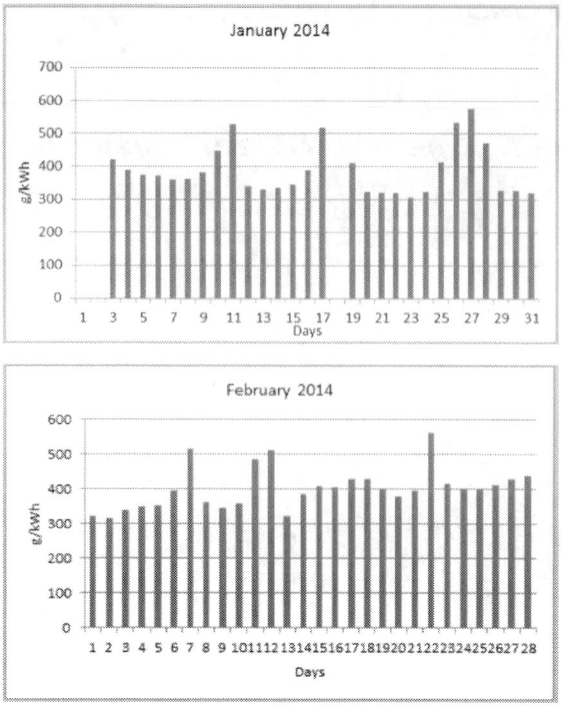

Figure 1. Sample SFR from Daily report

Obviously this kind of random variations in ships fuel efficiency cannot be attributed to hull condition or machinery. Most likely the cause of high SFR fluctuation could be some external factor such as the sea environment or operating procedures. Hence, to achieve accurate SFR estimates, the adverse effect of these factors on the fuel consumption needs to be filtered out. The author in this paper aims to answer this question by proposing a suitable data filter.

2 Data Acquisition and Management

An automatic data monitoring system is installed in the ship to record 24 hourly average value of following operational parameters which is then sent to the main office as daily report [3,4].

(i) Ship speed in knots
(ii) Duration of operation at specific speed in hours
(iii) SHP for duration of specific speed
(iv) Shaft rpm for duration of specific speed
(v) Liquid fuel consumption for the duration of specific speed, tons
(vi) Gas fuel consumption for the duration of specific speed, tons
(vii) Total equivalent fuel consumption for the duration of specific speed, tons
(viii) Distance covered by the ship at specific speed by ship's Log, nautical miles
(ix) Distance covered by the ship at specific speed by GPS, nautical miles
(x) Relative wind speed in knots and direction with respect to ship's heading, degree

Daily report entries also include summary of energy performance estimates, propeller slip and total power produced by the main engine and generators for the daily voyage. Fuel consumed during maneuvering and in port stay are entered separately. Typical daily report sample is shown in table 1.

3 Energy Efficiency Estimation

The shipboard performance monitoring system estimates ship EE by measuring the total fuel consumed by the ship in 24 hours (converted in grammes), Equation (1) and then dividing it by the kilo-watts average over the same period as in Equation (2). Algebraically, EE is expressed as in Equation 3.

Total fuel consumed during 24 hour voyage,

$$Q = \sum_{i=1}^{N} q_i \tag{1}$$

$$SHP_{average} = \frac{\sum_{i=1}^{N} P_i \times t_i}{\sum_{i=1}^{N} t_i} \tag{2}$$

$$\text{Ship SFR} = \frac{Q}{SHP_{average}} \tag{3}$$

The ship SFR in voyage leg method is calculated using equation 4 &5.

$$SFR_{Leg(i)} = \frac{q_i}{P_i} ,$$

$$SFR_{E(i)} = SFR_{Leg(i)} \times \frac{t_i}{\sum_{i=1}^{N} t_i} \tag{4}$$

Ship daily energy efficiency performance ,

$$\text{Ship SFR} = \sum_{i=1}^{N} SFR_{E(i)} \tag{5}$$

P_i= (Shaft power + Generator power) for specific voyage leg, kW

t_i -Voyage leg duration, hour
q_i -Voyage leg fuel consumption, grams
N=Number of voyage legs in a 24 hour sea passage

4 Ship Energy Efficiency Analysis [2,4,5]

The daily report data of the ship from January-December 2014 has been considered for analysis. The benchmark SFR of 300 g/kWh from the original sea acceptance trial report conducted by the shipbuilder has been used for comparison of ship energy efficiency in this analysis. KYMA system recorded SFRs for both months in fig 1 are above benchmark acceptable level. For instance, on days 11,17 in January and 12, 22 in February the SFR is over 500 g/kWh while on many days during this intervals the SFR is very close to the benchmark of 300 g/kWh. But the physical nature of the machinery and hull aging process should show a gradual change in the SFR with time. Therefore, this kind of abrupt variation in SFR readings cannot be used to predict actual condition of the machinery or growth in underwater hull fouling which is the chief objective of ship energy efficiency monitoring.

5 Voyage Leg Analysis

To determine actual SFR of the ship from daily report data, voyage leg analysis method has been used which works as follows. A day's voyage of 24 hours is generally made of a number of short voyage legs of few hours each with ship sailing at different speeds and shaft power. Also each voyage leg of a particular day's sailing is entered separately in the daily report with corresponding fuel consumption data. Therefore, by using Equation 3, it is possible to determine the SFR of an individual voyage leg of a 24 hours voyage together with its actual contribution to the day's overall SFR. A voyage leg with small or insignificant contribution to the overall day's SFR may be then treated as spurious operating noise source which is not truly related to ship energy system performance and hence can be ignored without causing serious error to the day's overall SFR estimate. The voyage leg SFR analysis can be thus used as filter to remove all spurious and non-standard operating conditions which will then give more steady and accurate SFR estimate of a 24 hours day's voyage. This estimation method has been used by the author to determine true SFR of the ship. Table 1 shows details of sample voyage legs, durations, shaft power and corresponding fuel consumption on four different days in January 2014 when recorded SFRs are high (ship's energy performance is low).

6 Results and Discussions [2,3,5]

Result of analysis using voyage leg method to estimate actual effective SFR of 24 hour voyage for January and February 2014 in figure 1 is discussed in this section. Day 11 in January had three distinct voyage legs with corresponding SFR and their impact on the effective SFR as shown in figure 2(a). The impact of voyage leg 1 on the overall effective SFR is very small which is indicative of a non-standard ship operation that does not really relate to actual machinery or hull condition. Therefore, the true ship effective SFR for 11 January 2014 voyage may be quite accurately estimated from the sum of only voyage legs 2 and 3 SFRs i.e 430.8 g/kWh as compared to the SFR of 528 g/kWh recorded by the KYMA energy performance monitoring system. This gives SFR estimation error of 97.2 g/kWh which for an $SHP_{average}$ of 8290 kW amounts to 1.85 tons extra fuel consumption by the ship.

Table 1(a)– Vessel SFR Data January 2014

Voyage Legs	Day 11			Day 17		
	SHP kW	T, hr	F.O, ton	SHP kW	T, hr	F.O, Ton
Leg 1	8903	0.1	0.3	3624	4.7	9
Leg 2	6899	1.9	7.1	6147	9.3	29.5
Leg 3	8404	0.3	1.2	--	--	--
Leg 4	--	--	--	--	--	--
Leg 5	--	--	--	--	--	--
Leg 6	--	--	--	--	--	--
Leg 7	--	--	--	--	--	--
Leg 8	--	--	--	--	--	--
Leg 9	--	--	--	--	--	--
Leg 10	--	--	--	--	--	--
Leg 11	--	--	--	--	--	--
Leg 12	--	--	--	--	--	--
SHPavg	8290			6602		

Table 1(b)– Vessel SFR Data January 2014

Voyage Legs	Day 27			Day 25		
	SHP kW	T, hr	F.O ton	SHP kW	T, hr	F.O ,ton
Leg 1	8419	0.1	0.	5795	1.1	4.2
Leg 2	8979	0.9	3.6	5345	0.7	2.4
Leg 3	--	--	--	12398	2	10.5
Leg 4	--	--	--	14126	14	77.5
Leg 5	--	--	--	16940	2.1	12.6
Leg 6	--	--	--	--	--	--
Leg 7	7070	0.4	1.5	6343	3.9	13.6
Leg 8	9196	0.8	3.8	5320	0.1	0.3
Leg 9	--	--	--	--	--	--
Leg 10	--	--	--	--	--	--
Leg 11	--	--	--	--	--	--
Leg 12	--	--	--	--	--	--
SHPavg	9893.54			12629		

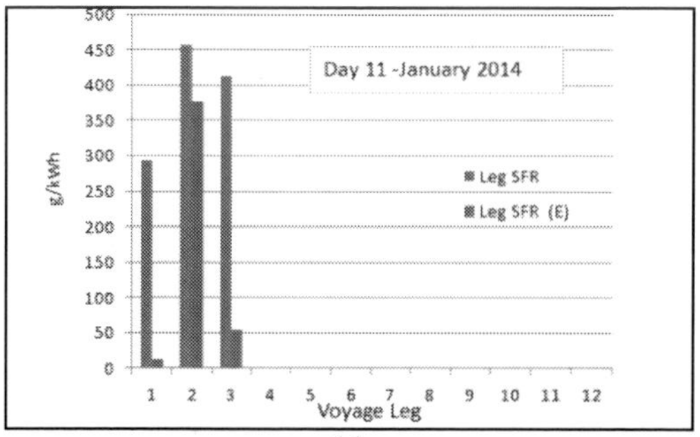

(a)

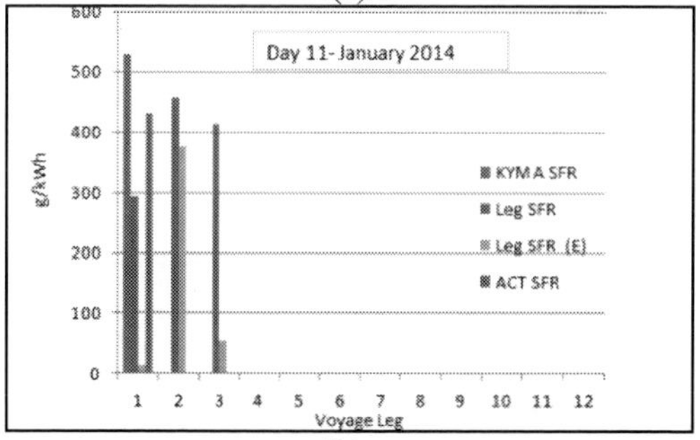

(b)

Figure 2 SFR Recorded versus estimated

For a voyage duration of 2.3 hours this is a significant estimation error in the KYMA energy efficiency monitoring system.

The voyage leg SFR, effective voyage Leg SFR(E), effective SFR and KYMA recorded SFR for day 11 in figure 2(b) clearly indicate that the actual effective SFR estimate from voyage leg analysis is more accurate than the SFR recorded by KYMA EE monitoring system. Similar analysis results of the daily report data for day 17 January is shown in figure 3(a,b). As expected, the day's effective SFR estimate is lower than the recorded KYMA SFR. The reason why voyage leg analysis method gives lower SFR values is obvious because it filters out fuel consumption of non-standard operations from the day's total fuel consumption and improve EE estimates.

Though, filtering spurious and non-standard operations by using voyage leg analysis method give lower and more accurate effective SFR estimates as shown in figures 2&3 but still in absolute terms the values on most days in January/February are above the sea trial benchmark of 300 g/kWh. This indicates that removal of spurious and non-standard operating conditions from the daily report data in figures 2&3 has not been fully achieved and further filtering of daily report data is still necessary to obtain ship actual SFR which truly reflects the hull and machinery conditions.

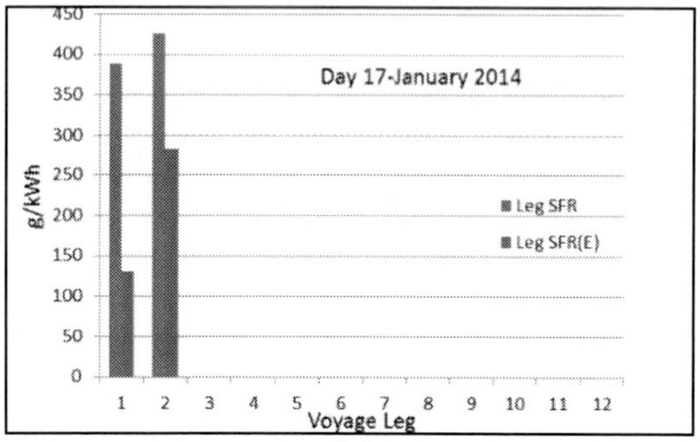

(a)

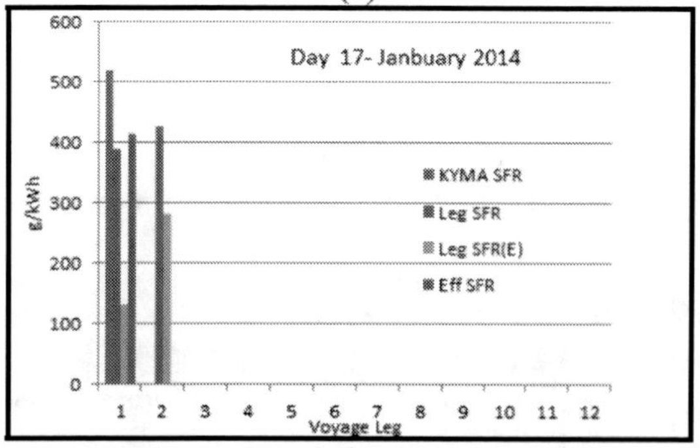

(b)

Figure 3 SFR recorded versus estimated

To improve data filtering method further, daily reports of 22,23,24 and 30 January 2014 have been analyzed. As the KYMA SFRs recorded on those days closely agree with the sea trials benchmark of 300 g/kWh, this analysis is expected to reveal underlying physics of improved filtering. Table 2 shows daily report data which are used for estimating effective SFR for those days.

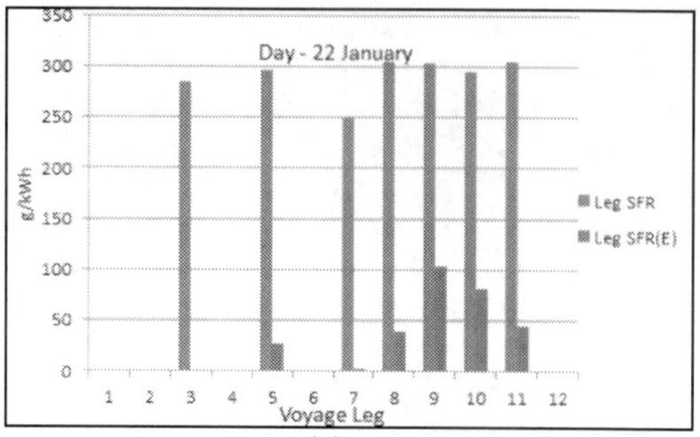

(a)

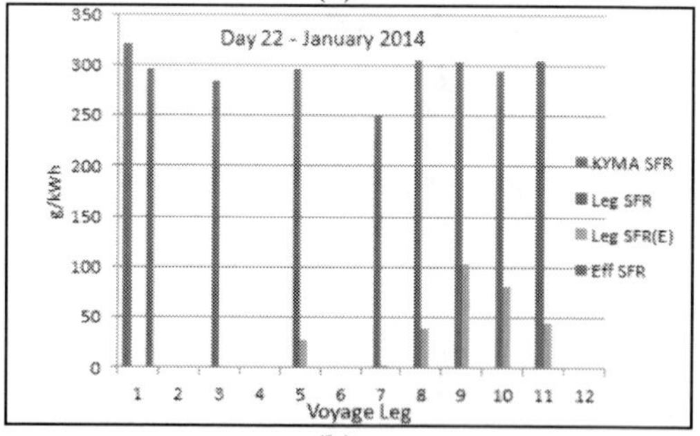

(b)

Figure 4 SFR recorded versus estimated

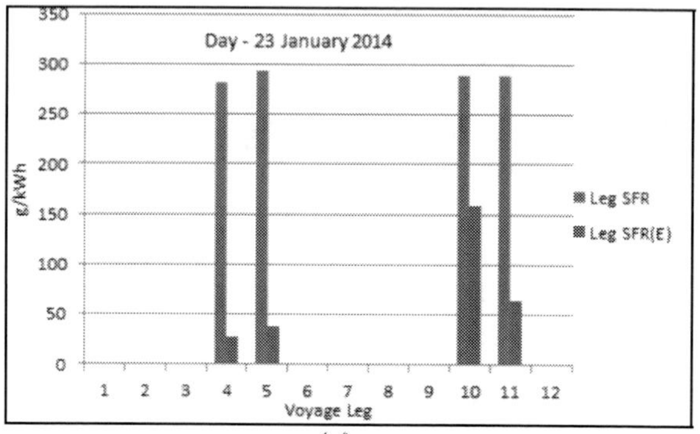

(a)

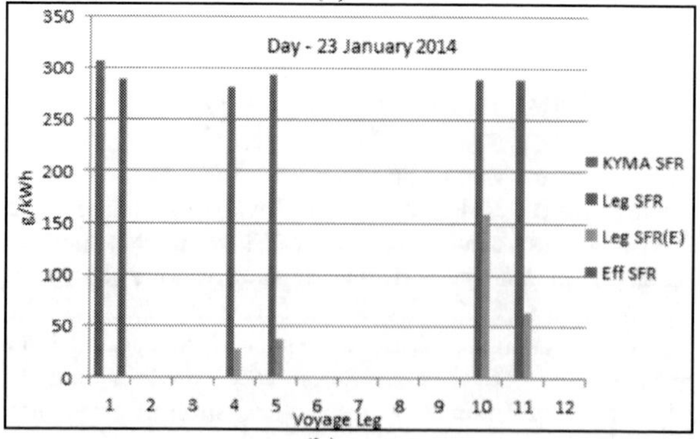

(b)

Figure 5 SFR recorded versus estimated

On day 22, figure 4, voyage legs 3 and 7 have negligible contribution to the overall day's SFR and hence can be safely ignored. The actual SFR for that day may be the sum of the remaining voyage legs which adds to 297 g/kWh as against the KYMA recorded SFR of 321 g/kWh. Contrarily, on day 23, all four voyage legs make significant contribution, figure 5 which indicates absence of any spurious or non-standard operation. The effective SFR is thus the total sum of all voyage legs.

Table 2 – Vessel SFR Data January 2014

Voyage Legs	Day 22			Day 23		
	SHP, kW	T, hr	F.O ton	SHP , kW	T, hr	F.O ton
Leg 1	--	--	--	--	--	--
Leg 2	--	--	--	--	--	--
Leg 3	19616	0.1	0.6	--	--	--
Leg 4	--	--	--	21785	2.2	14.4
Leg 5	21139	2.2	14.7	21667	3.0	20.3
Leg 6	--	--	--	--	--	--
Leg 7	18424	0.2	1.0	--	--	--
Leg 8	18537	3.1	18.9	--	--	--
Leg 9	19344	8.2	51.7	--	--	--
Leg 10	21272	6.6	44.3	21758	12.7	84.9
Leg 11	21281	3.5	24.3	21722	5.1	34.1
Leg 12	--	--	--	--	--	--
SHPavg	21679			23166		

Table 3(a) SFR Estimated versus KYMA

Voyage	Day – 11 January 2014			
	Leg SFR	Leg SFR(E)	Eff SFR	SFR KYMA
Leg 1	294.30	12.80	430.73	528
Leg 2	456.27	376.92	-	-
Leg 3	412.60	53.81	-	-
Leg 4	0	0	-	-
Leg 5	0	0	-	-
Leg 6	0	0	-	-
Leg 7	0	0	-	-
Leg 8	0	0	-	-
Leg 9	0	0	-	-
Leg 10	0	0	-	-
Leg 11	0	0	-	-
Leg 12	0	0	-	-
SHPavg	8290			

Table 3(b) SFR, Estimated versus KYMA

Voyage	Day – 17 January 2014			
	Leg SFR	Leg SFR(E)	Eff SFR	SFR KYMA
Leg 1	388.64	130.47	413.30	518.60
Leg 2	425.77	282.83	-	-
Leg 3	0	0	-	-
Leg 4	0	0	-	-
Leg 5	0	0	-	-
Leg 6	0	0	-	-
Leg 7	0	0	-	-
Leg 8	0	0	-	-
Leg 9	0	0	-	-
Leg 10	0	0	-	-
Leg 11	0	0	-	-
Leg 12	0	0	-	-
SHPavg	6602			

Table 4(a) SFR, Estimated versus KYMA

Voyage	Day – 22 January 2014			
	Leg SFR	Leg SFR(E)	Eff SFR	SFR KYMA
Leg 1	0	0	296.92	321.00
Leg 2	0	0	-	-
Leg 3	284.61	1.20	-	-
Leg 4	0	0	-	-
Leg 5	295.60	27.2	-	-
Leg 6	0	0	-	-
Leg 7	251.38	2.10	-	-
Leg 8	304.80	39.53	-	-
Leg 9	303.00	103.95	-	-
Leg 10	295.20	81.52	-	-
Leg 11	305.23	44.70	-	-
Leg 12	0	0	-	-
SHPavg	21679			

Table 4(b) SFR, Estimated versus KYMA

Voyage	Day – 23 January 2014			
	Leg SFR	Leg SFR(E)	Eff SFR	SFR KYMA
Leg 1	0	0	288.45	307
Leg 2	0	0	-	-
Leg 3	0	0	-	-
Leg 4	282.00	26.97	-	-
Leg 5	293.00	38.21	-	-
Leg 6	0	0	-	-
Leg 7	0	0	-	-
Leg 8	0	0	-	-
Leg 9	0	0	-	-
Leg 10	288.34	159.21	-	-
Leg 11	288.84	64.04	-	-
Leg 12	0	0	-	-
SHPavg	23166			

The summary of voyage leg analysis for days 11,17, 22 and 23 January 2014 are shown in tables 3&4. The voyage leg analysis reveals, when $SHP_{average}$ is much lower than the sea trials condition i.e SHP of 24124 kWh then the day effective SFR is much higher than the sea trials benchmark of 300 g/kWh. But as the value of $SHP_{average}$ increases and reach closer to the sea trials condition the day's effective SFR also improves and moves close to the benchmark value of 300 g/kWh, as shown in tables 3&4. Therefore apart from voyage duration, strong wind and optimum trim the $SHP_{average}$ is another critical parameter of the filter to estimation of accurate SFR from the daily report data.

The difference in effective SFR estimates from voyage leg analysis and KYMA EE monitoring system records arises mainly due to inclusion of non-standard operating conditions in the KYMA monitoring system which are filtered out in the voyage leg analysis method. The selected filter parameters are shown in table 5.

Table 5 Data filter Parameters [2,6]

S.No	Parameter	Filter limit
1	Voyage Leg Duration	<0.5 hour
2	SHPaverage	>21000 kW
3	Trim	$<\pm 0.5$ m
4	Wind speed	<15 knots
5	Beaufort Sea State	<3

7 Conclusions

SFR estimate obtained from KYMA energy efficiency monitoring system is based on total fuel consumed during normal and also non-standard short operational voyages. Because of that it does not represent true performance of the main machinery and underwater hull condition. In contrast the effective SFR estimate from voyage leg method is lower than values recorded by the KYMA system because it has been obtained by using a data filter to eliminate the contributions of spurious and non-standard operations of the ship. The Data filter parameters, table 5, have been selected after analysing contributions of each voyage leg to the ship SFR for a 24 hour passage and those relating to non-standard operating conditions have been ignored or filtered out. This procedure therefore gives effective SFR of the ship coming from only standard operating conditions. The effective SFR can be then used for comparison with the ship benchmark performance index to assess actual condition of ship's underwater hull and machinery.

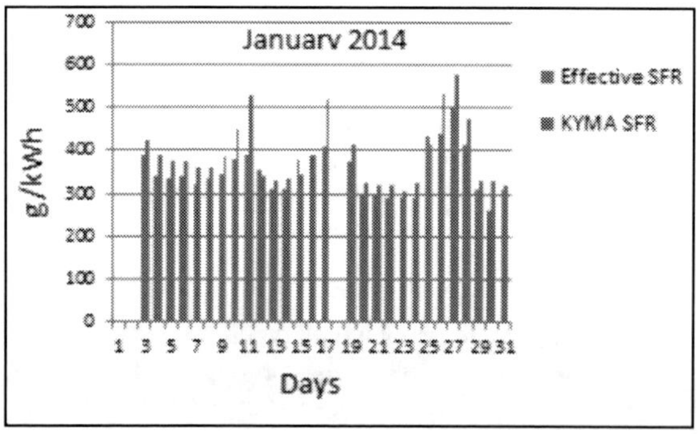

Figure 6 Ship SFR estimate versus KYMA

Comparison of energy efficiency i.e effective SFR of the ship based on voyage leg analysis and those recorded from the KYMA monitoring system for January 2014 in figure 6 indicates voyage leg analysis gives a lower and more accurate SFR estimates.

Glossary:

EE	Energy Efficiency
SHP	Shaft power, kW
LNG	Liquefied Natural Gas
KYMA SFR	Specific Fuel Rate calculated by the Energy Performance monitoring system
Leg SFR	Specific Fuel Rate of a Voyage Leg
Leg SFR(E)	Net SFR contribution of individual voyage leg on over all SFR of the day
Eff. SFR	Estimate of actual SFR after filtering contribution of nonstandard voyages

References

[1] Tadeusz Borkowski,Lech Kasyk, Przmyslaw Kawalak, Assessment of Ship Engine effective Power – Fuel Consumption and emissions using the vessel speed, Journal of Kones Powertrain and Transport, Vol 18, No2, 2011.

[2] Kevin P Logan, Using Ship Propeller for Hull Condition Monitoring, Presented at ASNE Intelligent Ship Symposium IX, 25 May 2011, Philadelphia, PA,USA

[3] Jonas Kallman, Ship Power Estimation for Marine Vessels Based on System Identification, Thesis, Department of electrical Engineering LIKOPINGS University , SF-581 83 Linkoping, Sweden.

[4] Sinha R P, Fuel Efficient Ship Operation – An Optimization Approach, 6th International Engineering Conference (ENCON 2013) 1-4 July 2013, Kuching, Malaysia

[5] Thomas Hellstrom, Optimum Pitch, Speed and Fuel Control at Sea, Journal of Marine Sciences and Technology , Vol 12, No2 pp 71-77(2004)

[6] Stefan Petursson, Predicting Optimal Trim Configuration of Marine Vessels with respect to fuel usage, Master Thesis, School of Engineering and Natural Sciences, University of Iceland, Reykjavik 2009

[7] Rashad A, Maihy A. EL , Energy and Exergy Analysis of Steam Power Plant in Egypt, International Conference on Aerospace Sciences & Aviation Technology, ASAT-13, May 26-28 2009, asat@mtc.edu.eg

[8] Sinha R P, Wan M Norsani Wan Nik, Energy Analysis of Steam Power Plant of a Medium Size LNG Tanker, International Conference on Maritime Technology , Kuala Terengganu, 2012.

[9] Zabi Bazari, Tore Longva, Assessment of IMO Mandated Energy Efficiency measures for International Shipping, L:\LED\INF\PRESS\2011\EEDI Study Presentation\Report.

MATEC Web of Conferences 74, 00028 (2016) DOI: 10.1051/matecconf/20167400028
ICMER 2015

Review on Recent Development Micro Gas Turbine -Trigeneration System and Photovoltaic Based Hybrid Energy System

MRR Chand[1,a], H.Ibrahim[1] , Z.Azran [2], A.Arshad [1], F.Basrawi [1]

[1]Faculty of Mechanical Engineering, University Malaysia Pahang, 26600 Pekan, Pahang Malaysia
[2]Faculty of Health Science, International Islamic University Malaysia, 26000, Kuantan, Pahang, Malaysia

Abstract. Research on distributed power generation as an alternative method to the conventional power generation system continue to be developed to improve its commercialization capabilities. The cogeneration system and trigeneration system are technological improved alternatives in distributed generation where they offer enhancement and reliability in term of efficiency, emission performances and economic benefits. However, it is more feasible to implement the trigeneration system for most commercial and domestic distributed generations as the cooling demand is deliberately high compared to heating demand especially in hot and humid climate locations. Moreover, micro gas turbine is observed to be a beneficial prime mover in cogeneration and trigeneration system based on several criteria such as ability on acquiring high heat to power ratio characteristic as well as lower greenhouse gas emission. On the other hand, the role photovoltaic in building integrated system provides opportunities for renewable energy system engagement in trigeneration based distributed generation systems. This paper emphasize on summarizing the research work perform on cogeneration system or trigeneration system in hybrid mode with photovoltaic. There are also preceding sections on overviewing the state of art of cogeneration system and the trigeneration system as well as photovoltaic technologies in power generation.

1 Introduction

The IEA report on the global CO_2 emissions in year 2014, reflects positive pronounced effect on the effort of mitigating the climate change and improvisation of power generation technologies. The emission levels were broad constant from serious exaggeration hiked to 32.3 billion tons as reported in year 2013 [1]. However, magnifying the permanent shortcomings of conventional power generation system such as its inefficiency power generations, high levels of emissions as well as transmission loses and costs , diverging into new power generation technologies such as smart grid based distributed system are much anticipated. Encountering the shortcomings through introducing integrated system called cogeneration system (CGS) and trigeneration system (TGS) had yielded in way of obtaining higher efficient system with less system loses, reduced emission and conveniences in applications wise[2, 3].

It was estimated that the CGS and TGS system would reduce the global CO_2 emissions by 10% that is 950 Mt/year by the year 2030 [4]. Technologies development also had enhanced the market capabilities of CGS and TGS system in mini and micro scaled , since the inventions of smart grid distributed generations [5]. Currently, the Germans lead the CGS and TGS utilization in power generation by 22%, followed by Poland and Demark where 46% of Denmark power generations were contributed by the CGS and TGS systems [6]. Implanting renewable energy based systems [7], integrating low emission prime movers in a way of reducing fuel consumptions, improvisation of current conventional system, instilling awareness as well as imposing policies and invectives are measures that able to assist further reduction on emission level in power generation sectors [8].

On the other hand, renewable energy system (RES) technology is also promoted as a sustainability solution on encountering the drawback of the conventional power system. Photovoltaic (PV) technology which converts solar energy into electricity is adopted as promising source of energy where it provide reliability in reducing the emission level in the absence of mechanical parts. Furthermore, the ability of PV to be building integrated provide suitability in distributed power generation application [9]. However, the intermittency issue of PV has led to research and development in hybrid based system where integration with CGS and TGS provide opportunity in developing system which is efficient and economical promising [10].

Hence, this paper emphasis on providing review on research work performed under the scheme of PV and CGS or TGS hybrid based system. System design, analysis and optimization as well as recommendation for future work for latter system are presented in this paper. There are preceding elaboration included on the state of art of CGS,

[a] Corresponding author: rizdwan@hotmail.com

TGS and PV where prime movers technology as well as waste heat utilization technology are discussed in detail.

2 Overview on CGS and TGS

CGS are engineered through integrating prime movers and exhaust heat recovery units, operating to provide both electrical power and thermal energy simultaneously. Henceforth, the thermal energy output are utilized for domestic water heating, space heating or in any industrial process which requires heat input such as pulp process in paper industry, food fermentation and brewing process in food industry, petroleum as well as in refining process [11, 12]. A CGS plant attains the ability to perform a high efficient energy conversion up to 80% of fuel energy compared to conventional power plant which converts an average of 30% to 35% of fuel energy[13, 14]. However, in most application heat requirement is fairly seen to be fluctuating and low depending on the climate. The shortcomings of CGS in term of its inadequacy in fulfilling the thermal requirement in most end use demand leads to advance energy utilization technology where conversion of thermal energy into cooling energy is seen to be more reliable.

Henceforth, the idea of including a thermally activated equipment to utilize exhaust heat for district cooling has assisted in the development of an advanced system called TGS. TGS is a system that able to provide power as well as heat and cooling energy simultaneously has broaden the application of on-site power generations especially for commercial buildings such as office buildings and hospitals as the requirement of space cooling are extremely high. Indeed, the TGS system reduce the pay back periods and improve the prime movers operations as there were more higher load demand through reducing partial operations[15]. From economic stand of view, the capital cost of TGS system is higher compared to any other power generations technologies, however, considering several factors such as electrical efficiency of prime movers, energy price and electrical price , TGS will be favourable in some cases [16]. Both CGS and TGS competent in providing an overall system efficiency up to 60% to 80% and 70% to 90% respectively. It is higher compared to conventional power generation system where the recorded efficiency is between 39% to 47% for coal based power plants, 38% to 44% for oil based power plants and 54% for combine cycle gas turbine (CCGT).

2.1 Prime Movers and Waste Heat Recovery Technologies

The majority of prime mover technologies are heat engines that convert chemical energy in fuel into electricity through combustion with the exception for fuel cell which converts fuel energy electrochemically. Common prime movers in distributed generations are reciprocating engines which include diesel generator and gasoline generator, MGT fuel cells and sterling engines. Table 1 display the comparison of several parameters involving technical performances, economics and emission characterises for TGS applications in between MGT, reciprocating engines and fuel cells. Reciprocating engines are the spark ignition Otto-cycle engines and the compression ignition Diesel-cycle engine. Reciprocating engines offer several advantages in power generation such as high efficiency, fast start up which is suitable for peak shaving and back up applications, good partial operation performances as well as operating on low pressure gas. However, narrowing the application particularly into CGS and TGS, reciprocating engines suffer several downsides such as low exhaust temperature which lead to limitation in heat utilization application as well as high emission and noise level.

Reciprocating engine such as natural gas engine and diesel engine releases 60% to 70% of waste heat during its operation where 45% to 50% are recovered through jacket cooling water and lube oil cooling system and the remaining in the form of exhaust gases. The temperature of waste heat ranges from 85°C to 100°C from the engine jacket coolant as well as the 380°C to 540°C contributed by the exhaust gases. However, only 110°C of hot water or low pressure steam up to 400psig able to extracted from the reciprocating engines. The common mechanism used for exhaust heat recovery are close loop cooling system, ebullient cooling system, heat exchanger and boiler. The Closed-loop cooling system and ebullient cooling system technologies are used to extract the waste heat from engine jacket coolant whereas heat exchangers are used for exhaust heat recovery. The waste heat are delivered in the form of hot water or steam depending on the targeted area of applications.

On the other hand, the type of acknowledged fuel cell technologies for CGS and TGS are PEMFC, PAFC, MCFC and SOFC where currently the total installed capacity of CGS employing fuel cell are 83.6MW. PEMFC and PAFC produce low temperature heat mostly below 100°C which is appropriate for space heating and water heating. Fuel cells have extremely low emission and high efficiency compared to other prime movers as there is no fuel combustion process. Other advantages of the fuel cell include low noise frequency, high partial load performances and fuel flexibility as well as modular design. However, economic wise, fuel cell technologies require higher capital cost than any other prime mover that further upsurge the payback period due to low-volume custom production as well as additional cost involved in power conditioning. In addition, the fuel cells requires longer startup period with at least 3 hours to 2 days period required where feasibility in peak shaving application and back up application are lower.

MGT is classified as a small gas turbine with power generation capability ranging from 25kW to 400kW , and which operates based on regenerated Brayton's Cycle with additional heat recovery heat exchanger called recuperator [17].The MGT units require lesser maintenance compared to reciprocating engines thanks to air bearing design that avoids any activities of oil lubrications and in addition there is no necessity. In additional cooling system as the atmospheric air provide the cooling needs [18]. Other merits of MGT over the reciprocating engines include less moving parts, lower emissions, small size and lightweight

Table 1. The comparison of various CGS and TGS prime movers

Parameters	DE	N-GE	MGT	Fuel Cell
Size (MW)	0.05-5	0.05-5	0.025-0.25	0.2-2
Electrical Efficiency (%)	30-50	25-45	20-30	40-70
CGS Efficiency (%)	77-80	77-80	63-70	55-80
CGS Output (Btu/kWh)	3400	1000-5000	4000-15000	500-3700
Start-up Time	10sec	10sec	60sec	3 hrs-2days
CGS installation Cost ($/kW)	800-1500	800-1500	500-1300	>3000
NO_x (lb/MWhr)	3-3.3	2.2-2.8	0.4-2.2	<0.02

fuel flexibilities, longer lifetime, less noise and higher heat outputs[19] . Correspondingly, MGT also exhibit better performance on economic prospect and operation start up time in contrast to fuel cell technologies. Application wise, MGT acquiring the ability to be implemented in peak shaving, grid connected, back up auxiliaries and in CGS and TGS [20].

In term of waste heat recovery, MGT delivers high quality exhaust gases with high oxygen content up to 15% which is suitable for direct utilization. Furthermore, the utilization of MGT in CGS and TGS are further enhanced by the fact that the system exhaust heat temperature ranges from 260°C to 350°C with CGS output as high as 4000Btu/kWh to 15000Btu/kWh which is suitable for direct fired absorption chiller as well as other industrial process Commonly, the exhaust heat is transferred to heat water by means of heat exchanger mechanism. Fig 1 shows the useable heat temperature range of various prime movers where the fuel cell acquires a wider range and highest temperature to be utilized in CGS and TGS. However, economically, MGT offers better useable heat temperature range compared to reciprocating engines and fuel cell. In both TGS and CGS the exhaust heat recovered can be utilized for various heating application. In both TGS and CGS the exhaust heat recovered can be utilized for various heating application. Moreover, the recovered waste heat also requires capability for application for thermally activated equipment for space cooling such as absorption chillers, adsorption chiller, liquid and solid desiccant cooling as well as ejector cooling [21].

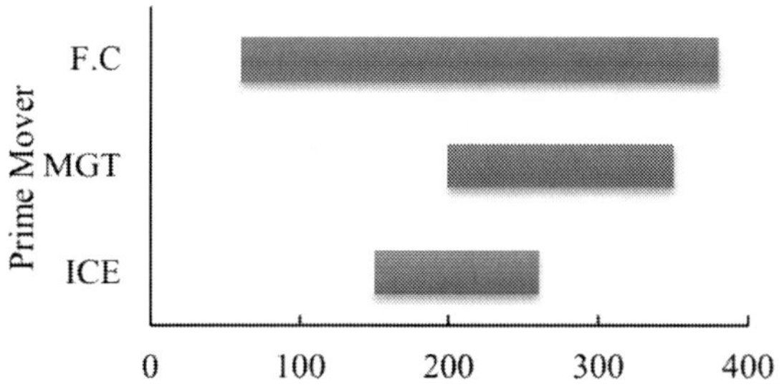

Figure 1. The range of useable exhaust temperature for several prime movers

Absorption chillers are well developed and mature technologies in the cooling industry compared to other cooling option [22]. Table 2 list features of several type of absorption chillers for integration in the TGS.

Through comprehensive view on the emission performances and economic benefits, MGT demonstrate better optimum prime movers for CGS and TGS. Even though, the efficiency is lower compared to both

Table 2. Features of Absorption Chiller Technologies [22]

Absorption Chiller Type	Heat Input	COP
LiBr- Water Single Effect	Hot water 70°C- 90°C ,Steam 2-3bar	~0.7
LiBr- Water Double Effect	Hot water 120°C- 170°C,Steam 4-8bar	~1.2
LiBr- Water Triple Effect	Hot water 200°C- 230°C	1.4-1.5
Water- NH_3 Single Effect	Hot water 80°C- 200°C,Steam 2-16bar	~0.8
Water- NH_3 Double Effect	Hot water 170°C- 220°C,Steam 8-16bar	0.8-1.2

[a] Corresponding author: rizdwan@hotmail.com

DOI: 10.1051/matecconf/20167400028

Even though, the efficiency is lower compared to both reciprocating engines and fuel cells, proper integration in CGS and TGS will result in better overall efficiency in addition to feasibility on economics and environmental indicators

3 Photovoltaic in Power Generation

Photovoltaic (PV) is made of positive layer (P-dope) and negative layer (N-dope) semiconductors sandwiched together. Both layers are separated by a junction developed by internal electrical field that permits electrons and holes movement. When sunlight waveforms strike the photovoltaic (PV) cells surface the atom ionization process take place where the electrons are excited to P-dope and holes to N-dope. This creates a voltage difference in between the layers and inducing an external circuit within which allows the electrons to move from P-dope to N-dope[23]. PV panels are categorized into 3 main type of technologies which are PV panels with silicon crystalline structure, thin film PV technologies and PV made of compound semiconductors [24].

Beforehand, the atomic structure of the silicon based PV were either manufactured in monocrystalline (Mono-PV) or in multi-crystalline (Poly-PV) cells. There are several types of thin film technologies which are differentiated though the semiconductor materials that includes amorphous silicon (a-Si), copper indium gallium selenide (CiGS) and cadmium telluride (Cd-Te). Thin films PV are manufactured through depositing semiconductors on glass and stainless steel by using a spurting tool. Currently, Cd-Te PV displayed dominance by as much as 50% in thin film industries mainly due to lower production cost compared to CiGS type which has slightly higher efficiency[23-25].

The efficiency of respective various PV cells and modules are listed in Table 3 where concentrated multifunction PV acquires the highest efficiency due to stacking of different band gaps cells as well as utilization of mirror lenses to concentrate the light by a factor of 50-1000.There are two significant parameters accounted on estimating the PV output power which are the solar irradiation and ambient temperature. The correlation in Eq.(1) shows that high solar irradiation increases the PV output however, high ambient temperature decreases the efficiency [26].

$$P_{PV}(t)=\left[P_{peak}\left(\frac{G(t)}{G_{STD}}\right)-\alpha_t\left(T_{c(t)}-T_{STD}\right)\right] \qquad (1)$$

$$T_{c(t)}-T_{amb(t)}=\frac{NOTC-20}{800}G(t) \qquad (2)$$

where, $P_{PV}(t)$ and P_{peak} are instantaneous power and peak power in W, $G(t)$ and G_{STD} are instantaneous and standard solar irradiation in W/m^2 respectively, T_{STD}, $T_{c(t)}$ NOTC are standard ambient temperature ,cell temperature and nominal cell temperature in K respectively, and α_t is temperature coefficient . Through an innovative intervention programs initiated by the government in recent years, the price of installing a $3kW_p$ -$10kW_p$ of PV arrays had reduced from Rm 55000/ kW_p as recorded in year 1990 to Rm 7000/ kW_p in year 2014

Table 3: The laboratory efficiency of respective PV cell and modules [27].

Type of PV Technology	Efficiency (%)	
	Cell	Module
Mono- Crystalline PV	25.0	22.9
Poly – Crystalline PV	20.4	18.5
Cd-TE	19.6	16.1
CIGS	19.8	15.7
a-Si	13.4	10.9
Multi-Junction (Concentrated)	44.7	36.7

The Germany emerged as the top leading country on the PV installation where the total generation 29,060GWh was by the end of 2012 followed by Italy, Belgium, Czech Republic, Greece and Australia and China is currently dominating the PV production industries [28, 29].

PV are interconnected forming an arrays for either standalone or grid tied applications. Narrowing the application to integrated building wise, PV had great potentials compared to other renewable energy system, based on criteria such as availability of energy source , convenience in integrated system and hybrid based system as well as due to the temporal correlations of electricity demand to the solar irradiations [30, 31]. Fig 3 illustrates the common configuration of PV in both standalone system and grid connected system (without back-up battery). In standalone system, batteries are utilized to store as well as distribute to the energy produced by the PV where charge controllers are installed to control the charging and discharging process. On the other hand, the simplest form of gird connected PV system without the battery storage system for back up purpose, only require inverters and metering units for system configurations. The end-users are allowed to inject clean power to the utility grid as well as consume electricity from the utility grid.

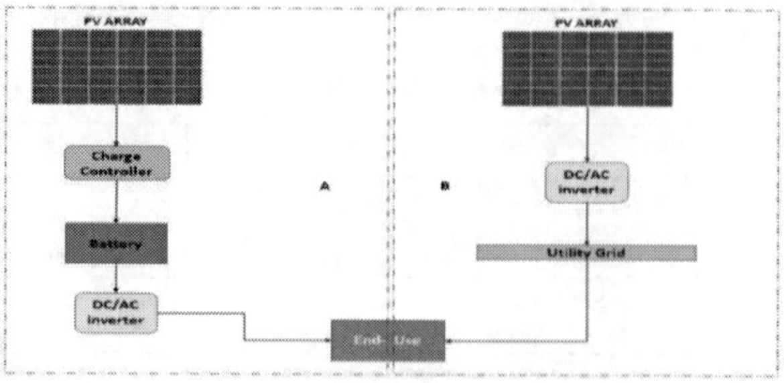

Figure 3: The figure shows the configuration of PV in standalone (A) mode and grid connected mode (B)

4 Photovoltaic in CGS/TGS

Foremost, even though PV is emerging as one of the attractive renewable energy technology, the issue intermittency of PV output are yet to be resolved. Irregular weather patterns and lower conversion efficiency are considered as key factor to unfavorable installation of PV. On the other hand, application of CGS and TGS suffered several obstacles such as unpredictable thermal and power demand on typically application. In certain application, the both thermal and power demand are high in which the prime movers of the respective CGS and TGS failed to satisfy the respective demand. Furthermore, too frequent partial operation of certain prime movers to satisfy variable demands causes incomplete fuel combustion which lead to lower efficiency and higher emission. In additions, increasing fuel price also raises the issue on the reliability of CGS and TGS in power generations [32]. Hence, the hybridization of PV in CGS and TGS are one of the several sustainable solution that can be implemented to resolve the above issues. This section was intentional to summarize the design, analysis and optimization work perform under the PV-CGS/TGS scheme

4.1 PV-CGS/TGS System Design

Pearce et al [33] had provided extensive explanation on the concept of expanding the photovoltaic penetration through designing a PV-CGS/TGS hybrid based system. In his paper, the author had briefly idealized four generation concept of utilizing PV and in CGS and TGS from the simplest form of grid connected PV with CGS unit as back up to the most advanced PV-thermal integration with TGS unit. In addition, Carmeli et al [34], had designed a PV-CGS consisting of an internal combustion engine (ICE) where several system topologies and power flow control strategies were further analyzed. The simulation results were then employed during an experimental setup in Delebio, Italy to analyze the dynamic performances. The system operated in such a way that it was connected to the grid for normal applications and was also able to operate in a standalone mode during emergency.

There was also an assessment performed to investigate the combination of photovoltaic and solar thermal (PVT) in an integrated TGS system that included a natural gas prime mover, exhaust heat recovery unit and hybrid cooling system as well as storage tank [35]. Life cycle assessment optimization method was developed in order to analyze the environmental impact in three categories that namely global warming potential, acidification potential and respiratory effect potential. The result indicate that the energy efficiency is 51.66% and better performances is obtained through thermal load following strategy

4.2 PV-CGS/TGS System Design

Nosrat et al [36] compared the PV-CGS hybrid system with PV-TGS hybrid system through simulation in a self-developed software, PVTOM. The results interpreted from the literature show that the PV-TGS system had better economic and environmental performances that the PV-CGS system. Derewonko et al [37] , studied on the feasibility of PV-CGS system where both rated power of PV and CGS were equivalent that is 1.2kW. The effect of climate and the operation time required by both power generation unit were analyzed extensively. The result shows that, PV only utilizes 50% of available solar energy due to losses suffered from cloudy weather.

In addition the author insisted on more studies to be performed on the PV-CGS system as the analysis displayed that the CGS prime mover were operated more than 20 hours daily to satisfy the load demand. The studies on economic analysis of PV-TGS for commercial office buildings as well as the optimization of the respective system were also studied by [38]. The analysis managed to summarize that the system was able to reduce the payback period compare to the separate power generation using CHS and PV as well as noting that the electric following strategy displayed better performances.

Ismail et al [39] had compared four utilization scenarios of PV and MGT in based on the Cost of Energy analysis (COE) for an application in Palestinian remote areas. The configurations compared were the stand alone PV system, MGT simple electricity generation, PV-MGT hybrid where MGT act as a backup power generator as well as a PV-MGT-CGS system. Through optimizing the tilt angle for PV, the results indicate that the PV-MGT-CGS system has lower COE as compared to other system configurations where the COE is RM 0.95/ kWh.

4.3 PV-CGS/TGS System Optimization

Yang et al [40], had design an optimum sizing method for renewable energy system which included a PV and wind energy system integration with the MGT .The author had equipped an optimum sizing with simulated annealing algorithm to perform the optimization process and subsequently the results obtained were positive. Bando et al [41], had performed optimization studies on PV-CGS plant which included a battery storage system. The objective function for the optimization studies was the operation and maintenances cost (O&M), primary energy consumption and the CO_2 emission where the manipulated variable used in the studies was the size of PV arrays and the CO_2 emission coefficient of the utility gird. Based on the simulation analysis, the PV-CGS system managed to reduce the CO_2 emission and the primary energy consumption as a high PV capacity installation was utilized. Basrawi et al [42] had studied on the economic by manipulating different operation strategies on a PV-MGT-TGS system that integrate auxiliaries component such as battery, thermal heat storage and boilers . The results show that power match strategies are economic as compared to the other operation strategies as well as the CCGT plant. However, the emission performances did not display positive result as compared to a CCGT plant due to lower efficiency during partial operation of MGT.

5 Conclusion

The simultaneous production of heating energy and power through CGS as well as cooling loads through TGS

[a] Corresponding author: rizdwan@hotmail.com

improve the overall system efficiency, reduce greenhouse gases emissions and offers economical energy generations in some cases. The selection of prime movers is crucial in obtaining top notch CGS and TGS where through literates and comparison, MGT is observe to serve better environmental and economic performances concurrently than other commercially available prime movers. The flaw existences in prime movers partial performances as well as varying energy demand lead to hydriziation of such CGS and TGS with renewable energy system. PV is one of the potential green technology which well fit in CGS and TGS hybrid based system as it is building integrated and resource abundancy. The integration of PV also assisted in resolving the intermittency problems of PV while improving the economic performances.

Based on the literature study, the integration of CGS or TGS and PV, require crucial attentions on the system topologies, sizing, weather conditions and as well as end-use demand where proper optimization should be perform. Analyzing a hybrid system with equipment of different working principles and individual properties, require proper control algorithm to optimize the energy flow. The control algorithm which is formulated based on dispatch strategy is important to avoid reduction in efficiency or system failure.

Hence, adequate optimization through the develop control algorithms should be perform prior to real life application. Extensive research work especially utilizing ICE as the prime movers were performed to design, analyze and optimize the hybrid system with proper dispatch strategy. Nevertheless, there are yet no specific control algorithm develop for CGS or TGS utilizing MGT in hybrid mode with PV. The development of algorithm for MGT-CGS or MGT-TGS and PV hybrid based system is essential as operation performance of MGT is highly dependent on ambient condition and energy demand. Hence, it is recommended on developing control algorithms for such system as it was simulated to provide better performances in both environmental and economical performances compare. Through the development, the reliability and flexibility of such system.

Acknowledgement

The author would like to acknowledge the authority of Malaysia Ministry of Education and University Malaysia Pahang upon the funding provide for the research under the grant (FGRS/2/2013/TK06/UMP/02/1) or RDU130133.

References

1. (IEA), I.E.A. *Global energy-related emissions of carbon dioxide stalled in 2014*. March 2015 [cited 2015 23 April]; Available from: http://www.iea.org/.
2. Ministry of Energy, C.a.M., *Policy Framework and the Promotion of Cogeneration in Malaysia*. 2013.
3. Cho, H., et al., *Evaluation of CCHP systems performance based on operational cost, primary energy consumption, and carbon dioxide emission by utilizing an optimal operation scheme*. Applied Energy, 2009. **86**(12): p. 2540-2549.
4. Cho, H., A.D. Smith, and P. Mago, *Combined cooling, heating and power: A review of performance improvement and optimization*. Applied Energy, 2014. **136**(0): p. 168-185.
5. Gelazanskas, L. and K.A.A. Gamage, *Demand side management in smart grid: A review and proposals for future direction*. Sustainable Cities and Society, 2014. **11**(0): p. 22-30.
6. Kuhn, V., J. Klemeš, and I. Bulatov, *MicroCHP: Overview of selected technologies, products and field test results*. Applied Thermal Engineering, 2008. **28**(16): p. 2039-2048.
7. Ahmad, S., M.Z.A.A. Kadir, and S. Shafie, *Current perspective of the renewable energy development in Malaysia*. Renewable and Sustainable Energy Reviews, 2011. **15**(2): p. 897-904.
8. Wong, S.L., et al., *Recent advances of feed-in tariff in Malaysia*. Renewable and Sustainable Energy Reviews, 2015. **41**(0): p. 42-52.
9. Singh, G.K., *Solar power generation by PV (photovoltaic) technology: A review*. Energy, 2013. **53**: p. 1-13.
10. Nehrir, M.H., et al., *A Review of Hybrid Renewable/Alternative Energy Systems for Electric Power Generation: Configurations, Control, and Applications*. Sustainable Energy, IEEE Transactions on, 2011. **2**(4): p. 392-403.
11. Holmberg, H., et al., *Allocation of fuel costs and CO2-emissions to heat and power in an industrial CHP plant: Case integrated pulp and paper mill*. Applied Energy, 2012. **93**(0): p. 614-623.
12. Tian, J., et al., *Measures and potentials of energy-saving in a Chinese fine chemical industrial park*. Energy, 2012. **46**(1): p. 459-470.
13. Onovwiona, H.I. and V.I. Ugursal, *Residential cogeneration systems: review of the current technology*. Renewable and Sustainable Energy Reviews, 2006. **10**(5): p. 389-431.
14. Bhatt, M.S., *Mapping of general combined heat and power systems*. Energy Conversion and Management, 2001. **42**(1): p. 115-124.
15. Chicco, G. and P. Mancarella, *Distributed multi-generation: A comprehensive view*. Renewable and Sustainable Energy Reviews, 2009. **13**(3): p. 535-551.
16. Schicktanz, M.D., J. Wapler, and H.M. Henning, *Primary energy and economic analysis of combined heating, cooling and power systems*. Energy, 2011. **36**(1): p. 575-585.
17. P.A.Pilavachi, *Power Generation with Gas Turbine System and Combined Heat and Power* Applied Thermal Engineering 2000.
18. Bruno, J. and A. Coronas. *Distributed generation of energy using micro gas turbines: polygeneration systems and fuel flexibility*. in *Proceedings of the international conference on renewable energy and power quality (ICREPQ'04)*. 2004.

19. Ismail, M.S., M. Moghavvemi, and T.M.I. Mahlia, *Current utilization of microturbines as a part of a hybrid system in distributed generation technology*. Renewable and Sustainable Energy Reviews, 2013. **21**(0): p. 142-152.
20. Yu, X., Z. Jiang, and A. Abbasi. *Dynamic modeling and control design of microturbine distributed generation systems*. in *Electric Machines and Drives Conference, 2009. IEMDC'09. IEEE International*. 2009. IEEE.
21. Steven Brown, J. and P.A. Domanski, *Review of alternative cooling technologies*. Applied Thermal Engineering, 2014. **64**(1–2): p. 252-262.
22. Jradi, M. and S. Riffat, *Tri-generation systems: Energy policies, prime movers, cooling technologies, configurations and operation strategies*. Renewable and Sustainable Energy Reviews, 2014. **32**(0): p. 396-415.
23. Meral, M.E. and F. Dinçer, *A review of the factors affecting operation and efficiency of photovoltaic based electricity generation systems*. Renewable and Sustainable Energy Reviews, 2011. **15**(5): p. 2176-2184.
24. El Chaar, L., L.A. lamont, and N. El Zein, *Review of photovoltaic technologies*. Renewable and Sustainable Energy Reviews, 2011. **15**(5): p. 2165-2175.
25. Parida, B., S. Iniyan, and R. Goic, *A review of solar photovoltaic technologies*. Renewable and Sustainable Energy Reviews, 2011. **15**(3): p. 1625-1636.
26. Khatib, T. and W. Elmenreich, *Novel simplified hourly energy flow models for photovoltaic power systems*. Energy Conversion and Management, 2014. **79**: p. 441-448.
27. ISE. *Photovoltaics Report*. 24 October 2014 [cited 2015 23 April]; Available from: www.ise.fraunhofer.de.
28. Kumar Sahu, B., *A study on global solar PV energy developments and policies with special focus on the top ten solar PV power producing countries*. Renewable and Sustainable Energy Reviews, 2015. **43**(0): p. 621-634.
29. Sener, C. and V. Fthenakis, *Energy policy and financing options to achieve solar energy grid penetration targets: Accounting for external costs*. Renewable and Sustainable Energy Reviews, 2014. **32**(0): p. 854-868.
30. Brandoni, C. and M. Renzi, *Optimal sizing of hybrid solar micro-CHP systems for the household sector*. Applied Thermal Engineering, 2015. **75**(0): p. 896-907.
31. Mohammed, Y.S., M.W. Mustafa, and N. Bashir, *Hybrid renewable energy systems for off-grid electric power: Review of substantial issues*. Renewable and Sustainable Energy Reviews, 2014. **35**(0): p. 527-539.
32. Immovilli, F., et al. *Solar Trigeneration for Residential Applications, a Feasible Alternative to Traditional Micro-Cogeneration and Trigeneration Plants*. in *Industry Applications Society Annual Meeting, 2008. IAS '08. IEEE*. 2008.
33. Pearce, J.M., *Expanding photovoltaic penetration with residential distributed generation from hybrid solar photovoltaic and combined heat and power systems*. Energy, 2009. **34**(11): p. 1947-1954.
34. Carmeli, M.S., et al. *Hybrid PV-CHP distributed system: Design aspects and realization*. in *Clean Electrical Power, 2009 International Conference on*. 2009.
35. Wang, J., et al., *Life cycle assessment (LCA) optimization of solar-assisted hybrid CCHP system*. Applied Energy, 2015. **146**(0): p. 38-52.
36. Nosrat, A.H., L.G. Swan, and J.M. Pearce, *Improved performance of hybrid photovoltaic-trigeneration systems over photovoltaic-cogen systems including effects of battery storage*. Energy, 2013. **49**: p. 366-374.
37. Derewonko, P. and J.M. Pearce. *Optimizing design of household scale hybrid solar photovoltaic + combined heat and power systems for Ontario*. in *Photovoltaic Specialists Conference (PVSC), 2009 34th IEEE*. 2009.
38. Jiang-Jiang, W., et al. *Economic analysis and optimization design of a solar combined cooling heating and power system in different operation strategies*. in *Industrial Electronics and Applications (ICIEA), 2012 7th IEEE Conference on*. 2012.
39. Ismail, M.S., M. Moghavvemi, and T.M.I. Mahlia, *Design of an optimized photovoltaic and microturbine hybrid power system for a remote small community: Case study of Palestine*. Energy Conversion and Management, 2013. **75**(0): p. 271-281.
40. Yang, Y., W. Pei, and Z. Qi. *Optimal sizing of renewable energy and CHP hybrid energy microgrid system*. in *Innovative Smart Grid Technologies-Asia (ISGT Asia), 2012 IEEE*. 2012. IEEE.
41. Bando, S., et al. *Optimal Operation Planning of a Photovoltaic-Cogeneration-Battery Hybrid System*. in *Power System Technology, 2006. PowerCon 2006. International Conference on*. 2006.
42. Basrawi, F., T. Yamada, and S.y. Obara, *Economic and environmental based operation strategies of a hybrid photovoltaic–microgas turbine trigeneration system*. Applied Energy, 2014. **121**(0): p. 174-183.

Sandwich Structure Based On Corrugated-Core: A Review

N. Z. M. Zaid[1,a], M.R.M. Rejab[2] and N.A.N Mohamed[1]

[1]*Faculty of Mechanical Engineering, Universiti Malaysia Pahang, 26600 Pekan, Pahang, Malaysia*

Abstract. Sandwich structures are commonly based on polymeric foam and honeycomb core material, for use in lightweight applications such as fuselage in aero plane, hull in marine construction and others. A review of sandwich structure based on corrugated-core is proposed and presented in this paper. Firstly, this paper aims to provide a means of comparing available sandwich structure in industries. Secondly, this paper aims to provide sandwich structure with corrugated-core for future research development efforts in field of sandwich construction. This paper starts with introduction of composite material such as sandwich structure, the advantages of sandwich structure was shown. After that these papers provide the structure of sandwich structure which includes the two faces and the cores. Furthermore, sandwich structure with different cores, which is honeycomb, foam core and corrugated core are discussed. At the end, the paper discussed more on corrugated-core for future research development.

1 Introduction

Advanced composites materials such as sandwich structure have remained progressively substituting traditional materials such as aluminum and steel in various industries fabricating all from small part to the key structural components of heavyweight vehicle such as airplane and marine ship[1].

The advantages of sandwich structures include great stiffness to weight ratios [2-5] , improved fatigue life, steadiness under compressive forces, good thermal and acoustic isolation properties. These benefits determine the fact that the use of composites has improved ominously in a wide range of structural applications and first of all in aerospace manufacturing [6-9].

Katzman et al. [10] said that sandwich structures are usually based on honeycomb core and polymeric foam materials may keep air and humidity. Moisture retention is one of the problems in aircraft sandwich construction. This problem may lead to growth in the whole weight of the sandwich construction and degrading of the core properties. To overcome problems, an open channel core material such as two-dimensional prismatic core is vent able in order to avoid moisture accumulation. In addition, the suitability of the corrugated-core as replacement core design structures in the sandwich construction will be as well serves the concept of sustainable manufacturing.

2 Sandwich Structure

Generally, single layer sandwich structure consists of three main parts that are a core and two face sheets. With an extra sheet, called internal sheet, inserted into the core, a two-layer sandwich panel is then formed. Sandwich structure comprise of two relatively thin, stiff and strong faces divided by a relatively thick lightweight core, for example, honeycomb core, foam core and cellular metal. Stiff and simultaneous light component is very important in nowadays industries. To achieve lightweight and high stiffness component, sandwich structure construction is develop [1-4].

Depending on the determination of the materials can differ, nevertheless the most significant characteristics for sandwich constructions are, they are lightweight compared to metallic, high stiffness and cost effective compared to other composite structures. For specific applications using core material for example in aerospace, automotive, marine transportation, satellites, truck structures, containers, tanks, body parts, rail cars and wind energy systems, the construction technology used for the sandwich construction is significant [1, 5].

Sandwich structures are being considered for application to airplane main structures, durability and damage tolerance is a first rank contemplation, thus, understanding the opposing effect of in-service events. In fact, development of composite structure to sensitive fields, where high reliability is required, such as domestic flight, was so far restricted by the poor information of their behavior under intricate dynamic loads. Then, the structure needs to be assessed in order to verify that damage occurring during the service life will not lead to failure or extreme structural deformation until the damage is detected. The understanding of their static and fatigue behavior are essential in order to use the material in different application and a better considerate of the

[a] Corresponding author: nzakiahmz@gmail.com

various failure mechanisms under static and fatigue loadings situations is essential and extremely necessary [5].

Because of their high specific strengths and stiffness, sandwich structures are extensively using in lightweight construction especially aerospace industries. Sandwich panel comprise of a lightweight core cover by two thin face sheets. Every face sheet may be an isotropic material or a fiber-reinforced composite laminate while the core material may either be of metallic/polymeric foam or metallic/aramid honeycomb [11]. The main profits of using the sandwich concept in structural components are the high stiffness and low weight ratio. These structures can carry out-of-plane and in-plane loads and show good steadiness under compression, possession excellent strength to weight and stiffness to weight characteristics. There are many benefits of sandwich constructions, the expansion of new materials and the essential for great performance and low-weight structures cover that sandwich structures will continue to be in demand [1, 8].

Polymers, wood, aluminum and composites are used for the core. These are to minimize weight that are used in form of foams, corrugated and honeycombs construction [12]. Sandwich structures with cellular core materials compromise great definite strength and an interest energy absorbing capacity. These sandwich structure properties make them a good answer for the protection of aircraft structure from impacting unknown objects. For instance, such panels are frequently used ahead of airplane to avoid unintended bird strikes, which can effect major damages to equipment and thus affect their safety. This shield is to avoid damage of panels which can cause the depressurization of aircraft [13].

2.1 Faces/Skin

The faces adhesively stuck to the core to gain a load transfer among the components. This method the properties of each detached component are utilized to the structural advantages of the whole assembly leading to a very high stiffness-to-weight and high bending strength-to-weight ratios. Typically, the facing layer realized by high pressure laminates, aluminum plates, and glass fiber reinforced plastics. Basically, the skins are thin, stiff and very strong. Figure 1 shows the example of skin part in sandwich structures.

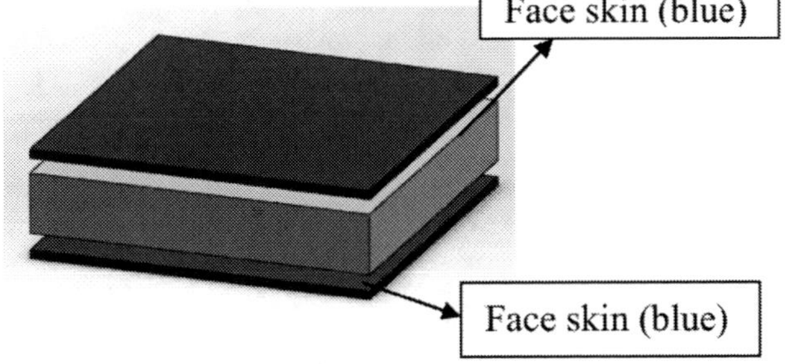

Figure 1: The skin of sandwich structure (Konka et. al 2012) [6]

2.2 Core

The sandwich structure core is the structure that placed between two thin faces. The material of the core is usually low strength material but its higher thickness provides the sandwich composite with high bending stiffness with overall low density. The purpose of the core is to backing the thin skin so that it does not deform and stay fixed relative to each other.

In recent years, there is several type of sandwich structures with better quasi-static and dynamic properties have been introduced, comprising those based on various foam, honeycomb cores, origami cores and truss cores [14].

2.3 Honeycomb

Aktay et al. [15] said that honeycomb is well-known core used to build sandwich structure. The name comes from the structure of honeycombs made by bees to store honey. Honeycombs and flex cores are used in many applications, for example chassis of modern cars use honeycomb sandwich structures. The honeycomb, flexcores and nomex are sandwiched between two carbon skins with the purpose of making a very stiff and strong structure that offers shield to the driver in case of a simple crash.

In aerospace applications, honeycomb materials have been broadly used among core materials. Various research have been conducted to know the mechanical response of honeycomb structures under dissimilar loadings [16].

Normally honeycomb has a uniform shape of hexagonal construction defined by the cell size, material, cell wall thickness and bulk density like Figure 2 below. The main materials for the core are glass fiber reinforced plastic, aluminum and aramid paper. Between them, aramid paper and aluminum are usually used in engineering application.

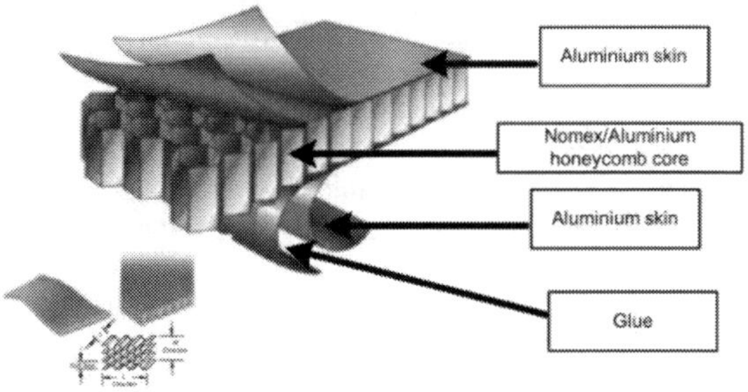

Figure 2: Example of Honeycomb sandwich structure (Abbadi et al. 2009)[17]

2.4 Foam Core

A big number of core materials and core configurations have been proposed these days. The most often used core materials are foams and honeycomb. The foam cores are preferably used when the waterproof, sound and heat insulation qualities of cores are required. Moreover, the foam cores are the cheapest among core materials and can offer some benefits in sandwich manufacture [18].

The stuffing of honeycomb cells with foam can be considered as the enhancement of debonding resistance and ability to produce new types of sandwich cores. This concept combines the benefits of honeycomb and foam cores. The increased adhesive area of foam-filled honeycomb cells is only one of them. On the other hand, the filling leads to changes of the dynamic properties of the honeycomb sandwiches [18].

Adding foam in the honeycomb core causes adequate decrease of the magnitudes of the natural frequencies. This effect is enlarged by the density of foam fillers, which trivially increases the stiffness and total mass of the filled sandwich plates. Filling with the foam promotes the slight increases of buckling loads and the insignificant decreases of imperfection sensitivity of foam-filled sandwich plates. Instead, adding the foam can make changes in stress distributions in the core to sheet faces interface [18].

The sandwich constructions have become more widely used thanks to the use of cellular foams as a structural element and load bearing component. Now day the main effort in design is put into face materials, but as the demands for more improved structures are raised, the research in the field of core materials is increased [19].

The situation of aluminum alloy foam with cold-worked aluminum face sheets have been studied experimentally and theoretically for plastic collapse modes of sandwich beams. Plastic collapse is by three contending mechanisms: face yield, indentation and core shear, with the active mechanism provisional upon the choice of geometry and material properties [20].

Recently, Chen et al. [20] explained that a range of metal foams have been established with a relative density ρ (defined by the ratio of foam density to the density of the cell walls) in the range 0.05 to 0.3. Generally are based on aluminum alloys, though steel foams and creep-resistant nickel alloy foams are also under progress. Metal foams compete favorably with polymer foams as the lightweight cores of sandwich beams, plates and shells, due to the higher stiffness and high-temperature ability.

Crupi et al. [21] discussed the aluminum foam sandwiches (AFS), gained by joining metal face sheets with a lightweight metal foam core, it have low specific weight, efficient capacity of energy dissipation, high impact strength, acoustic and thermal insulation, high damping, etc., that made them interesting for a number of practical applications, such as the realization of lightweight structures with high mechanical strength and good capacity of energy dissipation under impacts. When compared to traditional honeycomb panels, Aluminum Foam Sandwich offers several advantages. It can be prepared into curved shapes and with integral skins (without adhesive bonding), allowing higher working temperature and higher resistance to damage from water intrusion, which could be important for marine applications. Aluminum foams have potential to change polymer foams in sandwich panel applications due to their improved specific stiffness and higher temperature ability [22].

Usually, the low density cores of the sandwich structure are made of polymeric foams of numerous relative densities. Foam materials are however bending controlled structures, and the material is thus not fully developed, which results in a lower weight specific performance compared to a stretching controlled structure [23].

2.5 Corrugated-core

A corrugated-core sandwich structure is embraced of a corrugation sheet between two thin surface sheets. The important feature of this structure is its high strength-to-weight ratio. The corrugated-core keeps the face sheets apart and stabilizes them by resisting vertical deformations, and also enables the whole structure to act as a single thick plate as an asset of its shearing strength. This second feature gives better strength to the sandwich structures. Furthermore, unlike soft honeycomb shaped cores, a corrugated-core opposes bending and twisting as well to vertical shear. Then, corrugated-core sandwich panels, due to their extremely high flexural stiffness-to-weight ratio are usually used in aeronautics, aerospace, civil engineering and other applications, where weight is a significant design issue. This structure approach to form a sandwich plate may be defined as 'structurally composite', since its behavior characteristics are defined by the composite action of its components [24].
Rejab et al. [14] showed that corrugated cores when tested in the longitudinal direction offer shear strengths that are comparable with square honeycombs and significantly greater than those exhibited by diamond cores and more traditional foam cores. The second different feature of a corrugated-core is its ability to gives outstanding ventilation characteristics, avoiding problems related with humidity retention that is common in cellular core materials (e.g. polymeric foams and honeycombs). Humidity-retention can be a problem in many aerospace structures, e.g. aluminum honeycombs, but the adoption of corrugated, origami-type and truss core structures can minimize this problem.

Corrugated-cores with metal sandwich panels are an interesting industrial solution as structural components this is because their high stiffness-to-mass ratio. However, using detailed finite element models for numerical computation of their properties leads to large models and long solution time, specifically for acoustic simulations. Then, decrease of the complex shaped core to an equivalent homogenous material is usually used [25].

Amongst all sandwich panels, corrugated-core structures are an interesting alternative that is being progressively used in the transportation industry. For these panels there are dissimilar core shapes, such as truss-type corrugations (i.e. triangular), circular shape or trapezoidal cores [25].

Cote et al. [26] explain that prismatic, such as the Y-core and NavTruss, are chosen in naval sandwich structure for two reasons (i) they are direct to construction on large length scales by a welding route and (ii) the high longitudinal elongating and shear strength of the cores makes them ideal for application in sandwich beams.

Hou et al. [27] aims to investigate the effects of the key shape and dimensional parameters on the crashing behaviors of corrugated sandwich structures and improve the sandwich cores with the trapezoidal and truss-type corrugation configurations for crashworthiness standards. It will also compare the improved crashworthiness of these two different corrugated sandwich structures, thus given some guides to design of sandwich structures. Corrugated metal sandwich cores have verified excellent shock resistant properties, generally due to their high longitudinal stretching and shear strength [26].

2.5.1 Triangular

For triangular shaped corrugated-core, several authors already discussed the strength and properties. For example, Buannic et al, [28] studies the homogenization of corrugated sandwich panels. In this studies several shapes of corrugated-cores were selected including triangular shaped. Rejab et al, [14]conducted series of experimental investigations and numerical analyses is presented into the compression response, and subsequent failure modes in corrugated-core sandwich panels based on an aluminum alloy, a glass fiber (GFRP) and a carbon fibre reinforced plastic (CFRP). Tian et al. [29] experimented compression panel comprised of different type of corrugated-core including triangular. Figure 3 below shows the shape of triangular corrugated-cores which is triangle shape.

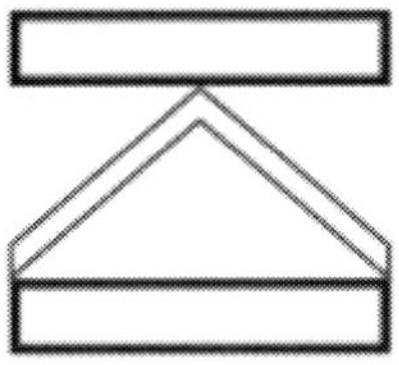

Figure 3: Triangle corrugated-cores (Bartolozzi et al. 2014) [30]

2.5.2 Sinusoidal

Bartolozzi et al. [25] investigated sinusoidal corrugated-cores aluminum sandwich structures. The main field of application of these aluminum structures is the transport area, e.g. in the automotive industry, where energy conservation, lightweight manufacture and recycling are critical requirements. Also marine internal applications are common, since these panels provide good structural performance with small thicknesses and can also be simply supplied in semi-finished components. The properties of the equivalent material are determined both analytically and numerically for the chosen Reissner–Mindlin orthotropic representation. The two derived models are compared in a comprehensive parametric study to validate the computationally much cheaper analytical formulation. Furthermore, a validation of the equivalent models is done based on the bending stiffness per unit width of the sandwich panel. Finally, the acoustic behavior of the structure is investigated comparing the reduced layered model with the fully detailed 3D model. Figure 4 shows sinusoidal corrugated-cores that have radius in the shape.

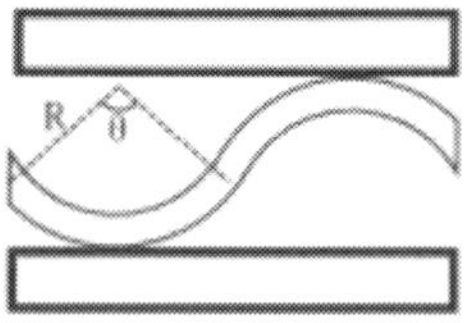

Figure 4: Sinusoidal corrugated-cores (Bartolozzi et al. 2014)[30]

2.5.3 Trapezoidal

The finite element modeling (FEM) and the impact responses of stacked trapezoidal corrugated aluminum core and aluminum sheet interlayer sandwich structures was studied by Kiliçaslan et al. [31]. Thill et al. [32] investigated about trapezoidal corrugated aramid/epoxy laminates under huge tensile displacements transverse to the corrugation direction. Figure 5 shows trapezoidal corrugated-cores that have trapezium in shape with the angle. Kiliçaslan et al. [33] do numerical studies and experimental on the quasi-static and dynamic crushing responses of multilayer trapezoidal aluminum corrugated sandwiches.

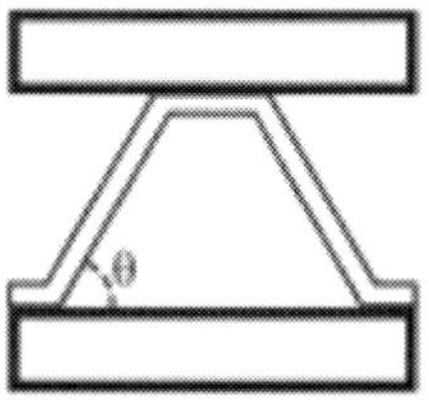

Figure 5: Trapezoidal corrugated-cores (Bartolozzi et al. 2014) [30]

3 Conclusions

Sandwich structures are commonly based on polymeric foam and honeycomb. Sandwich structure based on corrugated-core with different shape of corrugated-core, which is trapezoidal, triangle and sinusoidal was discussed in this paper. The important features of this corrugated-core sandwich structures is its high strength to weight ratio. Recent researchers already make several experiments to investigate the strength of the corrugated-core sandwich structures. Less of the research focus on trapezoidal sandwich structure which offer high strength in compression test and shear test. Furthermore, due to their high flexural stiffness to weight ratio, the structure usually used in making aircraft, marine ship, bridge and other application. From previous study, it showed that corrugated-core offer shear strengths when tested in the longitudinal direction which can be compare with square honeycombs and significantly greater than diamond cores and more traditional foam cores.

Acknowledgements

The authors would like to be obliged to University Malaysia Pahang for providing laboratory facilities and financial assistance under project no. RDU140116.

References

[1] D. G. Luchinsky, V. Hafiychuk, V. Smelyanskiy, R. Tyson, J. L. Walker, and J. L. Miller, "High-fidelity modeling for health monitoring in honeycomb sandwich structures," in *Aerospace Conference, 2011 IEEE*, (2011), pp. 1-7.

[2] Y. A. Bruno Castanié, Christophe Bouvet, Jean-Jacques Barrau, "Core crush criterion to determine the strength of sandwich composite structures subjected to compression after impact 2008 Composite Structures.pdf," *Composite Structures,* pp. 243–250, (2008).

[3] H. N. Wadley, "Multifunctional periodic cellular metals," *Philos Trans A Math Phys Eng Sci,* **vol. 364**, pp. 31-68, Jan 15 (2006).

[4] A. Mamalis, D. Manolakos, M. Ioannidis, P. Kostazos, and D. Papapostolou, "Axial collapse of hybrid square sandwich composite tubular components with corrugated core: numerical modelling," *Composite structures,* **vol. 58**, pp. 571-582, (2002).

[5] S. Belouettar, A. Abbadi, Z. Azari, R. Belouettar, and P. Freres, "Experimental investigation of static and fatigue behaviour of composites honeycomb materials using four point bending tests," *Composite Structures,* **vol. 87**, pp. 265-273, (2009).

[6] H. P. Konka, M. A. Wahab, and K. Lian, "On Mechanical Properties of Composite Sandwich Structures With Embedded Piezoelectric Fiber Composite Sensors," *Journal of Engineering Materials and Technology,* **vol. 134**, p. 011010, (2012).

[7] L. Librescu and T. Hause, "Recent developments in the modeling and behavior of advanced sandwich constructions: a survey," *Composite structures,* **vol. 48**, pp. 1-17, (2000).

[8] S. Heimbs, "Virtual testing of sandwich core structures using dynamic finite element simulations," *Computational Materials Science,* **vol. 45**, pp. 205-216, (2009).

[9] F. Abbassi, S. Mistou, and A. Zghal, "Failure analysis based on microvoid growth for sheet metal during uniaxial and biaxial tensile tests," *Materials & Design,* **vol. 49**, pp. 638-646, (2013).

[10] H. A. Katzman, R. M. Castaneda, and H. S. Lee, "Moisture diffusion in composite sandwich structures," *Composites Part A: Applied Science and Manufacturing,* **vol. 39**, pp. 887-892, (2008).

[11] C. C. Foo, G. B. Chai, and L. K. Seah, "Mechanical properties of Nomex material and Nomex honeycomb structure," *Composite Structures,* **vol. 80**, pp. 588-594, (2007).

[12] A. Petras and M. Sutcliffe, "Failure mode maps for honeycomb sandwich panels," *Composite Structures,* **vol. 44**, pp. 237-252, (1999).

[13] H. Zhao, I. Elnasri, and Y. Girard, "Perforation of aluminium foam core sandwich panels under impact loading—An experimental study," *International Journal of Impact Engineering,* **vol. 34**, pp. 1246-1257, (2007).

[14] M. R. M. Rejab and W. J. Cantwell, "The mechanical behaviour of corrugated-core sandwich panels," *Composites Part B: Engineering,* **vol. 47**, pp. 267-277, (2013).

[15] M. Yamashita and M. Gotoh, "Impact behavior of honeycomb structures with various cell specifications—numerical simulation and experiment," *International Journal of Impact Engineering,* **vol. 32**, pp. 618-630, (2005).

[16] L. Aktay, A. F. Johnson, and B.-H. Kröplin, "Numerical modelling of honeycomb core crush behaviour," *Engineering Fracture Mechanics,* **vol. 75**, pp. 2616-2630, (2008).

[17] A. Abbadi, Y. Koutsawa, A. Carmasol, S. Belouettar, and Z. Azari, "Experimental and numerical characterization of honeycomb sandwich composite panels," *Simulation Modelling Practice and Theory,* **vol. 17**, pp. 1533-1547, (2009).

[18] V. N. Burlayenko and T. Sadowski, "Analysis of structural performance of sandwich plates with foam-filled aluminum hexagonal honeycomb core," *Computational Materials Science,* **vol. 45**, pp. 658-662, (2009).

[19] M. Burman and D. Zenkert, "Fatigue of foam core sandwich beams—1: undamaged specimens," *International journal of fatigue,* **vol. 19**, pp. 551-561, (1997).

[20] C. Chen, A. Harte, and N. Fleck, "The plastic collapse of sandwich beams with a metallic foam core," *International Journal of Mechanical Sciences,* **vol. 43**, pp. 1483-1506, (2001).

[21] V. Crupi and R. Montanini, "Aluminium foam sandwiches collapse modes under static and dynamic three-point bending," *International Journal of Impact Engineering,* **vol. 34**, pp. 509-521, (2007).

[22] A.-M. Harte, N. Fleck, and M. Ashby, "The fatigue strength of sandwich beams with an aluminium alloy foam core," *International Journal of Fatigue,* **vol. 23**, pp. 499-507, (2001).

[23] S. Kazemahvazi, R. Ben, D. Vikram, and D. Zenkert, "Impact properties of corrugated composite sandwich cores," in *International Conference on Sandwich Structures, ICSS9,* (2010).

[24] W.-S. Chang, E. Ventsel, T. Krauthammer, and J. John, "Bending behavior of corrugated-core sandwich plates," *Composite Structures,* **vol. 70**, pp. 81-89, (2005).

[25] G. Bartolozzi, M. Pierini, U. Orrenius, and N. Baldanzini, "An equivalent material formulation for sinusoidal corrugated cores of structural sandwich panels," *Composite Structures,* **vol. 100**, pp. 173-185, (2013).

[26] F. Côté, V. S. Deshpande, N. A. Fleck, and A. G. Evans, "The compressive and shear responses

of corrugated and diamond lattice materials," *International Journal of Solids and Structures,* **vol. 43**, pp. 6220-6242, (2006).

[27] S. Hou, S. Zhao, L. Ren, X. Han, and Q. Li, "Crashworthiness optimization of corrugated sandwich panels," *Materials & Design,* **vol. 51**, pp. 1071-1084, (2013).

[28] N. Buannic, P. Cartraud, and T. Quesnel, "Homogenization of corrugated core sandwich panels," *Composite Structures,* **vol. 59**, pp. 299-312, (2003).

[29] Y. S. Tian and T. J. Lu, "Optimal design of compression corrugated panels," *Thin-Walled Structures,* **vol. 43**, pp. 477-498, (2005).

[30] G. Bartolozzi, N. Baldanzini, and M. Pierini, "Equivalent properties for corrugated cores of sandwich structures: A general analytical method," *Composite Structures,* **vol. 108**, pp. 736-746, (2014).

[31] C. Kılıçaslan, M. Güden, İ. K. Odacı, and A. Taşdemirci, "The impact responses and the finite element modeling of layered trapezoidal corrugated aluminum core and aluminum sheet interlayer sandwich structures," *Materials & Design,* **vol. 46**, pp. 121-133, (2013).

[32] C. Thill, J. A. Etches, I. P. Bond, K. D. Potter, P. M. Weaver, and M. R. Wisnom, "Investigation of trapezoidal corrugated aramid/epoxy laminates under large tensile displacements transverse to the corrugation direction," *Composites Part A: Applied Science and Manufacturing,* **vol. 41**, pp. 168-176, (2010).

[33] C. Kılıçaslan, M. Güden, İ. K. Odacı, and A. Taşdemirci, "Experimental and numerical studies on the quasi-static and dynamic crushing responses of multi-layer trapezoidal aluminum corrugated sandwiches," *Thin-Walled Structures,* **vol. 78**, pp. 70-78, (2014).

Speeding driving behaviour: Age and gender experimental analysis

Teo Sir Hiang[1] and Gan Leong Ming[1]

[1]Faculty of Mechanical Engineering, Universiti Malaysia Pahang, 26600 Pekan, Pahang, Malaysia

Abstract. Road safety is a substantial issue worldwide. Speeding contributed to one-third of all the fatal crashes reported from year 2002 to 2011with young drivers reported to have the highest fatality and injury rates. This paper studied on the speeding driving behavior of 10 teenagers and 10 adults, from both genders. The aim was to investigate the relationship between age and gender with speeding driving behavior. The drivers were required to drive within an enclosed compound by using a test car. Results showed young and male drivers averagely travelled at higher velocity before entering the roundabout and at the same time accelerate to higher velocity upon exiting the roundabout compared to old and female drivers.

1 INTRODUCTION

In a study carried out by World Health Organisation (WHO) [1] in the year 2013, about 1.24 million people involved in fatal road accidents. This trend however is worsening, from the ninth in 1990 to the third most common cause of death expected by the year 2020 [2] if no action were to be taken. Road accidents studies in Malaysia was reported as the fifth cause of death where most accidents occurred emerged from driver's behaviour [3][4].

Though there are many types of driving behaviour which lead to the high rate of road accidents, studies showed speeding was the contributing factor to one-third of all the fatal crashes reported from year 2002 to 2011 [5]. Teen drivers was reported to have the highest fatality and injury rates [6][7], especially young male drivers who are known to have greater affinity with speed than other drivers [8]. This suggests the necessity to understand the nature of speeding driving behaviours for this particular set of drivers.

In order to identify and understand the differences between these road accident risk potentials and meanwhile reduce the frequency of road accidents instrumented vehicle is widely used [9]. It is a naturalistic data collection equipment in studying driver behaviour and vehicle dynamics analysis, allowing quantitative assessments of driver performance under real driving events. Today's technology advancements and incorporations of sophisticated sensors in the instrumented vehicle have expanded the capabilities of naturalistic observations [10], correlating driver behaviour, vehicle dynamics performance and the road profile conditions [11].

This paper studied on the speeding driving behavior of 20 volunteer drivers, which are 10 teenagers aged from 20 to 25, with driving experience less than 10 years and 10 adults aged from 35-40, with at least 15 years of driving experience, from both genders, by utilising a test car. The aim was to investigate the relationship between age and gender with speeding driving behavior. With the equipped sophisticated sensors, the vehicle dynamics performance of the test car shall reveal the driving behavior of the drivers.

2 EXPERIMENTAL SET UP

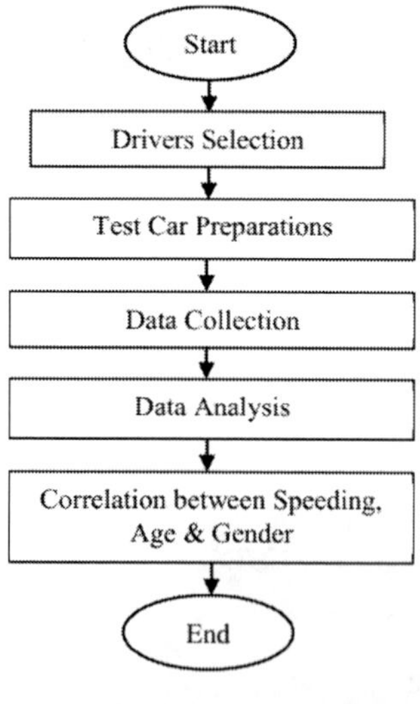

Figure 1. Standard Operating Procedures (SOP).

2.1 Drivers Selection

The drivers were selected via convenience sampling. Ten teenagers aged from 20 to 25 and ten adults aged from 35-40, from both genders.

2.2 Test Car Preparations

Before data collections, the sensors were calibrated and set to ensure proper and accurate data collection. The sensors involved in data collection were tabulated as shown below.

Table 1. Test Car and List of sensors involved.

Type	Description
Test Car	The test car is an instrumented vehicle, equipped with multitude of sensors for vehicle dynamics and performance testing.
Brake force sensor	To measure the brake force applied and displacement of the brake pedal upon car deceleration.
Throttle Pedal Displacement Sensors	Measures the displacement of the throttle pedal when the driver accelerates.
GPS sensor	Shows the direction and position of the test car and meanwhile obtain the velocity and course travelled.

2.3 Data Collection

The experiments were carried out in the morning under clear blue sky to ensure accurate GPS data collection. In order to minimize biased results due to test car unfamiliarity, the drivers were given the opportunity to drive along the test location for a few laps. Meanwhile, data collection was conducted when there was an absence of other vehicles on the test location when the experiment was carried out. The test location was within an enclosed compound to ensure the convenience of the drivers. The test location is as shown in Figure 2 below.

Figure 2. Test location.

2.4 Data Analysis

The location which was analysed in this study was further minimized to only around the roundabout 1 as labelled in Figure 3. The roundabout was further divided into 3 sections; section A: before entering roundabout; section B: roundabout; section C: exiting roundabout. GPS sensor was generally used to analyse on the travelling speed the drivers were in throughout the experiment. The brake force sensor was specifically used to study the brake force applied by the drivers upon reaching the roundabout (Section A) while the throttle pedal displacement sensor was specifically used to analysed the vehicle acceleration upon exiting the roundabout (Section C).

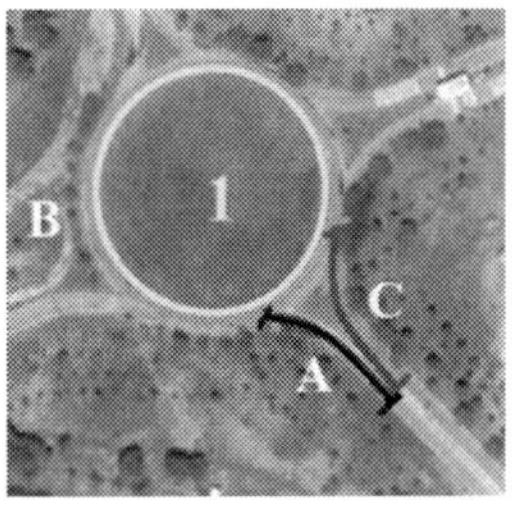

Figure 3. A closer look at roundabout 1.

2.5 Correlation between Speeding, Age & Gender

Travelling at higher speed increases the risk, severity and frequency of road accidents. With the vehicle dynamics results obtained from the test car, the relationship between the four groups of drivers namely male teens, male adults, female teens and female adults with speeding driving behaviour shall be revealed.

3 RESULTS AND DISCUSSION

The location which was analysed in this study was further minimized to only around the roundabout 1 as labelled in Figure 3 since the test car showed more obvious fluctuations in terms of vehicle dynamics performance in this section of road compared to the other two roundabouts.

3.1 Age analysis

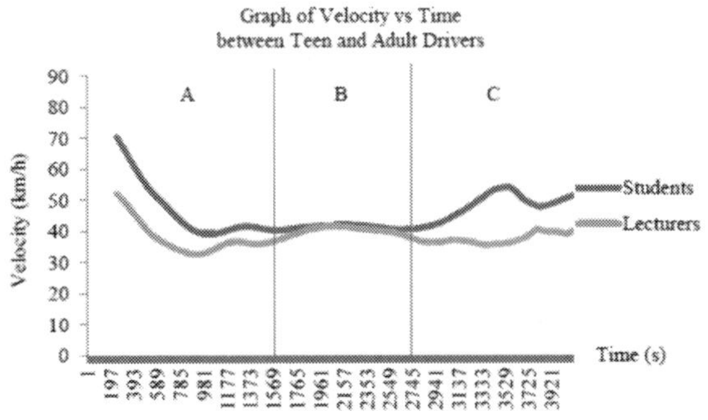

Figure 4. Graph of Velocity vs Time between Teen and Adult Drivers.

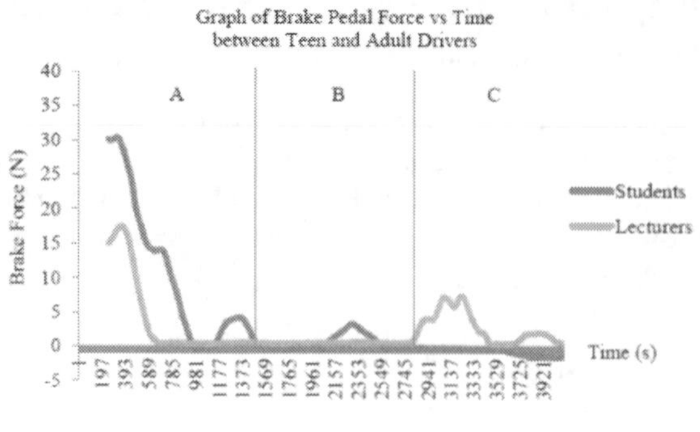

Figure 5. Graph of Brake Force vs Time between Teen and Adult Drivers.

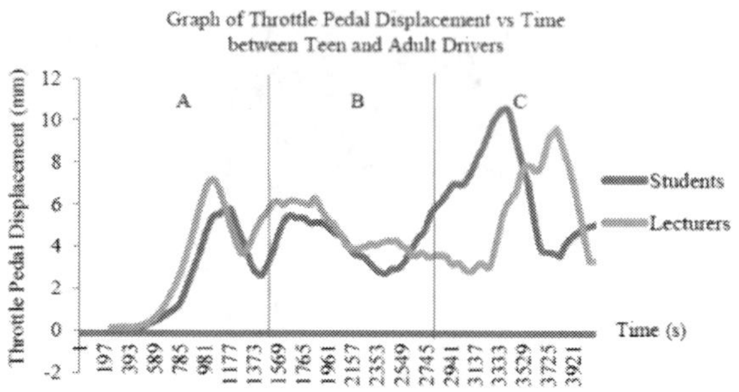

Figure 6. Graph of Throttle Pedal Displacement vs Time between Teen and Adult Drivers.

The age analysis was shown from Figure 4 till 6. From the roundabout inlet and outlet analysis, it can be seen that younger drivers prefer travelling at higher speed (Figure 4) compared to the older drivers in this study. Upon reaching the inlet of the roundabout, the teens applied higher brake force (Figure 6) where their travelling velocity dropped in a more drastic manner (Figure 4) and upon the exit of roundabout, the teens sped up again as shown by Figure 9. While travelling around the roundabout, which is shown by the middle section of all the graphs from Figure 4 till Figure 6, there was no significant difference among the teens and adults. This suggests that both the teens and adults travel at a considerably same manner.

3.2 Gender Analysis

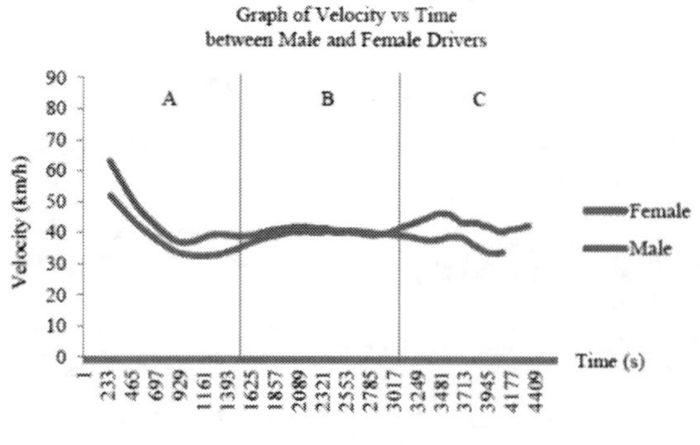

Figure 7. Graph of Velocity vs Time between Male and Female.

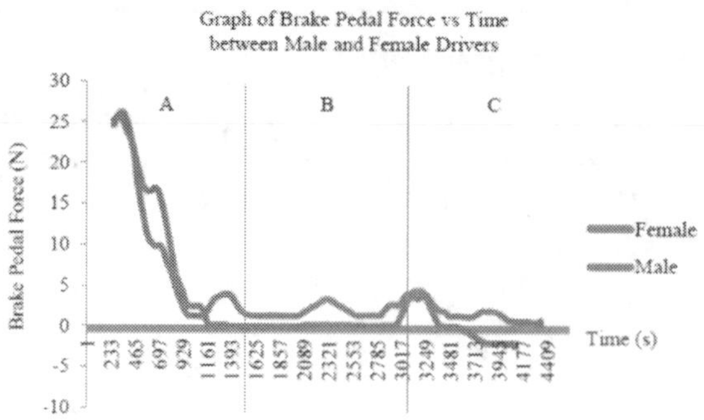

Figure 8. Graph of Brake Pedal Force vs Time between Male and Female.

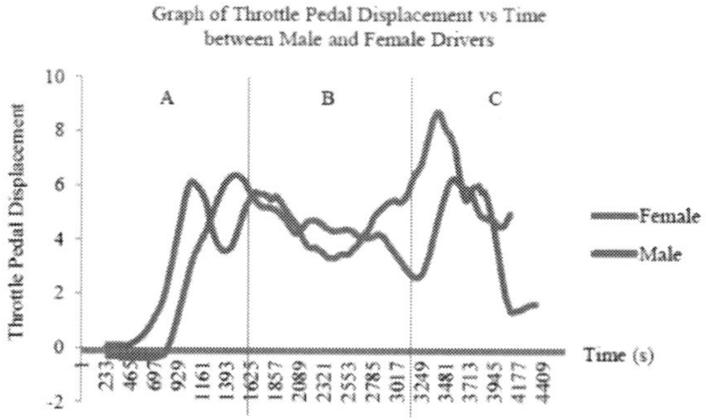

Figure 9. Graph of Throttle Pedal Displacement vs Time between Male and Female.

The gender analysis was shown from Figure 7 till 9. From the roundabout inlet and outlet analysis, it can be seen that male drivers prefer travelling at higher speed (Figure 4) compared to the female drivers. Upon reaching the inlet of the roundabout, even though the female drivers travel at lower speed, they applied a considerable equal amount of brake force as the male drivers. This indicates that the female drivers were more cautious upon entering the roundabout which explains the drop of velocity at the beginning section of Figure 7. Upon the exit of roundabout, the male drivers sped up faster, which is indicated by the throttle pedal displacement peak in Figure 9. While travelling around the roundabout, which is shown by the middle section of all the graphs from Figure 7 till Figure 9, there was no significant difference among the drivers. This suggests that both the male and female drivers travel at a considerably same manner.

4 CONCLUSIONS

Younger drivers practice speeding more often or to a greater extent than the older drivers and young male drivers have greater affinity with speed than other drivers as shown by the results obtained in this study.

5 FUTURE STUDY

The future study shall include more drivers and involve more sophisticated sensors in order to further verify the results obtained in this study. The traffic psychology of the drivers shall be included to enhance this driving behavior study.

ACKNOWLEDGEMENTS

The authors would like to be obliged to Universiti Malaysia Pahang for providing laboratory facilities and financial assistance under project no. RDU130130. Meanwhile, I would like to personally thank the drivers and everyone who has directly or indirectly involved in helping me in completing this study.

7 REFERENCES

1. WHO | Road traffic injuries, http://www.who.int/violence_injury_prevention/road_traffic/en/ (2013)
2. S. M. R. Ghadiri, J. Prasetijo, A.F. Sadullah, M. Hoseinpour & S. Sahranavard, Intelligent speed adaptation: Preliminary results of on-road study in Penang, Malaysia. *IATSS Research,* **36**(2), 106-114. doi: http://dx.doi.org/10.1016/j.iatssr.2012.08.001 (2013)
3. M.R. Othman, Z. Zhong, I. Takashi and M. Tetsuo, A study of analysis method for driver features extraction (2008)
4. N.M. Nor, A. Wahab, N. Kamaruddin and H. Majid, Pre-post accident analysis relates to pre-cursor emotion for driver behavior understanding. *Proceedings of the 11th WSEAS international conference on Applied computer science*: 152-157 (2011)
5. P. Schroeder, L. Kostyniuk & M. Mack, *2011 National Survey of Speeding Attitudes and Behaviors*. (Report No. DOT HS 811 865). Washington, DC: National Highway Traffic Safety Administration (2013)
6. A. Nayum, *The Role of Personality and Attitudes in Predicting Risky Driving Behavior.* Master Thesis. University of Oslo, Norway (2008)
7. J.Q. Wang, K.Q. Li and X.Y. Lu, Chapter 5 - Effect of human factors on driver behavior, in advances in intelligent vehicles. 111-157. ISBN 9780123971999, http://dx.doi.org/10.1016/B978-0-12-397199-9.00005-7 (2014)
8. C. Ferguson, L. Cohen, J.A. Pooley and A. Guilfoyle, The Theory of Planned Behaviour (TPB) and speeding behaviour of young drivers (2006)
9. A. Sukardi et al, Enhancement of instrumented car design for driver behaviour research. *Journal of Asian Scientific Research.* **4**(7): 314-319. Retrieved from http://www.aessweb.com/pdf-files/jasr(icetsr)-2014-4(7)-314-319.pdf (2014)
10. B.E. Porter, *Handbook of traffic psychology.* Elsevier (2011)
11. M. Rizzo, J. Jermeland and J. Severson, Instrumented Vehicles and Driving Simulators. **1**(4): 291-296 (2002)

ANALYSIS OF MODIFICATIONS ON A SPARK IGNITION ENGINE FOR OPERATION WITH NATURAL GAS

D. Ramasamy[1,2], M.M. Noor[1,3], K. Kadirgama[1], M.M. Rahman[1], W.G.B. Horizon[2]

[1]*Faculty of Mechanical Engineering, Universiti Malaysia Pahang, 26600 Pahang, Malaysia*
[2]*Focus Applied Technologies, 6305 Highland Pl, Black Forest CO 80908 USA.*
[3]*Department of Mechanical Engineering, University of Southern Queensland, Australia*
E-mail: deva@ump.edu.my / muhamad@ump.edu.my, contact: + 609 4246359

Abstract. Transportation is one of the key contributors to petroleum usage and emissions to the atmosphere. According to researchers, there are many ways to use transport by using renewable energy sources. Of these solutions, the immediate solution which requires less modification to current engine technology is by using gaseous fuels. Natural gas is the fuel of choice for minor modification to current engines. As it can be derived from anaerobic digestion process, the potential as a renewable energy source is tremendous, especially for an agricultural country such a Malaysia. The aim in the future will be operating an engine with natural gas only with pipelines straight to houses for easy filling. The fuel is light and can be easily carried in vehicles when in compressed form. As such, Compressed Natural Gas (CNG) is currently used in bi-fuel engines, but is mostly not optimized in term of their performance. The focus of the paper is to optimize a model of natural gas engine by one dimensional flow modeling for operation with natural gas. The model is analyzed for performance and emission characteristics produced by a gasoline engine and later compared with natural gas. The average performance drop is about 15% from its gasoline counterpart. The 4% benchmark indicates that the modification to ignition timing and compression ratio does improve engine performance using natural gas as fuel.

1 Introduction

Future energy needs is demanding and the use of petroleum as one of the main source of energy is leading to fuel depletion in many countries in the world [1-3]. Malaysia being an oil producing country has slowly reducing reserves as shown in Figure 1 [4, 5]. Consequently, petroleum price in Malaysia increased in price due to high demand in the world market. As vast amount of petroleum is used in the transportation sector, many alternatives have been introduced for the past few decades to reduce petroleum consumptions as well as reducing emissions. Of these alternatives only natural gas in compressed form and liquefied petroleum gas (LPG) appear to be more economic for worthy implementation [6, 7]. Since LPG is totally derived totally from petroleum, natural gas is a better choice of fuel as it can be obtained from non-petroleum sources [8]. Natural gas is derived from biogenic (transformation of organic matter) and abiogenic (from earth crust) processes. The gas is abundant and world production prediction is reaching to 155.5 trillion cubic feet (tcf) from 106.6 tcf from year 2007 to year 2035[4].

The important of natural gas to the engine combustion was applied to several applications [9]. In Malaysia the natural gas production, consumption and exports shows gradual production increase as compared to petroleum production. Figure 2 shows the current trend in natural gas usage in Malaysia [4]. Malaysia has the 8th largest reserves of natural gas with 83 tcf and producing 2.3 tcf as of year 2007. This ranking dropped in one year to 14th even though the reserve stood at 88 tcf [10]. More diversified methods are required for the gas production to be increased. The low demand of gas consumption compared to future production indicates the country is lacking gas fuelled technologies. However, more usage in vehicles is required to reduce the petroleum consumption.

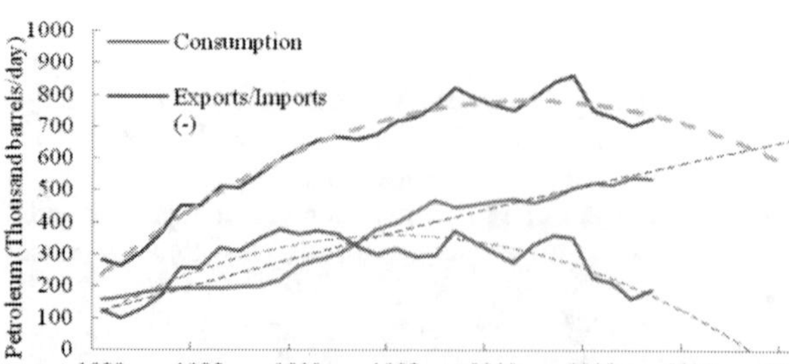

Figure 1: Petroleum Production in Malaysia (Ghosh and Prelas 2009)

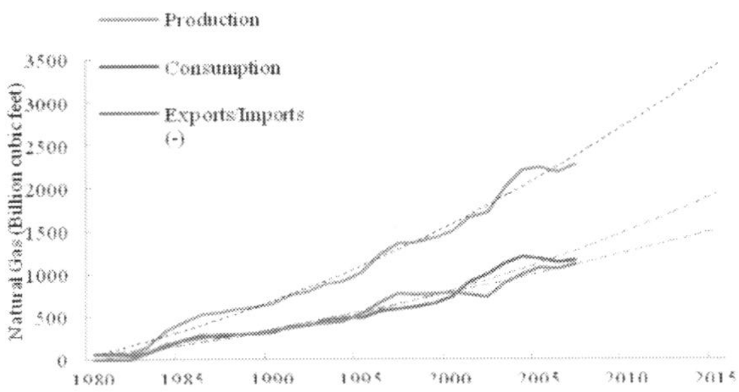

Figure 2: Malaysian Natural Gas Usage

To promote natural gas vehicles is a future target to reduce cost of operation of buses and cars. The country is an agricultural based country, planting crops such as oil palm, rubber, rice, mixed horticulture and coconut [11]. The waste products gathered from the plants is capable of producing natural gas from biogenic process. Biochemical methane potential (BMP), from crops and food using anaerobic digestion (AD) can reach up to 161-240 m3 CH4 /ton [12]. As Malaysia is producing 15000 tons of solid waste per day the amount of Methane generation will be around 3000000 m3. This source of methane is feasible to be used in vehicular usage.

Research for gas powered vehicles is required to reduce dependency to petroleum. This can be very beneficial to future energy cost in the country. This paper outlines engine testing with natural gas for a common car engine in Malaysia. A one dimensional approach shows how much power and emission is generated by natural gas usage. The result is compared with a gasoline counterpart simulation to get a benchmark of power loss due to fuel exchange. A minor optimization to the engine's compression ratio and ignition timing is done with natural gas so that the energy of the fuel is harnessed fully.

2 Engine Specification

The engine targeted for testing is a Perodua Myvi car engine named K3-VE, it has specification as in Table 1 below.

Table 1: Myvi Engine Specification

Engine Type	4-Stroke, EFI
Cylinders	4 inline
Bore x Stroke (mm)	72.0 x 79.7
Displacement (cc)	1298
Compression Ratio	10:1
Max. Output (kW/rpm)	63/600
Maximum torque (Nm/rpm)	120/3200
Valve timing	
Intake valve open	BTDC, 30° to (–12)°
Intake valve close	ABDC, 10° to 52°
Exhaust valve open	BBDC, 30°
Exhaust valve close	ATDC, 2°

Table 2 shows the specification of the fuel that was used in the simulation. As natural gas has higher octane rating it has the capability to be compressed and ignite earlier. In doing this the higher energy content of the fuel is targeted to be fully optimized. 1-Dimensional engine simulation software, was used. The simulation software provides the guidelines to modify the engine. The simulation results will provide a strong example of how the engine behaves after modification. With the simulation results, decisions of which compression ratio and ignition timing map to be use can be made easily. The general setup of the software is shown in Figure 3 below.

Table 2 Fuel Spécifications (Heywood 1988, S. Maji 2000, Ramasamy, Aik Soon et al. 2013)

Fuel	Gasoline	CNG
Phase (25°C)	Liquid	Gas
Stoichiometric A/F Ratio	14.7	14.5
Energy Content, MJ/Kg	44	50
Composition	Complex Hydrocarbon	92.73 % Methane
Octane Rating	80-90	120
Flame Speed in air at stoichiometric, mm/s	390	375
Autoignition Temperature	480~550 °C	537 °C

3 Methodology

As discussed earlier in, the capabilities of the natural gas to be ignited earlier and compressed is tested out in the 1 dimensional simulation. Two criteria were identified for the simulation which are the ignition timing advancing for more fuel energy harvesting and compression ratio modification for increasing the fuel compression. The ignition timing of the engine has to be advanced to compensate the slow burn rate of the compressed natural gas. The ignition timing will be varied from 60° before top dead center (BTDC) to 30° after top dead center (ATDC). Engine power, torque, and combustion data will be taken with varying the ignition timing at fixed engine speed. With the simulation data, the ignition timing can be optimized for maximum brake power and torque.

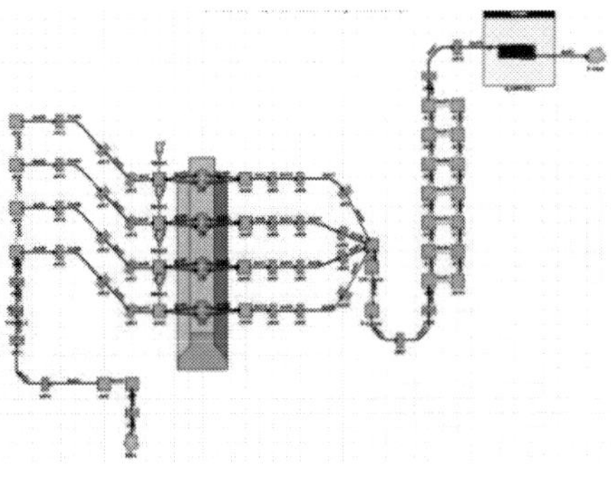

Figure 3: 1-Dimensional Simulation

The compression ratio was focus in this research. As compression ratio increase, the thermal efficiency of the engine will increase and this will increase the overall efficiency of the engine. This is shown in Figure 4. The octane number of compressed natural gas is 120. This allows the engine to run on higher compression ratio (Debnath, Sahoo et al. 2013). However higher compression ratio will induce knock to the engine. This can be simulated in the simulation software. The power curve and torque curve will be collected at different compression ratio (Korakianitis, Namasivayam et al. 2011). The compression ratio will be varied from 10 to 18. The knock plot will also be collected.

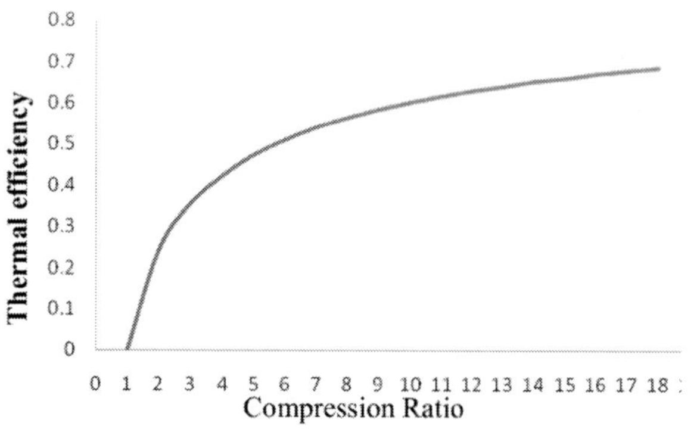

Figure 4: Thermal efficiency against different compression ratio

4 Results And Discussion

The result of the knock intensity at different compression ratio and ignition timing were presented. As compression ratio increases the knock intensity at low speed increases dramatically. This is due to the air fuel mixture has enough time to be influenced by the temperature of the combusted mixture. Similar result was found by [8, 18].Therefore at low speed the engine ignition timing needs to be advanced. Figure 5 to 8 shows the knock intensity at different engine speed and at different mass burn ratio of 50% crank angle.

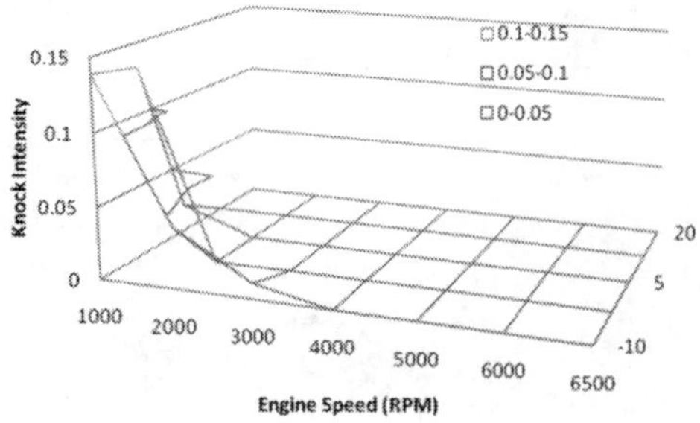

Figure 5: Knock intensity at compression ratio 12 at various different mass burn ratio of 50% crank angle

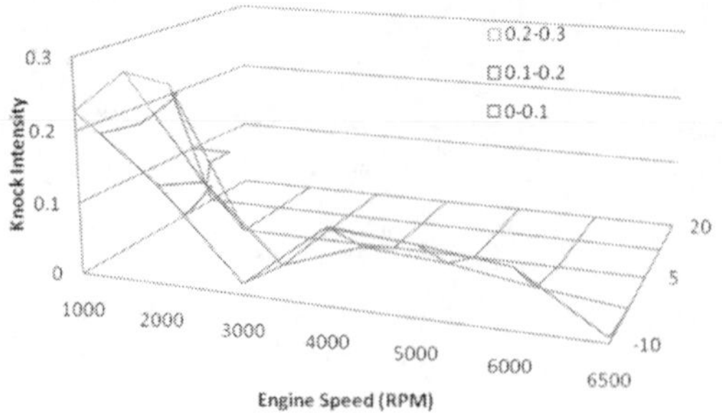

Figure 6: Knock intensity at compression ratio 14 at various different mass burn ratio of 50% crank angle

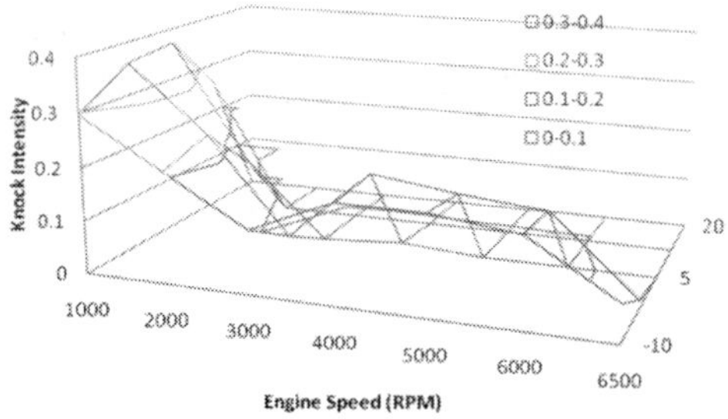

Figure 7: Knock intensity at compression ratio 16 at various different mass burn ratio of 50% crank angle

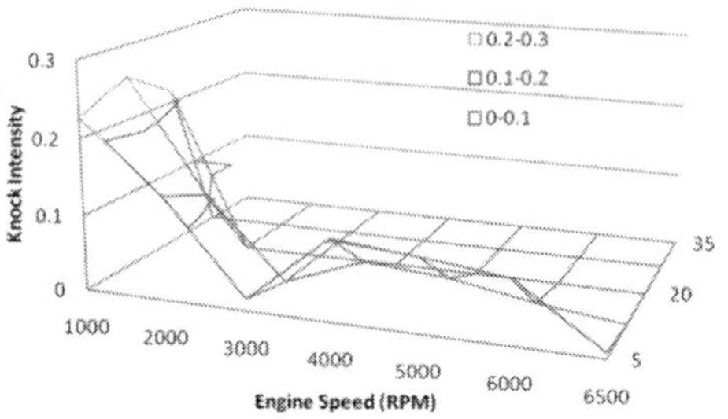

Figure 8: Knock intensity at compression ratio 18 at various different mass burn ratio of 50% crank angle

The power and torque curve at different compression ratio and ignition timing was discussed. Figure 9 to Figure 12 below shows the power and torque curve at different mass burn ratio of 50% crank angle. The effect of compression ratio to the power and torque curve diminishes as compression ratio increases (Caris and Nelson 1959, White, Steeper et al. 2006, Ramasamy, Zainal et al. 2014). Also, at low engine speed that is below 3000 rpm the ignition timing has to be advanced. However, above 3000 rpm the ignition timing needs to be retarded. The ignition timing is analyzed for 5 points from the baseline CNG. They are 10°CA BTDC, 2.5°CA BTDC, 5° CA ATDC, 12.5° CA ATDC and 20° CA ATDC. The negative denotes the piston position before TDC. Compression ratio 18 was ruled out due to the varying angle is too much, and it can compromise engine stability.

DOI: 10.1051/matecconf/20167400031

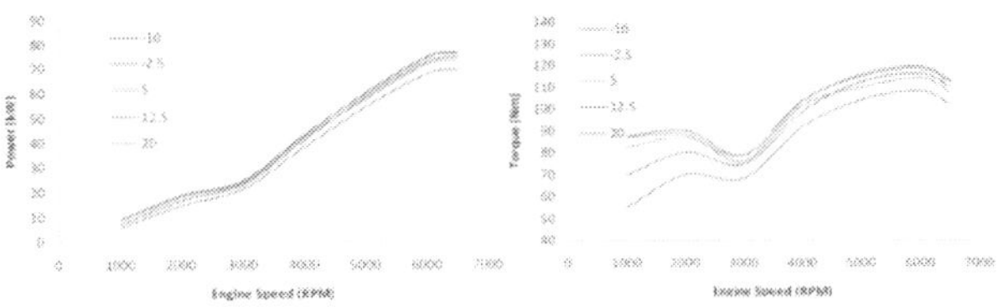

Figure 9: Power and torque curve at compression ratio 12 at various different mass burn ratio of 50% crank angle

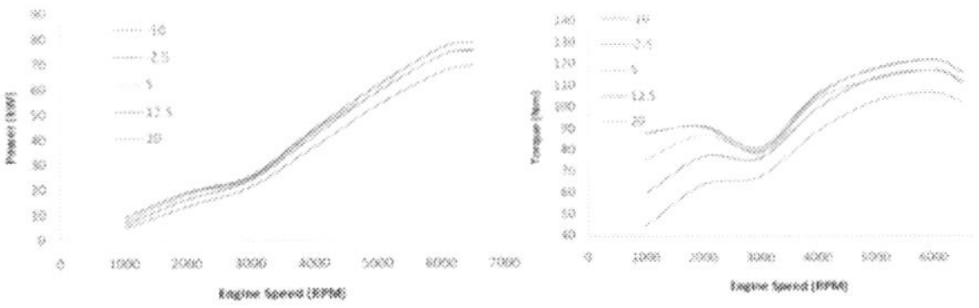

Figure 10: Power and torque curve at compression ratio 14 at various different mass burn ratio of 50% crank angle

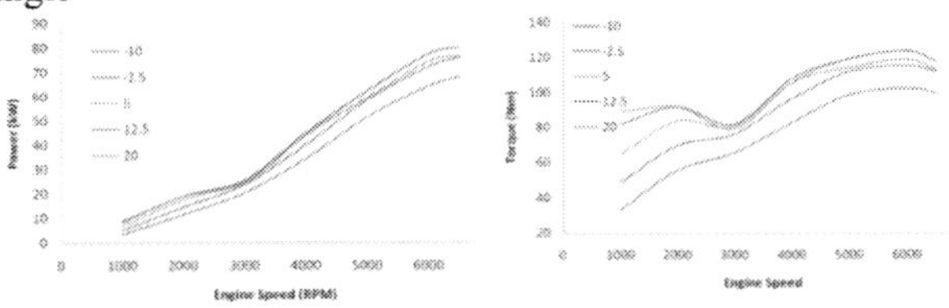

Figure 11: Power and torque curve at compression ratio 16 at various different mass burn ratio of 50% crank angle

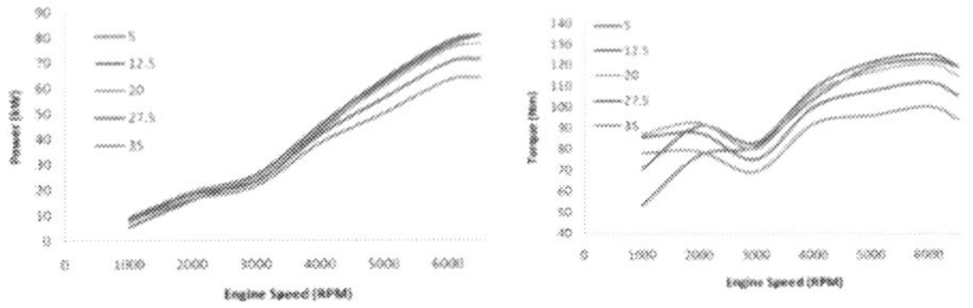

Figure 12: Power and torque curve at compression ratio 18 at various different mass burn ratio of 50% crank angle

Figure 13 shows the power and torque curve of gasoline and CNG without any modification. The average performance drop is about 19% from its gasoline counterpart (Anderson 2012, Vijayakumar, Ponnalagi et al. 2013, Ramasamy, Zainal et al. 2014). Figure 14 shows the power and torque curve of gasoline and CNG with compression ratio and ignition timing optimization achieved from the modification of compression ratio and ignition timing. The improvement can be compared by higher gradients in Figure 14 as compared to Figure 13. The optimized values shows improvement for the engine.

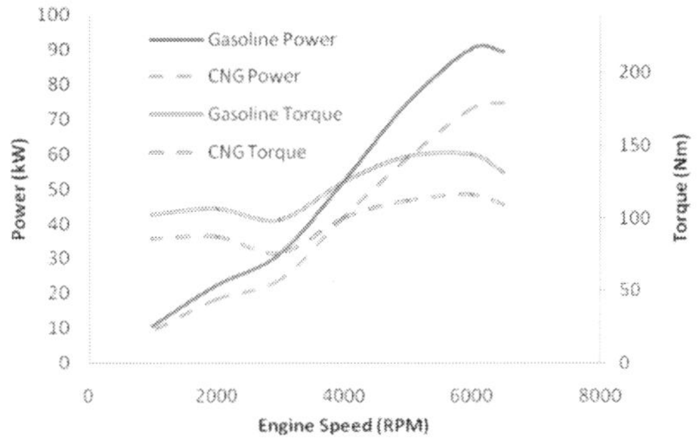

Figure 13: Power and torque curve of gasoline and CNG without modification.

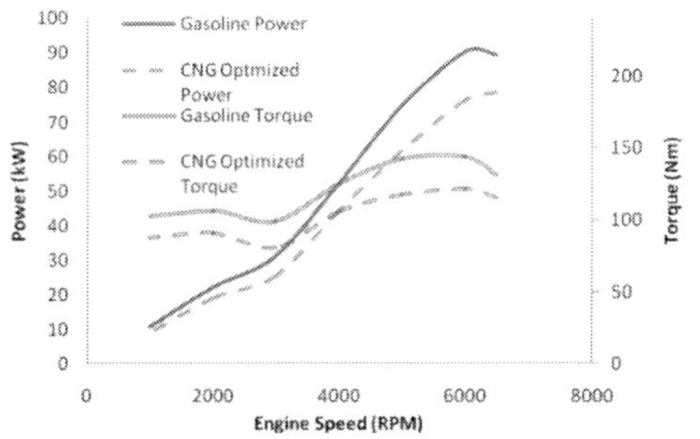

Figure14: Power and torque curve of gasoline and CNG with optimization

5 CONCLUSION

Performance drop is about 15% from its gasoline counterpart. The 4% benchmark indicates that the modification to ignition timing and compression ratio does improve engine performance using natural gas as fuel. However, the modification of the engine using compression ratio and ignition timing can be further studied. The only way to increase the power of engine to its original gasoline specification is to apply forced induction system to the engine. A recommendation is that a turbocharged system with slightly lower compression ratio and optimized ignition timing will give the best efficiency.

ACKNOWLEDGEMENTS

The authors would like to express their deep gratitude to Universiti Malaysia Pahang (UMP) for provided the laboratory facilities and financial support under MOHE FRGS project no. RDU140125 and Automotive Excellence Centre (AEC).

References

1. Shafiee, S. and E. Topal, *When will fossil fuel reserves be diminished?* Energy Policy, 2009. **37**(1): p. 181-189.
2. Noor, M.M., A.P. Wandel, and T. Yusaf, *Mild combustion: the future for lean and clean combustion.* International Review of Mechanical Engineering, 2013. **8**(1): p. 251-257.
3. Al-lwayzy, S. and T. Yusaf, *Chlorella protothecoides Microalgae as an Alternative Fuel for Tractor Diesel Engines.* Energies, 2013. **6**(2): p. 766-783.
4. Ghosh, T.K. and M.A. Prelas, *NATURAL GAS.* Vol. 1. 2009, Columbia: Springer Science + Business Media B.V. 281-381.
5. Supeni, E., et al., *Experimental and modeling investigation of the performance characteristics on diesel HCCI with hydrogen additive-a review.* National Conference in Mechanical Engineering Research, 2010. **2**(1): p. 580-596.
6. Lapidus A.I, K.I.F.T.B.P., *Natural Gas as Motor Fuel.* Chemistry and Technology of Fuels and Oils, 2005. **41**(3): p. 165-174.
7. Hosseini, S.E. and M.A. Wahid, *Biogas utilization: Experimental investigation on biogas flameless combustion in lab-scale furnace.* Energy Conversion and Management, 2013. **74**: p. 426-432.
8. Yusaf, T., et al., *Effect of Compress Natural Gas Mixing on The Engine Performance and Emissions.* International Journal of Automotive and Mechanical Engineering, 2013. **8**: p. 1438-1451.
9. Hairuddin, A.A., T. Yusaf, and A.P. Wandel, *A review of hydrogen and natural gas addition in diesel HCCI engines.* Renewable and Sustainable Energy Reviews, 2014. **32**: p. 739-761.
10. Malaysian Gas, A., *Malaysia : Natural Gas Industry Annual Review 2012.* 2013. p. 25-25.
11. Ahmad, F. *Sustainable agriculture system in Malaysia.* in *Regional Workshop on Integrated Plant Nutrition System (IPNS), Development in Rural Poverty Alleviation, United Nations Conference Complex, Bangkok, Thailand.* 2001.
12. Yu, Z. and F.L. Schanbacher, *Production of methane biogas as fuel through anaerobic digestion*, in *Sustainable Biotechnology.* 2010, Springer. p. 105-127.
13. S. Maji, R.R.P.B.S., *Comparison of Emmisions and Fuel Consumption from CNG and Gasoline.* SAE Technical Paper, 2000(2000-01-1432).
14. Heywood, J.B., *Internal Combustion Engine.* Vol. 21. 1988, New York: McGraw-Hill, Inc. 219-219.
15. Ramasamy, D., et al., *Variation Of Airflow Pattern Through Dissimilar Valve Lift In A Spark Ignition Engine.* Journal of the Chinese Institute of Engineers, 2013. **36**(8): p. 1083-1096.
16. Debnath, B.K., N. Sahoo, and U.K. Saha, *Thermodynamic analysis of a variable compression ratio diesel engine running with palm oil methyl ester.* Energy Conversion and Management, 2013. **65**: p. 147-154.
17. Korakianitis, T., A. Namasivayam, and R. Crookes, *Natural-gas fueled spark-ignition (SI) and compression-ignition (CI) engine performance and emissions.* Progress in Energy and Combustion Science, 2011. **37**(1): p. 89-112.
18. Yusaf, T., G. Najafi, and D.R. Buttsworth, *Theoretical and experimental investigation of SI engine performance and exhaust emissions using ethanol-gasoline blended fuels*, in *3rd International Conference on Energy and Environment (ICEE2009).* 2009, IEEE Power and Energy Society: Malacca, Malaysia.
19. Ramasamy, D., et al. *Mass fraction burn comparison of compressed natural gas and gasoline.* in *Applied Mechanics and Materials.* 2014.
20. White, C., R. Steeper, and A. Lutz, *The hydrogen-fueled internal combustion engine: a technical review.* International Journal of Hydrogen Energy, 2006. **31**(10): p. 1292-1305.
21. Caris, D. and E. Nelson, *A new look at high compression engines.* 1959, SAE Technical Paper.
22. Anderson, S.T., *The demand for ethanol as a gasoline substitute.* Journal of Environmental Economics and Management, 2012. **63**(2): p. 151-168.
23. Vijayakumar, S., et al., *Microwave assisted synthesis of Co 3 O 4 nanoparticles for high-performance supercapacitors.* Electrochimica Acta, 2013. **106**: p. 500-505.

Thermal Fatigue Of Die-Casting Dies: An Overview

Hassan A. Abdulhadi[1,2a], S.N. Aqida[2], M. Ishak[1], G. R. Mohammed[1, 2]

[1]Faculty of Mechanical Engineering, University Malaysia Pahang 26600 Pekan, Pahang, Malaysia
[2]Baghdad-Institute, Foundation of Technical Education, 10001 Baghdad, Iraq

Abstract. Coupled studies by experimental and numerical simulations are necessary for an increased understanding of the material behaviour as related to the interaction between the thermal and mechanical conditions. This paper focus on the mechanisms of thermal fatigue in the failure of dies and cores used in the die casting of aluminum alloys. The thermal fatigue resistance is expressed by two crack parameters which are the average maximum crack and the average cracked area. Samples of various types of H13 steel were compared with a standard H13 steel by testing under identical thermal fatigue cycles. To determine the thermal constraint developed in the sample during the test, a finite difference technique was used to obtain the temperature distribution, based on temperature measurements at the boundaries. The resulting stresses and strains were computed, and the strain calculated at the edge or weakest point of the sample was used to correlate the number of cycles to crack initiation. As the strain at the edge increased, the number of cycles to failure decreased. The influence of various factors on thermal fatigue behavior was studied including austenitizing temperature, surface condition, stress relieving, casting, vacuum melting, and resulfurization. The thermal fatigue resistance improved as the austenitizing temperature increased from 1750 to 2050°F.

***Keywords*:**; Aluminium alloys, air entrapment; die molten metal, materials

INTRODUCTION

Die soldering is the result of an interface reaction between molten aluminium and the die material during the impact of the high velocity molten aluminium onto the die surface and the intimate contact between alloy and die at high temperature. Once molten aluminum gets into the die with high motion, it destroys the protective film (lubricant and coating) on the die surface, it reaches the primal die surface. Then, the iron dissolves into the molten aluminum and a layer of inter metallic phases appears. A soldering layer, which is difficult to prevent, is formed over this intermediate layer at an atomic level. The adhesion of the cast metal to the die surface or core is recognized as soldering,which happens by different changing reactions along with the solidification phase, because of the dependence on die layer and the aluminum alloy. The chemical reactions take place because of the formation of intermetallic layers at the die substrate. Soldering is the resulting buildup of the aluminum alloy which takes place at the interface. During expulsion casting, soldering will create some adhesive problems which can even worsen to the extent of sticking wear, depending on the separation of the casting from the die [1, 2].

Washout of aluminum die-casting dies is basically caused by erosive and corrosive wear, and soldering. They are the main reasons of damage to the die surface and excessiv washout will eventually lead to the catastrophic they die. Corrosive wear can be best defined as the resolution of the die material in the molten aluminum and the stabilization of intermetallic substrate [3-5]. Corrosion occurs from high die temperatures which caused surface oxidation, at which point the alloying elements are soluble in liquid aluminium. Thus, the intermetallic material can form at the die surface. The main mechanism of erosionis because of liquid-encroachment erosion. Initial Si particles, oxide corpuscles, and impurities or intermetallic particles happen by filling solid erosion in result of the impact of solidified particles. When there is a high-speed flow of molten metal related to the die surface, erosion takes place. It gets further severe in the presence of hard particles like primary solid silicon in 390 alloys or SiC particles in metal matrix composites. Chemical corrosion is known as the process of material penetration and dissolution by the malt as well as the interface layer formation, when there is a trivial relative motion between the solid materials and the melt Gross cracking is normally devastating which might lead to a complete cracking of the die. Cracking and die fracture take place when the die material is stressed over its fracture strength as shown in Figure1. This can happen even when the mechanical stress applied to the die is less than the yield stress. A compound of residual, thermal and mechanical stresses create the crack [6-8]. This kind of failure is attributed to the inherent resistance of the die material against the brittle fracture termed fracture toughness.

DIE-CASTING DIES FAILURE MECHANISM

Roughly half of the aluminium alloys castings which are produced all through the world by the use of gravitational die casting or high pressure die casting (HPDC) are used in different automotive parts and consumer goods[9-13]. One of the major concerns in die casting is the durability of die materials/surface when they are exposed to pressurised casting process during filling, high temperature molten aluminium flow, and solidification and die holding stages[14, 15]. In die
casting process, the molten aluminum alloy is injected into die cavity at high speed of 30-100 m/s, at temperatures between 670-710 °C , and injection pressure of 50-80 MPa[3, 4, 8, 16, 17]. Several failure modes appear on aluminum die casting surface such as soldering effe.cts, washout gross-cracking (cleavage cracking) and thermal fatigue cracking which happen in effect of heat checking [18-20] .

[a]Email: h.shamary@gmail.com

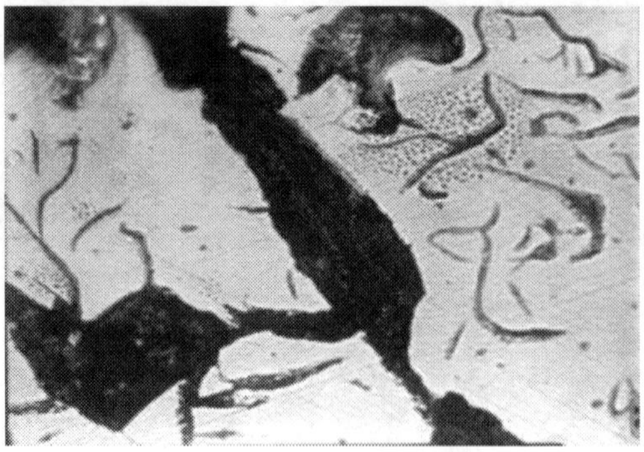

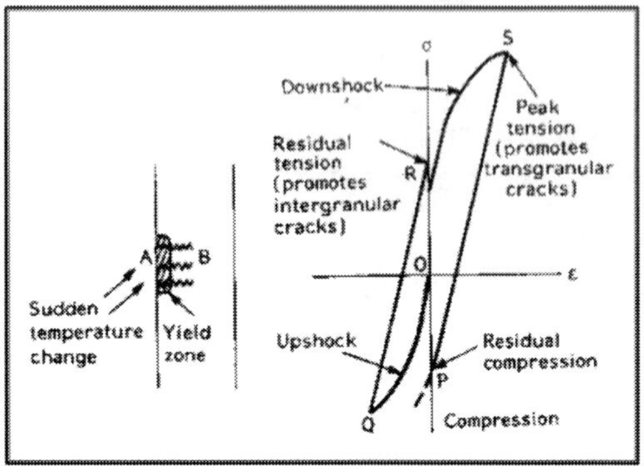

Figure 1. Hysteresis loop at the surface of a material subjected to cyclic heating cooling [18]

Moreover, die-casting dies are prone to high mechanical and thermal loads. Thermal fatigue cracking of dies which is caused by thermal cycling might considerably reduce the die lifetime. Cracks reduce the surface quality of dies and consequently the surface of castings will decline too. Previous studies have analysed thermal fatigue cracking of dies during the process of die casting aluminium alloys [7, 8]. During the process cracks are identified and their size and location are measured. Thermal and mechanical loads produce high local stresses which make the surface to crack. First cracks emerge right after 2000 cycles and spread progressively with cycles as illustrated in Figure1. [7, 18, 20, 21]

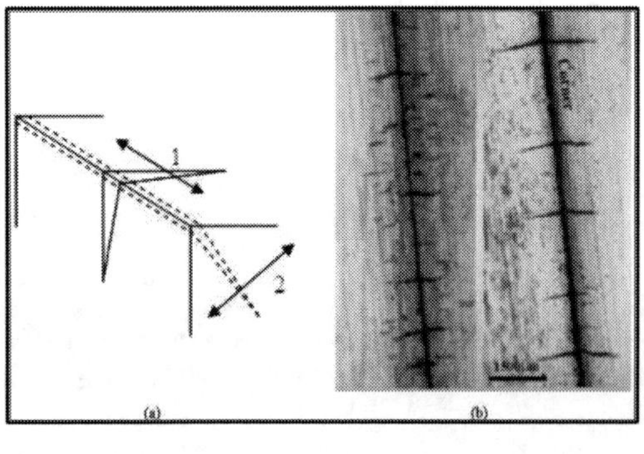

Figure 2. Direction of crack [18]

Moreover, die-casting dies are prone to high mechanical and thermal loads. Thermal fatigue cracking of dies which is caused by thermal cycling might considerably reduce the die lifetime. Cracks reduce the surface quality of dies and consequently the surface of castings will decline too. Previous studies have analysed thermal fatigue cracking of dies during the process of die casting aluminium alloys [7, 8, 22-24]. During the process cracks are identified and their size and location are measured. Thermal and mechanical loads produce high local stresses which make the surface to crack. First cracks emerge right after 2000 cycles and spread progressively with cycles as illustrated in Figure 2.[7, 18, 20, 21, 25].

THERMAL FATIGUE IN DIE-CASTING DIES

The thermal fatigue resistance of tool steels can be studied through differe tests. However, a majority of the studies involve the alternating immersion of samples in water and molten aluminium[26, 27] . This kind of experiment is like the actual high-pressure die-casting process, though it lacks repeatability. This is attributed to the sticking feature of the molten aluminum on the sample during testing which makes it the main weakness. Other researchers applied induction heating of samples surface followed by water cooling [28]. or gas cooling [26]. Researchers have also carried out tests by the applied conduction heating of samples, followed by water cooling of the tested surface [29]. Based on the findings of the previous studies, as shown in Table 1, the thermo-mechanical fatigue causes heat checks on the surface. The mechanical properties of the material grow instable in result of heating the die material. This problem and the surface treatment will be addressed in the following sections.

THERMAL STRESS IN DIES

The thermal stresses, which take place in the die, develop from the thermal gradient across the die area. The thermal gradient is made in he result of the heating and cooling of the surface during the ejecting, injecting, and lubricant spraying stages of the casting cycle. When the molten aluminum is injected into the die, the die surface heats up remarkably as opposed to the cooler underlying mass of the die. This sets up an initially steep thermal gradient. In the result, the surface enlarges more than the interior and because the interior is further massive, it does not allow the surface to expand. In consequence of this limitation, the surface undergoes compression. With an increase in the temperature, the yield strength of the material is lowered, and the compressive strains might grow plastic. The surface temperature decline quickly once a flow of heat is conducted to the lower layers. When the casting is ejected, cooling the surroundings from the surface, and the spray of die lubricants help to further decrease the surface temperature.[28, 30]. If the surface cools more swiftly than the interior, the compressive strains are released and tensile strains might be produced. Figure 3 shows schematic diagrams of the temperature, stress and strain distributions experienced by casting dies both on heating (during casting) and cooling (after the casting has been extracted from the die) [28, 31, 32]. Each half of the casting cycle is portrayed by three schematic drawings. Figure 3 (a) through 3 (c) show the effect of increased surface temperature on the stress and strain distribution from surface to interior of the die material. Figure 3(d) through 3 (f) illustrate the subsequent stress and strain conditions which exist during the second or cooling period of the casting cycle. It becomes clear that the thermal gradient is a function of the thermal conductivity of the die material. Increased thermal conductivity will result in a lower therm al gradient and thus low er stresses and strains.

Table 1. List of previous works conducted to prolong die life cycle.

Ref	Materials	Fatigue properties	Finding
[33]	AISI H13 tool steel	Thermal fatigue wear test	Improve thermal fatigue properties and reduce crack in semi-solid processing.
[14]	AISI H13 tool steel	The thermal fatigue cracks of aluminium alloy die casting is analyzed. Immersion test apparatus. A Finite Element Analysis (FEA) of immersion test was completed by ABAQUS software.	The results showed a good thermal fatigue resistance of the well hardened H13 hot work tool steel. The surface cracks growth increases by surface oxidation.
[1]	AISI H13 tool steel	Previously, the standard thermal fatigue immersion test was used to compare steels. The thermal stresses were kept equal for all the 2"x2"x7"samples.	Thermal fatigue samples have been modified to incorporate stress concentration factors typical to dies; their effect of thermal fatigue cracking is being assessed.The experimental data will be implemented to predict crack in die casting.
[34]	AISI H13 tool steel	The wear test result shows that O-CrN-H13 specimens possess the lowest coefficient of friction. In addition, SEM observation shows that only H13 specimens have plough cracks on the surface and other specimens show a very shallow wear and tear on the surface instead of Fe oxides wear test.	Based on the results, the wear resistance of AISI H13 can be effectively improved by different surface treatments. The lowest coefficient of friction was also noticed in the duplex surface treatment of CrN coated with oxynitriding. The coefficient of friction was found 0.28.
[1]	C40 and AISI H13 steels	The thermal fatigue test involved immersion of samples into molten aluminium, and quenched in ionised water emulsion at 17 °C.	High surface roughness of laser treated surface captured oxides and carbides which might begin the thermal fatigue failure. Cracks and heat checks took place on the carbides affected the area and soldered surface. Laser treatment parameter settings helped to enhance thermal fatigue properties during semisolid processing.

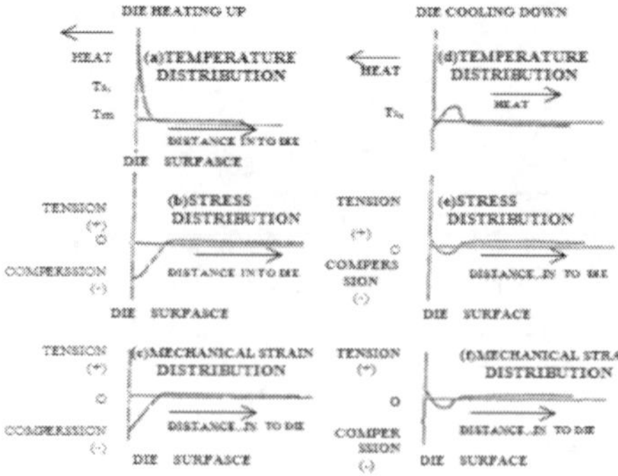

Figure3. Temperature, Stress and Strain Distributions when Dies are Exposured to Molten Metal [28]

During heating of the injection half of the die casting cycle, th e surface attempts to expand putting it in com pression with respect to the die interior, as shown in Figure 3(b). The corresponding compression strain is shown in Figure 3(c). The magnitude of the die surface stress is determined by the temperature gradient and the coefficient of thermal expansion of the die material. At first, the surface deform ation (strain) takes place within the die material elastic capabilities. However, if the integration of the temperature gradient and the widespread thermal expansion are high enough, the compression stress developed w ill exceed the elastic lim it of the die material as can seen in Figure 3(b), by the chang e in slope o f the stress plot n ear the die surface[28].

DIE MATERIALS ENHANCEMENT

AISI H13 tool steel is commonly used for die casting dies, due to its good resistance against temparature, thermal fatigue, tense hardness, corrosion and wear. Extension of the die casting die life, and often apply different thermal and chemical treatments on the surface of the die. A previous study found that samples oxynitriding could pose a negative film, which has anti-corrosion properties contributed to better through corrosion test. Moreover, improving the surface properties of alloys and metals without any change in the bulk properties, like oxidation , hardness, corrosion, and corrosion resistance, and can show types of metal ions with an adjustable flow of ions and energy Indicates that the surface treatments of CrN coating also deposited with the planting of metal and carbon ions lead to condensation and phase shift in the near-surface system to improve the corrosion resistance, abrasion resistance and fatigue strength significantly.Previous studies have suggested several methods to extend the service time of the casting dies such as nitriding [35], PVD/CVD coatings[36, 37], and laser surface engineering [10, 27, 38, 39]. The findings of studies carried out by Lin et al. [40], Koneshlou et al. [37] and Bronfin et al. [26] recommended the use of four different layers to change the properties of metal in the best way possible [19]

CONCLUSION

This study identifies and exemplifies the die casting process and the defects which appear during the casting. Detail is then given of ferrous casting alloys and non-ferrous casting alloys. Also, this paper focuses aluminium alloys, corrosion its corresponding mechanisms in the industry such as erosion, stress corrosion, and thermal fatigue has been expressed. At the same time, the protection of aluminium alloy corrosion (surface treatment/alteration, coating, etc.) is also presented. Because of the well-known characteristics of aluminium and its alloys (highly vulnerable to corrosion) and the extending challenges, the check of the working condition or surface treatment improved dies including cast as finer grain. New phases should be developed in a way to make the protection systems more practical, reliable, sustainable and cheaper and higher techniques need to be created to match with the practical applications. Furthermore, the prediction of defects represents the culmination of solidification modeling. It enables models to make practical contributions to real commercial processes, but it requires incorporating together and augmenting the models of almost every other aspect of casting simulation. Hot-tear, crack prediction requires accurate thermal and mechanical analysis, combined with criteria for embrittlement. With an advance in the computing power and software tools for computational mechanics, efficient analysis of fluid flow, strain, stress, temperature, deformation, and related phenomena in real casting processes has become highly possible. Computations are still affected and hampered by the limitations caused by mesh resolution, particularly for realistic three-dimensional geometries and defect analysis. Regarding the thermal fatigue damage, the most important factor to initiate cracks is the steel softening. Once the conditions embolden at the lower temperature on the surface, less thermal fatigue damage are observed and consequently the die steel hardness and strength is preserved more. It seems like one of the conditions for the extension of the thermal fatigue cracking damage is a decrease in strength the crack. In addition, the maximum temperature in die-casting applications happens in thin sections where the capacity of the material to absorb and transfer the heat from the surface is extremely different. From another view, high temperature - long resident time conditions are important, since the significant with the die casting of large components, when the die is exposed to raised temperature for the longer duration of time. Previous studies have reported an important decrease of cracking when the cooling condition and lubricant is good. Furthermore, the results of the present study show that the existence of a temperature threshold, under which the thermal fatigue defect is least. Finally, controlling parameters like velocity melting metal, the temperature of the surface and surface treatment lead to reducing thermal fatigue defects.

REFERENCES

1. Collot, J., *Review of New Process Technologies in the Aluminum Die-Casting Industry.* Materials and Manufacturing Processes, 2001. **16**(5): p. 595-617.
2. Avalle, M., et al., *Casting defects and fatigue strength of a die cast aluminium alloy: a comparison between standard specimens and production components.* International journal of Fatigue, 2002. **24**(1): p. 1-9.
3. Han, Q. and S. Viswanathan, *Analysis of the mechanism of die soldering in aluminum die casting.* Metallurgical and Materials ransactions A, 2003. **34**(1): p. 139-146.
4. Schwam, D., et al., *Effect of Design Factors on Thermal Fatigue Cracking of Die Casting Dies.* Work Performed Under Contract DE-FC07-00ID138486, US Department of Energy Assistant Secretary for Energy Efficiency and Renewable Energy, Washington, DC, October 2004, 2004.

5. Fash, J., et al., *Damage development during multiaxial fatigue of unnotched and notched specimens*. ASTM special technical publication, 1988(942): p. 874-898.
6. Jhavar, S., C.P. Paul, and N.K. Jain, *Causes of failure and repairing options for dies and molds: A review*. Engineering Failure Analysis, 2013. **34**: p. 519-535.
7. Sakhuja, A. and J.R. Brevick. *Prediction of thermal fatigue in tooling for die-casting copper via finite element analysis*. in *AIP Conference Proceedings*. 2004. IOP INSTITUTE OF PHYSICS PUBLISHING LTD.
8. Shi, Q., *Prediction of thermal distortion and thermal fatigue in shot sleeves*. 2002, The Ohio State University.
9. Obiekea, K.N., S.Y. Aku, and D.S. Yawas, *Effects of Pressure on the Mechanical Properties and Microstructure of Die Cast Aluminum A380 Alloy*. Journal of Minerals and Materials Characterization and Engineering, 2014. **02**(03): p. 248-258.
10. Otarawanna, S., *High Pressure Die Casting of Aluminium and Magnesium Alloys: Formation of Microstructure and Defects*. 2009.
11. Bai, Y. and H. Zhao, *Tensile properties and fracture behavior of partial squeeze added slow shot die-cast A356 aluminum alloy*. Materials & Design, 2010. **31**(9): p. 4237-4243.
12. ElTobgy, M.S., E. Ng, and M.A. Elbestawi, *Finite element modeling of erosive wear*. International Journal of Machine Tools and Manufacture, 2005. **45**(11): p. 1337-1346.
13. *<232 Materials_for_Automobiles14.pdf>*.
14. Klobčar, D., et al., *Thermo fatigue cracking of die casting dies*. Engineering Failure Analysis, 2012. **20**: p. 43-53.
15. Abed, D.E.J., *<42 The Influence of Different Casting Method on.pdf>*. International Journal of Engineering & Technology IJET-IJENS 2011. **Vol: 11 No: 06**: p. 8.
16. Lus, H.M., *Effect of casting parameters on the microstructure and mechanical properties of squeeze cast A380 aluminum die cast alloy*. ovove Materialy, 2012. **50**(4): p. 243-250.
17. Sadowski, T. and P. Golewski, *The influence of quantity and distribution of cooling channels of turbine elements on level of stresses in the protective layer TBC and the efficiency of cooling*. Computational Materials Science, 2012. **52**(1): p. 293-297.
18. David Schwam, P. and X. Zhu Sr, *Energy Saving Melting and Revert Reduction Technology: Improved Die Casting Process to Preserve the Life of the Inserts*. 2012, Advanced Technology Institute.
19. Fazarinc, M., et al., *Thermal fatigue properties of differently constructed functionally graded materials aimed for refurbishing of pressure-die-casting dies*. Engineering Failure Analysis, 2012. **25**: p. 238-249.
20. Koutiri, I., et al., *High cycle fatigue damage mechanisms in cast aluminium subject to complex loads*. International Journal of Fatigue, 2013. **47**: p. 44-57.
21. Hussein, A.M., et al., *A review of forced convection heat transfer enhancement and hydrodynamic characteristics of a nanofluid*. Renewable and Sustainable Energy Reviews, 2014. **29**: p. 734-743.
22. Kadiri, H.E., et al., *Identification and modeling of fatigue crack growth mechanisms in a die-cast AM50 magnesium alloy*. Acta Materialia, 2006. **54**(19): p. 5061-5076.
23. Kowanaga, S., T. Konaga, and M. Nakamura, *Initiation and Propagation of Heat Checking and Variation of Residual Stress in Aluminum Die Casting Dies*. J. Jpn. Die Cast. Assoc., 1989(90): p. 1-9.
24. Naimi, S. and S.M. Hosseini, *Tool Steels in Die-Casting Utilization and Increased Mold Life*. Advances in Mechanical Engineering, 2015. **7**(1): p. 286071-286071.
25. Heim, D., F. Holler, and C. Mitterer, *Hard coatings produced by PACVD applied to aluminium die casting*. Surface and coatings Technology, 1999. **116**: p. 530-536.
26. Bronfin, B., et al. *Metallurgical background to the development of creep resistant gravity casting magnesium alloys*. in *Magnesium Technology 2005, Proceedings of the Symposium Sponsored by the Magnesium Committee of the Light Metals Division (LMD) of TMS with the International Magnesium Association*. 2005. Minerals, Metals & Materials Society Warrendale.
27. Anderson, B., *Die Casting Engineering: a hydraulic, thermal and mechanical process*. 2005, Marcel Dekker NY.
28. Wallace, J., *Thermal Conditions in the Die*. Foundry, 1968. **96**(10): p. 176-179.
29. Totten, G.E., K. Funatani, and L. Xie, *Handbook of metallurgical process design*. 2004: CRC press.
30. Verran, G.O., R.P.K. Mendes, and M.A. Rossi, *Influence of injection parameters on defects formation in die casting Al12Si1,3Cu alloy: Experimental results and numeric simulation*. Journal of Materials Processing Technology, 2006. **179**(1-3): p. 190-195.
31. Doehler, H.H., *Die casting*. 1951: McGraw-Hill.
32. Abdelatif, A.O., J.S. Owen, and M.F.M. Hussein, *Modelling the prestress transfer in pre-tensioned concrete elements*. Finite Elements in Analysis and Design, 2015. **94**: p. 47-63.
33. Aqida, S.N., et al., *Thermal fatigue properties of laser treated steels*. International Journal of Material Forming, 2010. **3**(1): p. 797-800.
34. Chang, S.-H., K.-T. Huang, and Y.-H. Wang, *Effects of thermal erosion and wear resistance on AISI H13 tool steel by various surface treatments*. Materials Transactions, 2012. **53**(4): p. 745-751.
35. Beeley, P., *Foundry technology*. 2001: Butterworth-Heinemann.
36. Blair, M. and T.L. Stevens, *Steel castings handbook*. 1995: ASM International.
37. Koneshlou, M., K.M. Asl, and F. Khomamizadeh, *Effect of cryogenic treatment on microstructure, mechanical and wear behaviors of AISI H13 hot work tool steel*. Cryogenics, 2011. **51**(1): p. 55-61.

38. Parashar, B.N. and R. Mittal, *Elements of manufacturing processes*. 2002: PHI Learning Pvt. Ltd.
39. Fisher, K. and W. Kurz, *Fundamentals of solidification*. Trans Tech Publications, 1986.
40. Lin, J., et al., *Design methodology for optimized die coatings: The case for aluminum pressure die-casting: Invited paper B7-1-1, ICMCTF, presented Monday May 2nd, 2005, San Diego*. Surface and Coatings Technology, 2006. **201**(6): p. 2930-2941.

NATURAL FIBER REINFORCED COMPOSITES: A REVIEW ON POTENTIAL FOR CORRUGATED CORE OF SANDWICH STRUCTURES

A.F. Jusoh[1, a], M.R.M. Rejab[1], J.P Siregar[1], D. Bachtiar[1]

[1]*Faculty of Mechanical Engineering, Universiti Malaysia Pahang 26600 Pekan, Pahang, Malaysia*

Abstract. Natural fibers, characterized by sustainability, have gained a considerable attention in recent years, due to their advantages of environmental acceptability and commercial viability. In this paper, the characterization of natural fibers including the mechanical properties and alkalization of fibers is presented. Most recent study had gone through the mercerization process to improve the toughness of natural fibers; which is a well-known hydrophilic material. Traditional reinforcement method was commonly used to fabricate a natural fiber composite such as hand lay-up and mold press due to its convenience in terms of time and cost. Also, different kind of matrix material used in different kind of natural fibers gave high impact on the tensile and flexural test result. By selecting appropriate chemical treatment, matrix material and fabrication method, the tensile and flexural test gives different results and findings. As most researchers tend to use metals to create corrugated cores for sandwich structure, it is possible to develop this structure using natural fibers such as kenaf, wood dust, and other natural fibers.

1 Introduction

Natural fiber encircle all forms of fiber from grasses, seeds, woody plants, fruits, wild plants, agriculture crops, water plants, palms, leaves, animal feathers and its skins. By-products such as banana, pineapple, sugarcane, sugar palm, rice, cotton, coconut, kenaf, hemp, abaca, sisal, oil palm, jute and bamboo are very well known to produce composites [1]. Natural fibers can also be classified into certain fiber categories such as bast, leaf, seed, fruit, grasses, reeds and woods as in figure 1. Some of the advantages of natural fibers are very environmentally friendly, fully biodegradable, a non-toxic material and sources of income for rural and agricultural community.

There are disadvantages for natural fibers such as poor compatibility with hydrophobic polymer matrix. These drawbacks need several treatments to improve compatibility, such as alkali treatment and coupling agent [2]. Moreover, alkali treatment increased the mechanical properties of the composites itself, resulting improved surface wettability of the fiber against the matrix [3, 4].

On the other hand, sandwich structures with various types of cores are widely used in aerospace, high-speed train, civil engineering and so on. The structures have becoming quite familiar and famous in manufacturing nowadays due to its properties of lightweight and high stiffness [5]. Many researchers have focused on the dynamic mechanical behavior of sandwich structures with corrugated cores. A variety of such structure of different metallic cores including stochastic foams, honeycombs, prismatic topologies, and lattice topologies have been investigated experimentally and numerically [6-8]. Between all sandwich structures, desirable solutions are the corrugated cores. The characteristic of corrugated core is providing good structural performance with very restricted thickness in total and small thickness compared to overall measurement which affect negatively of their modeling [9]. Most researchers were using metallic cores, stochastic foams, honeycombs, lightweight ceramic and polymeric foams to study the characteristic and the effect of sandwich panel with corrugated cores [10-14].

[a] Corresponding author: fahmyjusoh@gmail.com

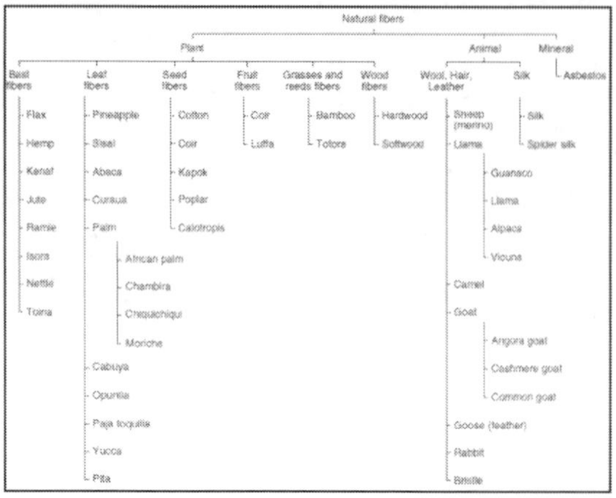

Figure 1. The classification of natural fiber [15].

2 Kenaf Fiber Composite

Kenaf was found to be comparative, much available and cheapest amongst other types of natural fiber reinforced material. Kenaf also labeled as industrial kenaf due to its great returns in manufacturing of industrial raw materials.

Kenaf fiber comes from species of Hibiscus cannabinus where its subdivision is Hibiscus and family Malvaceae obtained from plant's stems [16] which also includes cotton (Gossypium spp.) and okra (Abelmoschus esculentus L. Moench). The word kenaf itself originated from Persian which describing the plant having warm season, short day, and annually herbaceous plant, with the fiber diameter is 67.6 μm in average [17]. Kenaf is a tough, strong and hardy plant with a fibrous stalk, invulnerable to insect damages and needs relatively less of or no pesticides [18]. Kenaf fiber comprise of cellulose, hemicellulose, and lignin with the value of 56 – 64, 21 – 35, and 8 – 14 by weight percentage (Wt. %) [19, 20].

A long history recorded that kenaf fiber's cultivation was practiced in some areas in the world such as Malaysia, Thailand, Bangladesh, India, parts of Africa and Europe. The fiber has been mainly used in paper, cloth, coarse, twine and rope. However in today's market there is demand for kenaf reinforced for polymers. As encouraged by the local government of Malaysia, planting, cultivating and harvesting of kenaf plant has growth widely in rural area in order to replace the tobacco plant as well as to support local market in producing high-density fiberboard and then exported the products worldwide. Also, kenaf fibers received a lot of attention to support for multi-application including automotive parts, packaging and furniture as well as sports and leisures [21].

There are many types of research with regards in application of kenaf fiber as main material throughout the characterization and its properties. Mahjoub, R. et al. [17] recorded the tensile properties and elastic modulus of kenaf fibers are 930 MPa and 53 GPa, respectively. Compared to Mazuki, Davoodi and Summerscale et al. in Table 1, the tensile properties of kenaf is in range of 240 to 800 MPa and its elastic modulus range from 14 to 60 GPa [19, 20, 22]. The authors were using kenaf fiber with almost same density but the types of kenaf must be taken into consideration either short or long kenaf. A factor that affects the properties of kenaf fiber is the matrix material used in the process. Salleh et al. [16] using Polyester with 100 to 20 by weight as its matrix material and done the process by hand lay-up. Meanwhile, Meon et al. [21] used Polyethylene and Polypropylene with maximum 67% by weight of matrix but fabricated with hot press moulding, compared to Vijayakumar et al. [23] who used 90% by weight of Epoxy and then fabricated with hand lay-up process.

Table 1. Mechanical properties of kenaf fiber composite.

Density (g/cm^3)	Tensile Strength (MPa)	Elastic Modulus (GPa)	Elongation at break (%)	Ref.
1.44	393 – 773	26.5	1.5 – 1.8	[19]
1.4	284 – 800	21 – 60	1.6	[20]
1.193	240 – 600	14 – 38	-	[22]

3 BANANA, HEMP, WOOD DUST, AND JUTE FIBERS

Other selective natural fibers where some researchers used in determining its mechanical properties as in Table 2 are banana, hemp, wood dust, and jute. Gosh et al. [24] treated its banana fibers with 5% of NaOH and dried the banana fiber mat in 80°C for 24 hours, the same method used by Meon. The matrix material used for his experiment is Bisphenol Epoxy based and Vinyl Ester, fabricated using hand lay-up technique and the tensile properties is 122 MPa. Singh et al. [25] used hemp fibers in their study of tensile and flexural which gives a result of 17.39 MPa and 16.39 MPa for both tensile and flexural test respectively. The hemp fibers were mixed with 90% High Density Polyethylene (HDPE) and fabricated with injection molding machine after gone through mercerization process of 5% NaOH in 24 hours.

For wood dust fiber, Kumar et al. [26] managed to get the maximum of 28.29 MPa and 1335.20 MPa of tensile and flexural strength which through 15 % of Epoxy and Hardener using glass vacuum chamber as its fabrication method. But then, Kumar did not mention any mercerization process in the making of his composite material testing. The same goes to Mishra et al. [27] who study the mechanical properties of jute composed with Epoxy. However, Mishra recorded a 110 MPa and 55.8 MPa of tensile and flexural strength which is the results is higher than recorded by Singh for hemp fibers even though Mishra did not execute any mercerization process to his fibers.

Table 2. Other selective natural fibers reinforced composite.

Fiber	Matrix Material	Resin Comp.	Fabrication	Ref.
Kenaf	Polyethylene & Polypropylene	30, 67, 3 wt%	Hot press molding	[21]
Kenaf	Epoxy	90%	Hand lay-up	[23]
Banana	Bisphenol epoxy based Vinyl ester	-	Hand lay-up	[24]
Hemp	High Density Polyethylene HDPE	70%, 80%, 90%	Injection molding	[25]
Wood Dust	Epoxy + Hardener (10:8)	0, 2.5, 5, 7.5, 10, 12.5, 15 (wt%)	Vacuum glass chamber	[26]
Jute	Epoxy	100, 88, 76, 64, 52 (wt%)	Hand lay-up	[27]

4 Mercerization of Fibers

Another factor which affects the properties of kenaf is the mercerization process. Treating the fibers with Sodium Hydroxide (NaOH) also known as mercerization process helps in mending the interfacial bonding between the resin and fiber ensuing better mechanical properties. Several authors state that mechanical properties of fiber composites can be improved by chemically treating the fibers. It is well known that alkaline treatment is to clean the fiber surface off its impurities, modifies the surface structure and increase the fiber surface area. The moisture absorption by the fiber composites and its detrimental effect on the mechanical properties are a prime concern especially when these are used in marine application [24]. As an example, Meon and Vijayakumar did the mercerization process whereby the fibers were diluted and washed in NaOH solution in 4 to 24 hours and then dried in the fibers for the next couple days. As results, Meon recorded up to 15 MPa and 18 MPa for tensile and flexural strength but Vijayakumar managed to record higher which approximately 25 MPa and 35 MPa after both experimental gone through mercerization process.

However, there are disadvantages of natural fiber reinforced composites such as inadequate adhesion, poor wettability; high level of moisture absorption, and de-bonding but the alkali treatment of fibers significantly reduces these drawbacks. In order to increase elastic modulus of fiber by increasing molecular orientation, the alkali treatment needed to reduce the spiral angle by removing the lignin and hemicellulose content in the fiber. Also, to improve the mechanical properties of fibers, the improvement of interfacial bonding is crucial. The alkali treatment provides superior bonding, defects reduction, reduced moisture absorption, increasing crystallinity and finer fabric [28]. High moisture absorption and weak adhesion to hydrophobic matrices was resulted from the hydrophilicity of natural fibers. This set of drawback can be treated to improve and to increase the adhesion to matrix material [29].

Natural fibers also inherent high moisture absorption and it need to be removed by drying or conditioned before use. By removing or controlling moisture content, voids and poor fiber-matrix adhesion can be avoided. In addition, by applying more heat during production of composites could be enough to remove moisture. Rassman, et al. [30] found that by loading fibers into a preheated mold, placing the mold in the oven and drawing warm air through it can reduce moisture absorption but not all fibers could be dried in this method. Significant amount of moisture did not absorb when the fibers were dried in the oven and then sealed inside the mold for 2 min. Ramesh et al. [31] controlling the moisture content by drying the fibers under hot sun for 24 hours, compared to Srinivasababu et al. [32] who dried the fibers at ambient conditions until the moisture is removed by natural means. Boopathi et al. [33] dried the natural fibers with sunlight for only 3 to 5 hours and typically it depends on what fibers were used. However, the needs of mercerization and proper matrix material selected should impact the results of its mechanical properties. The results from each researcher are different as the purpose of study is differing.

5 Potential of Natural Fiber for Corrugated Cores

Corrugated core are a subset of two-dimensional periodic cores and most important feature is the corrugation shape, that can be varied for a fabricated design of the components [9]. A lot of theoretical, numerical, and experimental studies have been conducted to find its failure mode, properties of the design and the behavioral of corrugated core sandwich structure [6, 9-14, 34-37]. In Figure 2, Buannic et al. [35] reported that corrugated core sandwich structure can be categorized into four different conventional geometries; straight, hat-type, triangular and curvilinear. Then, comparison of the properties of these four geometries to a reference stiffened panel was made by the authors. The purpose of this comparison is to determine the pure bending characteristic of periodic plates and equivalent membrane.

Figure 2. Traditional corrugated core sandwich structure [35].

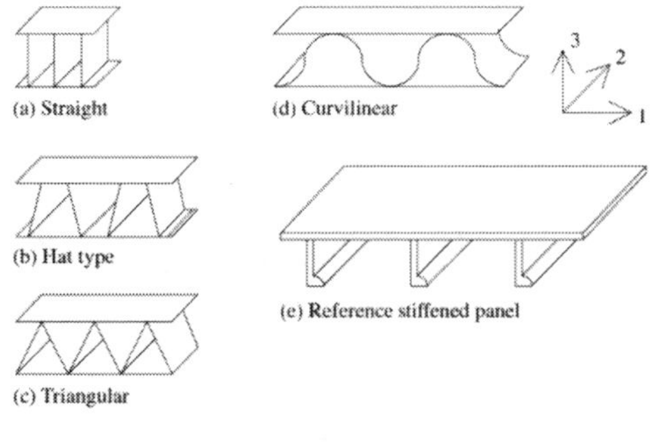

As in previous chapter, most researches uses sheet metal as such Bartolozzi et al. [9, 34] did reported the use of metal due to its main characteristic which is high stiffness-to-mass ratio, most likely under bending conditions. Zhang et al. [6, 10, 14] and Lim et al. [13] used stainless steel sheet to study the dynamic and buckling response of trapezoidal corrugated core sandwich structure panels. As compared to Yan et al. [12], the authors discussed more on bending points of sandwich beams using aluminum foam-filled and found out that foam insertions altered not only the failure mode of the corrugated sandwich but also increased dramatically its bending resistance. At this point, researchers tend to use metal in terms of sheets or foams and disregard the potential of natural fiber reinforced composite to replace the material use in experimental work even though that Wei et al. [11] was possible using Zr02 ceramic in studying of mechanical properties of corrugated core sandwich panel. Therefore, there are possibilities of using natural fiber reinforced composite to produce corrugated core sandwich structure and more studies need to be done to replace current material.

6 Conclusion

On top of all experimental disclosed in the literature, none authors fabricated a corrugated core sandwich structure using natural fibers. As most authors tend to use metals, it is possible to develop this structure using natural fibers such as kenaf, wood dust, and other natural fibers. Researchers may run several testing to study the characterization of corrugated core sandwich structure using only natural fiber reinforced composite. This is the gap in research field that require more experimental work to be done using natural fiber reinforced composite to fabricate corrugated core sandwich structure.

Also, different kind of matrix material used in different kind of natural fibers did gave an impact on the tensile and flexural test result. By selecting appropriate chemical treatment, matrix material and fabrication method, the tensile and flexural test gives different results and findings.

Acknowledgement

The authors are grateful to the Ministry of Education for funding this research (FRGS/1/2014/TK01/UMP/02/3).

References

1. Salit, M.S., *Tropical Natural Fiber Composite; Properties, Manufacture and Applications.* 2014. **1**.
2. Goulart, S.A.S., et al., *Mechanical Behaviour of Polypropylene Reinforced Palm Fibers Composites.* Procedia Engineering, 2011. **10**: p. 2034-2039.
3. Fiore, V., G. Di Bella, and A. Valenza, *The effect of alkaline treatment on mechanical properties of kenaf fibers and their epoxy composites.* Composites Part B: Engineering, 2015. **68**: p. 14-21.
4. Nirmal, U., J. Hashim, and M.M.H. Megat Ahmad, *A review on tribological performance of natural fibre polymeric composites.* Tribology International, 2015. **83**: p. 77-104.
5. Wu, Z.-J., F.-M. Li, and Y.-Z. Wang, *Vibration band gap behaviors of sandwich panels with corrugated cores.* Computers & Structures, 2013. **129**: p. 30-39.
6. Zhang, P., et al., *Experimental and numerical investigations on laser-welded corrugated-core sandwich panels subjected to air blast loading.* Marine Structures, 2015. **40**: p. 225-246.
7. Rejab, M. and W. Cantwell, *The mechanical behaviour of corrugated-core sandwich panels.* Composites Part B: Engineering, 2013. **47**: p. 267-277.
8. Rejab, M., K. Ushijima, and W. Cantwell, *The shear response of lightweight corrugated core structures.* Journal of Composite Materials, 2013: p. 0021998313514086.
9. Bartolozzi, G., N. Baldanzini, and M. Pierini, *Equivalent properties for corrugated cores of sandwich structures: A general analytical method.* Composite Structures, 2014. **108**: p. 736-746.
10. Zhang, J., et al., *Improving the bending strength and energy absorption of corrugated sandwich composite structure.* Materials & Design, 2013. **52**: p. 767-773.
11. Wei, K., et al., *Fabrication and mechanical properties of lightweight ZrO2 ceramic corrugated core sandwich panels.* Materials & Design, 2014. **64**: p. 91-95.
12. Yan, L.L., et al., *Three-point bending of sandwich beams with aluminum foam-filled corrugated cores.* Materials & Design, 2014. **60**: p. 510-519.
13. Lim, J.-Y. and H. Bart-Smith, *Theoretical approach on the dynamic global buckling response of metallic corrugated core sandwich columns.* International Journal of Non-Linear Mechanics, 2014. **65**: p. 14-31.
14. Zhang, P., et al., *Dynamic response of metallic trapezoidal corrugated-core sandwich panels subjected to air blast loading – An experimental study.* Materials & Design, 2015. **65**: p. 221-230.
15. L. Nicolais, A.B., *Wiley Encyclopedia of Composites - Natural Fibres.* 2012. **5 Vol Set - 2nd Ed.**
16. Salleh, Z., et al., *Fracture Toughness Investigation on Long Kenaf/Woven Glass Hybrid Composite Due To Water Absorption Effect.* Procedia Engineering, 2012. **41**: p. 1667-1673.
17. Mahjoub, R., et al., *Tensile properties of kenaf fiber due to various conditions of chemical fiber surface modifications.* Construction and Building Materials, 2014. **55**: p. 103-113.

18. Elsaid, A., et al., *Mechanical properties of kenaf fiber reinforced concrete.* Construction and Building Materials, 2011. **25**(4): p. 1991-2001.
19. Mazuki, A.A.M., et al., *Degradation of dynamic mechanical properties of pultruded kenaf fiber reinforced composites after immersion in various solutions.* Composites Part B: Engineering, 2011. **42**(1): p. 71-76.
20. Davoodi, M.M., et al., *Mechanical properties of hybrid kenaf/glass reinforced epoxy composite for passenger car bumper beam.* Materials & Design, 2010. **31**(10): p. 4927-4932.
21. Meon, M.S., et al., *Improving Tensile Properties of Kenaf Fibers Treated with Sodium Hydroxide.* Procedia Engineering, 2012. **41**: p. 1587-1592.
22. Summerscales, J., et al., *A review of bast fibres and their composites. Part 1 – Fibres as reinforcements.* Composites Part A: Applied Science and Manufacturing, 2010. **41**(10): p. 1329-1335.
23. Vijayakumar, S., et al., *Mechanical and Microstructure Characterization of Coconut Spathe Fibers and Kenaf Bast Fibers Reinforced Epoxy Polymer Matrix Composites.* Procedia Materials Science, 2014. **5**: p. 2330-2337.
24. Ghosh, R., et al., *Water Absorption Kinetics and Mechanical Properties of Ultrasonic Treated Banana Fiber Reinforced-vinyl Ester Composites.* Procedia Materials Science, 2014. **5**: p. 311-315.
25. Singh, S., et al., *Tensile and Flexural Behavior of Hemp Fiber Reinforced Virgin-recycled HDPE Matrix Composites.* Procedia Materials Science, 2014. **6**: p. 1696-1702.
26. Kumar, R., et al., *Study of Mechanical Properties of Wood Dust Reinforced Epoxy Composite.* Procedia Materials Science, 2014. **6**: p. 551-556.
27. Mishra, V. and S. Biswas, *Physical and Mechanical Properties of Bi-directional Jute Fiber Epoxy Composites.* Procedia Engineering, 2013. **51**: p. 561-566.
28. Prasad, V., et al., *Finite Element Analysis of Jute and Banana Fibre Reinforced Hybrid Polymer Matrix Composite and Optimization of Design Parameters Using ANOVA Technique.* Procedia Engineering, 2014. **97**: p. 1116-1125.
29. Sgriccia, N., M.C. Hawley, and M. Misra, *Characterization of natural fiber surfaces and natural fiber composites.* Composites Part A: Applied Science and Manufacturing, 2008. **39**(10): p. 1632-1637.
30. Rassmann, S., R.G. Reid, and R. Paskaramoorthy, *Effects of processing conditions on the mechanical and water absorption properties of resin transfer moulded kenaf fibre reinforced polyester composite laminates.* Composites Part A: Applied Science and Manufacturing, 2010. **41**(11): p. 1612-1619.
31. Ramesh, M., et al., *Processing and Mechanical Property Evaluation of Banana Fiber Reinforced Polymer Composites.* Procedia Engineering, 2014. **97**: p. 563-572.
32. Srinivasababu, N., J.S. Kumar, and K.V.K. Reddy, *Manufacturing and Characterization of Long Palmyra Palm/Borassus Flabellifer Petiole Fibre Reinforced Polyester Composites.* Procedia Technology, 2014. **14**: p. 252-259.
33. Bhoopathi, R., M. Ramesh, and C. Deepa, *Fabrication and Property Evaluation of Banana-Hemp-Glass Fiber Reinforced Composites.* Procedia Engineering, 2014. **97**: p. 2032-2041.
34. Bartolozzi, G., et al., *An equivalent material formulation for sinusoidal corrugated cores of structural sandwich panels.* Composite Structures, 2013. **100**: p. 173-185.
35. Buannic, N., P. Cartraud, and T. Quesnel, *Homogenization of corrugated core sandwich panels.* Composite Structures, 2003. **59**(3): p. 299-312.
36. Liu, P., Y. Liu, and X. Zhang, *Internal-structure-model based simulation research of shielding properties of honeycomb sandwich panel subjected to high-velocity impact.* International Journal of Impact Engineering, 2015. **77**: p. 120-133.
37. Magnucka-Blandzi, E., K. Magnucki, and L. Wittenbeck, *Mathematical modeling of shearing effect for sandwich beams with sinusoidal corrugated cores.* Applied Mathematical Modelling, 2015. **39**(9): p. 2796-2808.

Effect of Flux onto Intermetallic Compound Formation and Growth

Siti Rabiatull Aisha Idris[1a], Noor Farihan[1] and Hardinnawirda Kahar[1]

[1]*Faculty of Mechanical Engineering, Universiti Malaysia Pahang, 26600 Pekan, Pahang, Malaysia*

Abstract. In this study, the effect of different composition of no-clean flux onto intermetallic compound (IMC) formation and growth was investigated. The solder joint between Sn-3Ag-0.5Cu solder alloy and printed circuit board (PCB) was made through reflow soldering. They were further aged at 125oC and 150oC for up to 1000 hours. Results showed that fluxes significantly affect the IMC thickness and growth. In addition, during aging, the scallop and columnar morphology of IMC changed to a more planar type for both type of flux during isothermal aging. It was observed that the growth behavior of IMC was closely related to initial soldering condition.

1 Introduction

Nowadays, there are huge demand in multifunction of electronics products. To produce such products, it requires a larger number of input and output data which can be obtained through a good solder joint connection [1-3]. Nevertheless, the solder joint need to be controlled due to the existence of IMC, whereby excessive growth or unexpected morphological change of IMC may be detrimental to the solder joint performance which finally leads to failure [4, 5].

In general, reflow soldering process requires certain parameters to produce solder joint which includes flux. It is used to clean and activate the subtrate surface by forming a continuous film over the surface that inhibits access oxygen to the substrate. Thus, it can be wetted by the molten solder. In addition, flux also used to provide heat transfer and wetting enhancement among the solder material during solder making process. However, the solid solder or flux and molten solder/flux interfacial tensions differ with respect to flux type, composition and also temperature. This condition will also affect the type of IMC formed after reflow soldering since it encounters reaction between the solder and substrate [6, 7]. Most of the researchers were investigating on the wetting properties of solder alloy with respect to flux type or materials, but none of them were looking at the effect of flux type towards IMC formation and growth after reflow soldering and isothermal aging. The researchers were unaware of the importance of the IMC type produced by different type of flux. Different type of IMC leads to different solder joint strength.

Therefore, in this study the interfacial reaction between Sn-3.0Ag-0.5Cu solder alloy paste with different flux composition, and copper substrate were thoroughly investigated during soldering and isothermal aging with a focus on the IMC formation and growth.

[a] Corresponding author: author@ rabiatull@ump.edu.my

2 Experimental Set Up

The substrate used was FR4 (epoxy glass) material sandwiched between two electroplated copper layers. The substrate has the dimensions: width x length x thickness of 25 x 40 x 1 mm. Firstly copper substrates were polished and grind in order to remove oxide layer that formed onto it. Then, Sn-3.0Ag-0.5Cu in a paste condition was reflowed on copper substrate for 25 minutes at the temperature of 250°C. This was followed by cooling at room temperature. Since the composition of the flux remains confidential and disclosed by Electronic Packaging Research Society-Universiti Kebangsaan Malaysia (EPRS-UKM), thus it was labelled as Flux A and Flux B for this study. However, it is well known that both flux were epoxy based. The samples were then aged at 125°C and 150°C for up to 1000 hours. Cross sections from the samples were then prepared using standard metallographic steps and examined using optical microscopy as well as Scanning Electron Microscopy (SEM). The average thickness of the intermetallics formed is identified using Energy Dispersive Xray (EDX) and measured using imageJ software across the intermetallic layer on cross sections of the solder joints.

3 Results and Discussion

The results of the present work indicated that the interfacial characteristics of composite solder joints are affected significantly by the initial soldering condition since Flux A and Flux B produces different wetting characteristics onto copper substrate. This can be seen through Figure 1 whereby solder alloy with Flux B produced much larger grain and denser Cu_6Sn_5 IMC phase as compared to solder alloy with Flux A. The detail explanation of the wetting characteristics was explained in elsewhere [6, 7]. Nevertheless, this phenomena can be summarized as follow; whenever the flux was exposed to reflow temperature, some of Flux B begins to thermally decomposed due to its different composition, and thus activating the diffusion reaction between solder alloy and copper substrate. This will finally leads to producing IMC with larger and denser grain.

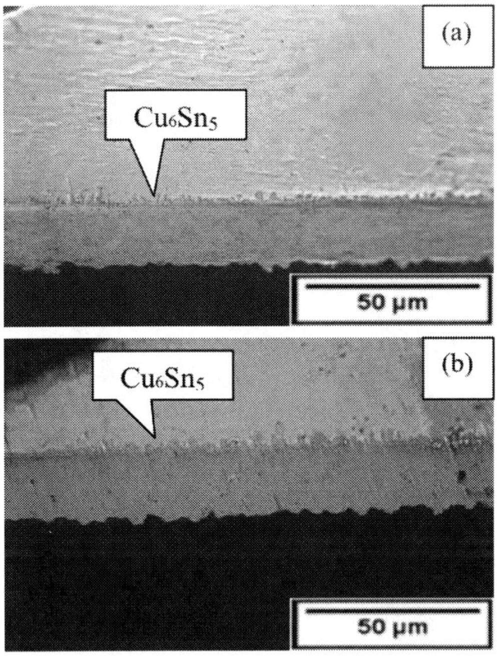

Figure 1. Interfacial reaction between Sn-3.0Ag-0.5Cu solder alloy and copper substrate after reflow soldering using; (a Flux A, and (b) Flux B.

In fact, after aging, the second layer of IMC produced by Flux B is also thicker as compared to Flux A for both aging temperature (Table 1). This explained faster interfacial reaction or diffusion rate for Flux B. However, no Cu_3Sn was detected for Flux A after 250 hours aging duration. It has happened because the temperature used is low enough to slow down the solid state diffusion. Thus, it took longer time to form Cu_3Sn. In addition, the IMC layer for both fluxes became thicker, compact and denser as aging duration increases.

Nevertheless, as the isothermal aging process started taking place, the IMC thickness for Flux B slightly lower than Flux A for both temperatures (Figure 2). It is believed that Flux A is suitable to be used for lower temperature of aging, whereas Flux B much more stable for both temperatures. It is suspected that its different composition or ingredient helps it maintain at that

Table 1. Cross sectional microstructure of the Sn-3.0Ag-0.5Cu using dissimilar flux for different aging temperature at 250 hours duration.

Flux	Aging temperature	Cross Section Image
A	125°C	
B	125°C	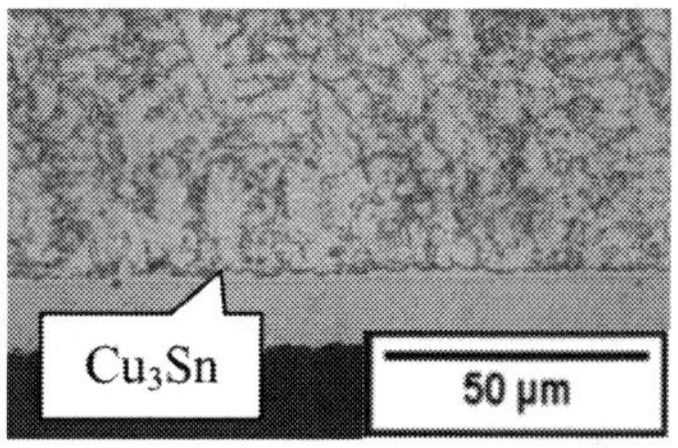
A	150°C	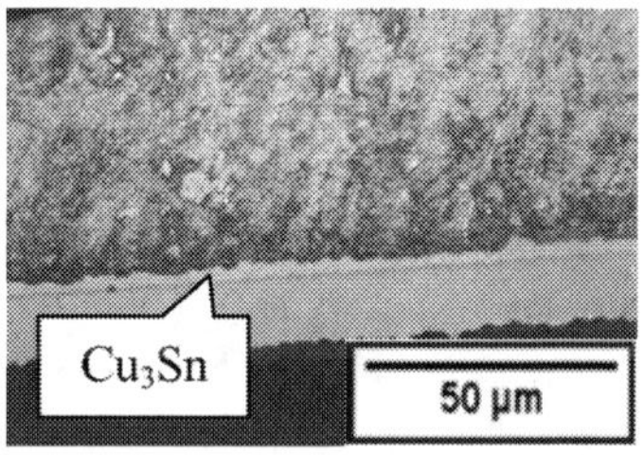
B	150°C	

temperature. Besides, it can also be seen that for isothermal aging temperature of 150°C, the total IMC for flux B became slightly lower than those IMC produced in Flux A at the duration of 500 and 1000 hours. This is happened because lack of Sn from the solder alloy that could diffuse towards the copper substrate due the existence of Cu_6Sn_5 between them. Therefore, the diffusion rate became slower. However, it was still increasing in thickness as the aging duration increased.

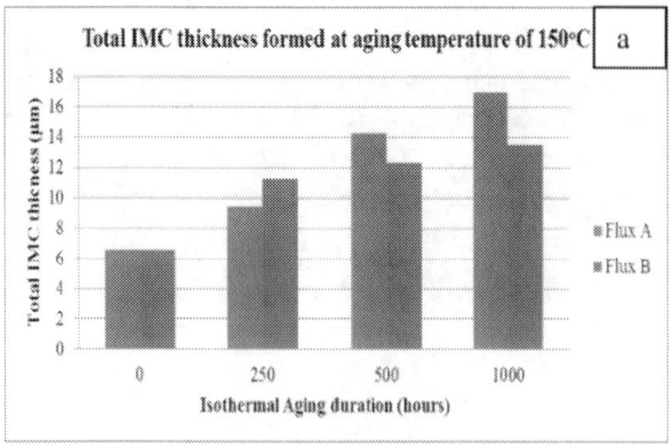

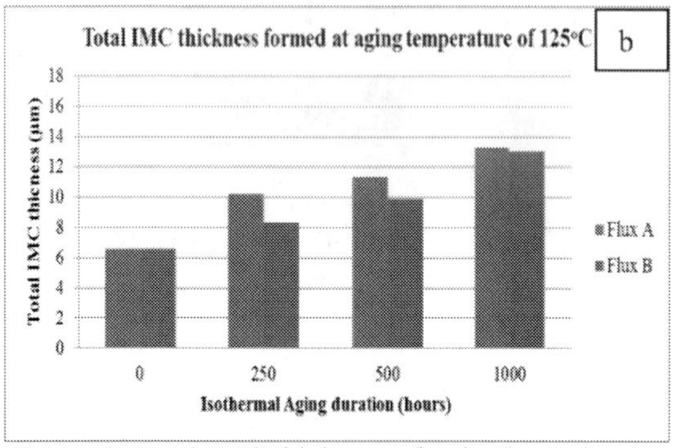

Figure 2. Total IMC thickness after isothermal aging at different temperatures; (a) 150°C, and (b) 125°C.

4 Conclusions

This paper presented the results of interfacial reaction between Sn-3.0Ag-0.5Cu solder alloy paste with different flux composition, and copper substrate with a focus on the IMC formation and growth. The results of this study show that different composition of flux significantly affect the IMC thickness and growth. The evidence from this study suggests that Flux B is the best candidate that could be used for reflow soldering and aging at high and low temperature because it is more stable. However, it is recommended that further research be taken to establish its optimum solder joint strength.

Acknowledgement

The authors would like to be obliged to Universiti Malaysia Pahang for providing laboratory facilities and financial assistance under project no. RDU130107.

References

1. W. Choi and H. Lee, *Effect of soldering and aging time on interfacial microstructure and growth of intermetallic compounds between Sn-3.5Ag solder alloy and Cu substrate.* Journal of Elec Materi, **29**(10): p. 1207-1213.(2000)
2. K.S. Kim, S.H. Huh, and K. Suganuma, *Effects of intermetallic compounds on properties of Sn–Ag–Cu lead-free soldered joints.* Journal of Alloys and Compounds, **352**(1–2): p. 226-236.(2003)
3. I.S.R. Aisha, A. Ourdjini, M.A. Hanim, and O.S. Azlina, *Effect of Reflow Profile on Intermetallic Compound Formation.* IOP Conference Series: Materials Science and Engineering, **46**(1): p. 012037.(2013)
4. H. Nishikawa and N. Iwata, *Formation and growth of intermetallic compound layers at the interface during laser soldering using Sn–Ag Cu solder on a Cu Pad.* Journal of Materials Processing Technology, **215**(0): p. 6-11.(2015)
5. L. Jiahui, M. Haitao, L. Shuang, S. Junhao, A. Kunwar, M. Wang, H. Jianjie, and B. Yanpeng. *The study of interficial reaction during rapidly solidified lead-free solder Sn3.5Ag0.7Cu/Cu laser soldering.* in *Electronic Packaging Technology (ICEPT), 2014 15th International Conference on*.(2014)
6. S.-P. Yu, C.-L. Liao, M.-H. Hon, and M.-C. Wang, *The effects of flux on the wetting characteristics of near-eutectic Sn-Zn-In solder on Cu substrate.* Journal of Materials Science, **35**(17): p. 4217-4224.(2000)
7. M. Arenas, M. He, and V. Acoff, *Effect of flux on the wetting characteristics of SnAg, SnCu, SnAgBi, and SnAgCu lead-free solders on copper substrates.* Journal of Elec Materi, **35**(7): p. 1530-1536.(2006)

Numerical Investigation on Effect of Immersed Blade Depth on the Performance of Undershot Water Turbines

Nor Fadilah Yah[1], Idris Mat Sahat[1] and Ahmed Nurye Oumer[1]

[1]Faculty of Mechanical Engineering, Universiti Malaysia Pahang, 26600 Pekan, Pahang, Malaysia.

Abstract. Energy, especially electricity, plays a vital role in global social and economic development. High annual rain rate in Malaysia seems a good potential for electricity generation especially through small hydro powers. Undershot water turbines are one of the hydropower turbines used for many years. However, the effect of blade depth immersed in the flowing water is not fully investigated. Therefore, the purpose of this paper is to study the effect of immersed blade depth for straight blade undershot water turbine in power generation by using Computational Fluid Dynamics (CFD) method. ANSYS CFX 15.0 was used to perform three dimensional analysis under steady state, incompressible, and non-isothermal conditions. The water wheel with number of blades of 6 and four different immersed depth was applied for each simulation. There are four different immersed depth was applied to each simulation, which are 20 mm, 40 mm, 60 mm and 80 mm. From the simulation result, it was found that the optimum immersed depth is 40 mm where the torque load and power generated were 0.264 N.m and 1.318 Watt respectively.

1 Introduction

Energy plays a vital role in global social and economic development. The demand of energy has increased as a result of the industrial, domestic and agricultural activities that have been grown rapidly in order to fulfill the user needs.

Hydropower is one of the renewable energy resources which is getting popular and raising attention of many researchers. There are three types of hydropower technologies for rural or off-grid application which are mini-hydro, micro-hydro and pico-hydro. These hydropower technologies differ based on their requirement of suitable flow rate and head [5].

Recently, Pico hydro become attractive because of its sustainability, cleanness and ability to generate power in low head area [6]. The water current turbine are very suitable for the rural electrification because of many advantages such as low head application, pollution free and economical [7]. The vertically mounted undershot wheel employs only the potential energy of the water as the driving force. The idea of utilizing potential energy instead of kinetic energy of the low head water in rivers was noticed by a French engineer, Poncelet [8].

Undershot water wheel could operate in very small head differences of 0.5m to 2.5m and large flow volume ranging from 0.5 to 0.95 m^3/s per m width. The water enters the wheel's blade. The wheel will rotate as a result of reaction between water and lowest paddles at the bottom of wheel. Undershot water wheel is suitable for flat country like Malaysia because it almost has no interest on the head and can be operates in shallow water. Besides that, it can also be installed on the floating platform and portable [9]. Compared to water turbine, undershot water wheel has simpler design and can be operated in river with very small head and speed. However, the effect of blade depth immersed in the flowing water is not fully investigated. Therefore, the purpose of this study is to study the effect of immersed blade depth for straight blade undershot water turbine in power generation by using CFD method. ANSYS CFX 15.0 was used to perform three dimensional analyses under steady state, incompressible, and non-isothermal conditions.

2 Numerical Model

2.1 Governing equations and Computational Domain

Numerical simulations were performed using finite volume software called ANSYS CFX in three dimensional steady flow analysis. Flow equations in ANSYS CFX consist of equation of continuity and the Navier-Stokes equation. In this section, the numerical models for the turbine blade and river together with the

boundary conditions imposed for different immersed blade depths simulations will be described.
The turbine blades considered for this study are straight type blades as shown in Figure 1 with overall dimensions 200 mm in diameter and width of 150 mm. Each paddle wheel has area of 70 mm (H) x 150 mm (W) with six number of paddles. The turbine is partially immersed in the flowing fluid (river) which is assumed to have constant flow velocity. The river flow field is considered to be incompressible, non-isothermal and behaves as Newtonian fluid (neglecting the effects of slurries inside the river). Therefore, the general forms of the mass and momentum equations can be written as:

$$\frac{\partial \rho}{\partial t} + \nabla \cdot (\rho \mathrm{u}) = 0 \quad (1)$$

$$\rho \frac{\partial}{\partial t}(\mathrm{u}) + \rho \nabla \cdot (\mathrm{uu}) = -\nabla p + \nabla \cdot \tau + \rho \mathrm{g} \quad (2)$$

Where u is the velocity vector, ρ is the density of water and p the hydrostatic pressure. The term τ represents the total shear stress due to viscosity and g is the gravity vector.

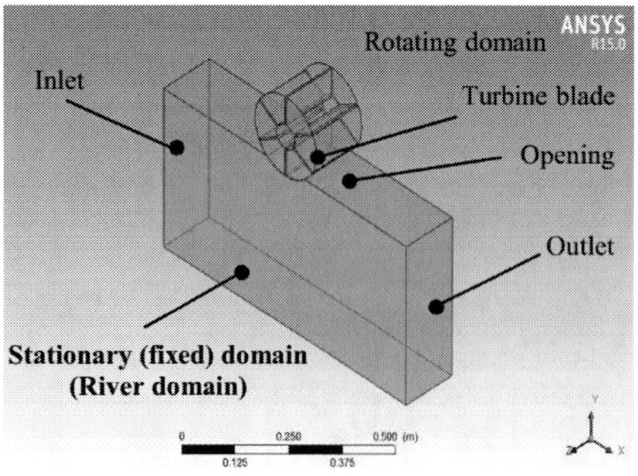

Figure 1: Schematic of turbine blade system considered in this study

2.2 Mesh Generation

The mesh generation was done by using Ansys-ICEM CFD. This software is not only effectively able to mesh large and complex models but also can provide advanced size function of the grid. Basically, computational mesh are classified into structured mesh and unstructured mesh [10]. Structured meshes are not preferred for complex geometries but it is more efficient in terms accuracy, CPU time and in most cases preferable. Unstructured mesh are preferable for complex geometries but the size difference between two adjacent elements can be large thus affect its numerical accuracy.

In this study, the simulation domain was divided in tow two parts; rotating and stationary parts. The stationary domain represents the domain for control surfaces of upstream and downstream region of the river. The rotating (turbine blade) domain meshed with advanced size function "proximity and curvature" and consist mainly of tetrahedral elements with 658948 numbers of elements. Table 1 summarizes the computational grids of both domains at various immersed blade depths. The stationary domain which represent the upstream and downstream regions were meshed with advanced size function turned off and consist mostly hexahedral elements with some tetrahedral, prismatic and wedges element type as shown in Figure 2.

Table 1: Computational grids for rotating and stationary domains at various immersed blade depths

Immersed depth (mm)	**Immersed radius ratio**	**Number of elements**	
		Rotating domain	Stationary domain
20	0.20	658948	6590
40	0.40	658948	6909
60	0.60	658948	6879
80	0.80	658948	7039

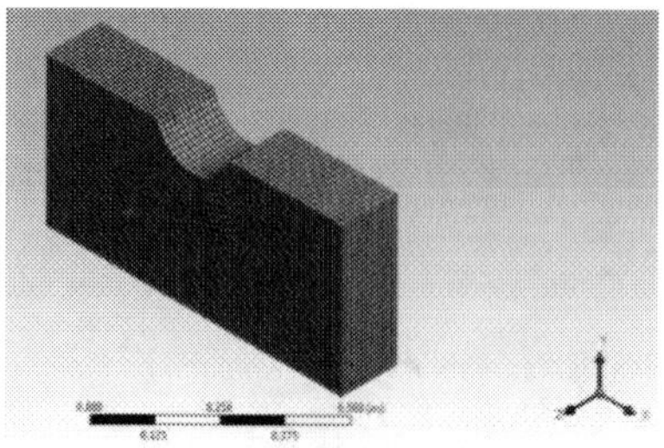

(a)

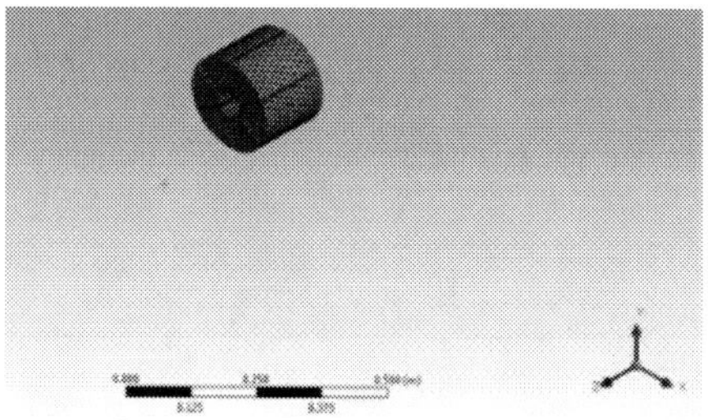

(b)

Figure 2: (a) Structured mesh for the stationary domain (b) Unstructured mesh for the rotating domain

2.3 Boundary Conditions

The simulation is for the straight blades turbine which consists of 6 blades and rotates at 5 rad/s. The inlet flow velocity of the river is considered to be 0.5 m/s (laminar flow). Control surfaces are established at a distance of twice the diameter of the turbine in the direction of upstream and downstream from the turbine center [1]. Besides that, the top surface of the computational domain was permeable to the atmosphere by applied with the entrainment with zero gradients. Meanwhile, no slip condition was applied to all other walls. The boundary of stationary and rotating domains were joined by using rotational periodicity method. The volume fraction of water (VFW) and volume fraction of air (VFA) were defined by using step functions as:

$$VFW = step\ ((y\text{-}h)/1\,[m]) \quad (3)$$

$$VFA = 1\text{-}VFW \quad (4)$$

Where, y is the coordinate of the height direction and h is the depth of the river water. Other conditions include, free outflow with a relative gauge pressure of 0 Pa for the outlet boundary and rotational speed was applied to the rotating domain.

3 Result and Discussion

3.1. Flow Patterns around Blade Surface

Figures 3 and 4 show the steady flow velocity vector and pressure contour for various immersed blade heights. The velocity vectors and pressure contours for each immersed depth was compared. From the simulation result, it can be clearly seen that the as the blade depth increases the number of blades in contact with the flowing water increases. At blade depth of 60 mm, surface area of the second blade in contact with the flowing fluid is higher that the recirculation contact area. This makes the back pressure to be higher the flow pressure for the second immersed blade. On the other hand, the effect of reverse pressure due to recirculation is negligible for 80 mm blade depth. The pressure difference between front and back surface of blade leads to the higher power generation. For blade depths 20 mm and 40 mm, since the velocity of blade region that immersed in the water are higher than the air region which have higher volume fraction of air, the effect of reverse pressure due to recirculation of air is negligible.

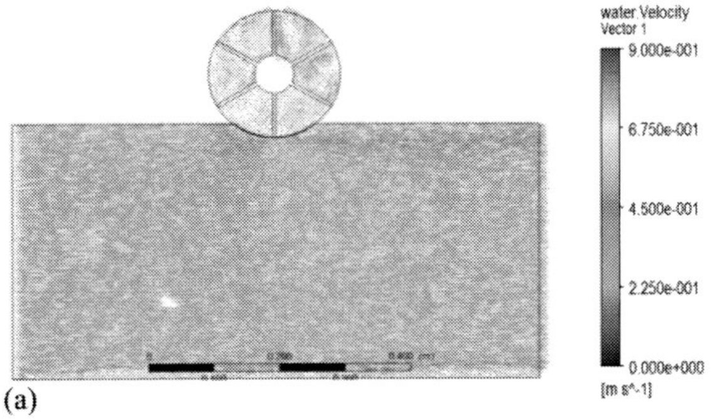

(a)

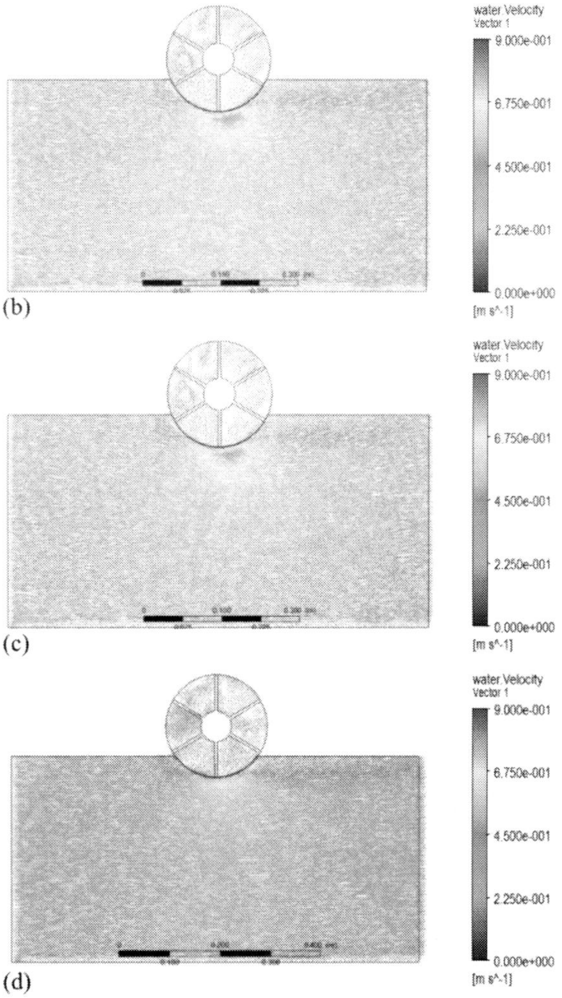

(b) (c) (d)

Figure 3: Velocity vector for different immersed blade depths (a) 80 mm (b) 60 mm (c) 40 mm (d) 20 mm

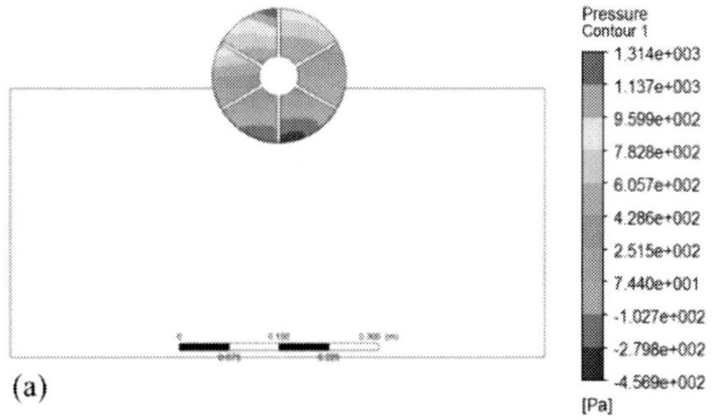

(a)

DOI: 10.1051/matecconf/20167400035

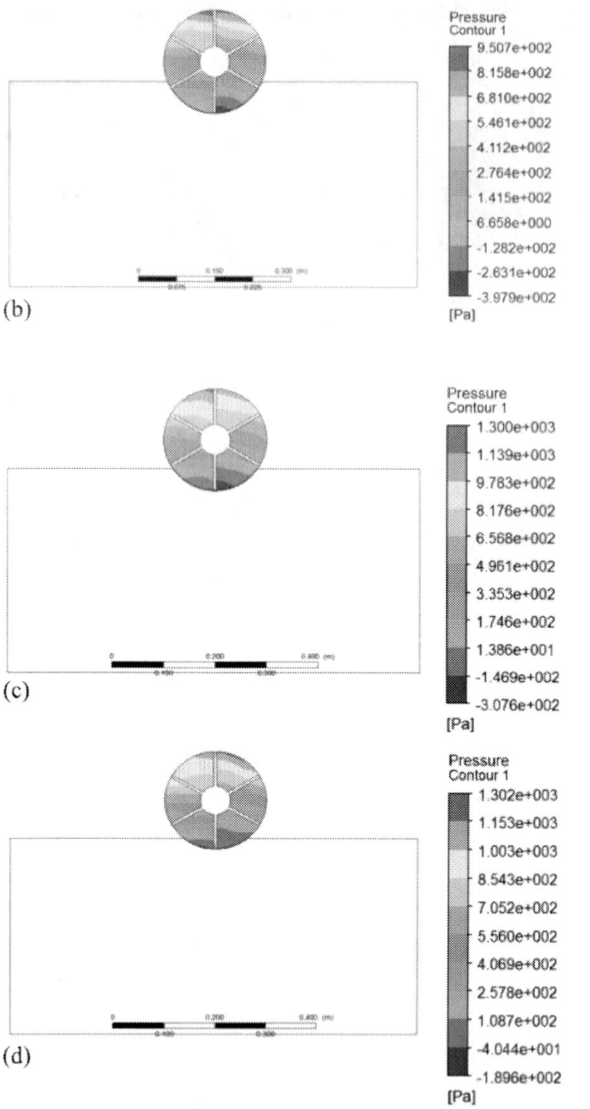

Figure 4: Pressure contours for different immersed blade depths (a) 80 mm (b) 60 mm (c) 40 mm (d) 20 mm

Figure 5 show the peak pressure, p acted on the blade immersed in the water. The value of peak pressure was increased when the immersed depth increased. As can be seen in Figures 6, power is varied with the immersed blade depth. It was found that the maximum power was generated when the immersed radius was 40 mm and the torque load at this depth was 0.264 N.m. When immersed depth increased, power also increased initially until optimum immersed depth which is 40 mm before decreasing. Besides that, the power coefficient for 40 mm immersed depth is the highest as shown in the Figure 7. This may happen due to the drag force generated when large area was immersed in the water. The obstacles due to the drag force generated are less in the 40 mm immersed radius.

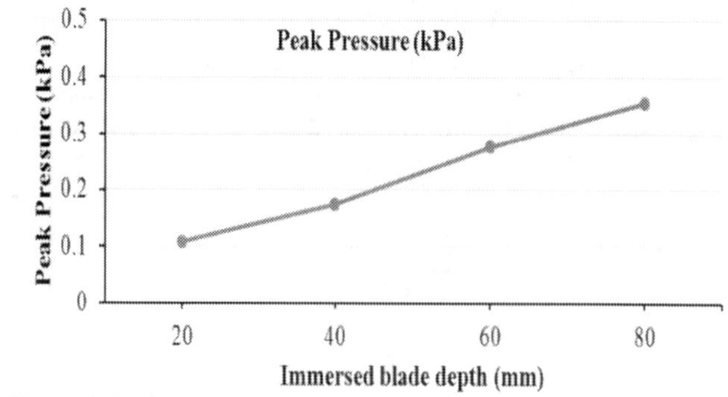

Figure 5: Peak pressure, p, for different immersed blade depths

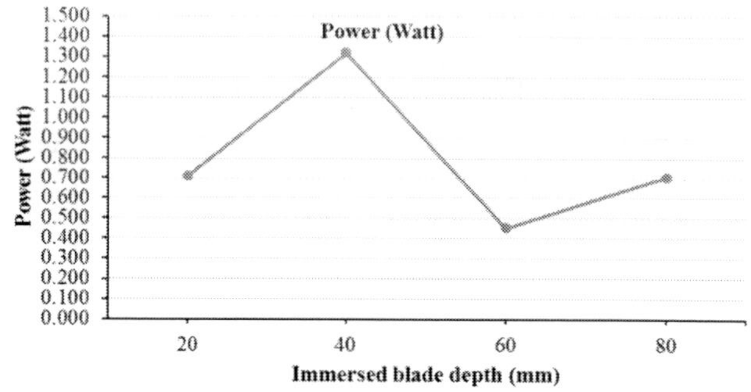

Figure 6: Variation of Power, P based on immersed depth

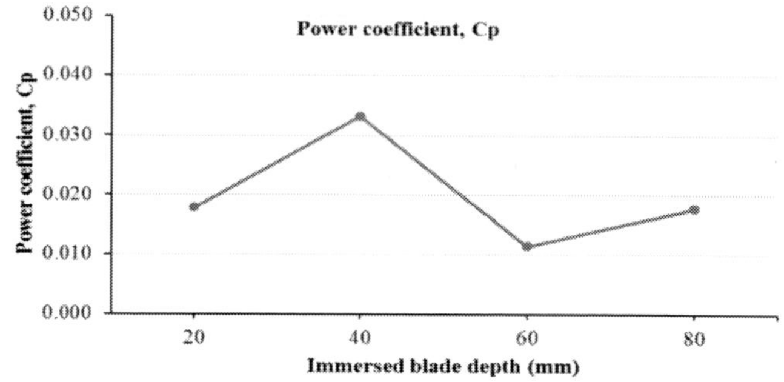

Figure 7: Power coefficient, Cp for different immersed blade depths

3 Conclusion

The performance of straight blade water turbine was numerically investigated. From the simulation, the performance of the undershot water turbine with different immersed depth was observed. The maximum power generation was generated when the immersed depth of the floating water turbine is 40 mm. The value of maximum power generated at this depth is 1.32 Watt. However, the generated power is very small to light up even a single house hold bulb. Therefore, a mechanism to increase the flow velocity of the river should be done in order to increase the power generation from the turbine.

Acknowledgement

The authors would like to thank Universiti Malaysia Pahang for providing necessary assistance and Ministry of Higher Education for providing financial support under FRGS project number RDU140127.

References

1. N.L. Panwar, S.C. Kaushik, and S. Kothari, *Role of renewable energy sources in environmental protection: A review*. Renewable and Sustainable Energy Reviews, 2011. **15**(3): p. 1513-1524.
2. K. Beckman, *Renewable Sources Provide Over 20% Of Global Power Production*. 2014; Available from: http://cleantechnica.com/2014/06/05/renewable-sources-global-power-production/.
3. I. Yüksel, *Hydropower for sustainable water and energy development*. Renewable and Sustainable Energy Reviews, 2010. **14**(1): p. 462-469.
4. A.T. Din, M.Y. Othman, and M.Y. Yaakob, *Harvesting Energy from Flood Mitigation Ponds using Water Wheel*. 2011: Solo Java.
5. L. Barelli, et al., *Mini-hydro: A design approach in case of torrential rivers*. Energy, 2013. **58**: p. 695-706.
6. L. Sule, I.N.G.W., R. Soenoko, S. Wahyudi, *Performance of a Straight Bladed Water Current Turbine*. Advances in Natural and Applied Sciences, 2013.
7. P. Maher, N.P.A. Smith, and A.A. Williams, *Assessment of pico hydro as an option for off-grid electrification in Kenya*. Renewable Energy, 2003. **28**(9): p. 1357-1369.
8. G. Muller, *Water Wheels as a Power Source*. The Queen's University of Belfast p. 1-9.
9. S.R. Turnock, G.M., R. F. Nicholls-Lee, S. Denchfield, S. Hindley, and a.S.S. R. Shelmerdine. *Development of a floating tidal energy system suitable for use in shallow water*. in *Proceedings of the 7th European Wave and Tidal Energy Conference,*. 2007. Porto, Portugal.
10. R. Iovanel, *Numerical simulation of the Flow Through Kaplan Draft tube*, in *Department of Engineering Sciences and Mathematics*. 2014, Luleå University of Technology.

Chromium Enrichment on P11 Ferritic Steel by Pack Cementation

F. A. Fauzi[1,a], T. Kurniawan[1], M. S. Salwani[1], Y. S. Bin[1], W. S. W. Harun[1]

[1]*Faculty of Mechanical Engineering, University Malaysia Pahang, 26600 Pekan, Pahang Darul Makmur, Malaysia*

Abstract. The future thermal power plant is expected to operate at higher temperature to improve its efficiency and to reduce greenhouse gas emission. This target requires better corrosion properties of ferritic steels, which commonly used as materials for superheater and reheater of boiler tubes. In this work, chromium enrichment on the surface of ferritic steel is studied. The deposited chromium is expected to become a reservoir for the formation of chromia protective layer. Chromium was deposited on the substrate of steel by pack cementation process for two hours at the temperature of 850°C, 950°C and 1050°C, respectively. XRD analysis indicated that chromium was successfully deposited at all temperatures. Somehow, SEM cross sectional image showed that continuous layer of chromium was not continuously formed at 850oC. Therefore, this research clarify that chromium enrichment by pack cementation may be conducted at the temperature above 950°C.

1 Introduction

Nowadays, world is facing environmental impact caused by the greenhouse gasses emission from fossil fuel combustion. The awareness of public on global climate change leads to the growing concern on the need to reduce emissions of carbon dioxide and reductions in the range 60-80% may be needed [1]. The reductions can be done by using low- or zero-carbon energy sources such as renewables energy: wind, hydro and others [2], but for the current stage, fossil fuel power plant is very important in power generation. The fossil fuels that used to generate power included coal, oil and gas mainly covered the energy consumption[3, 4]. Therefore, in order to reduce the emissions of greenhouse gases from fossil fuel power plant, increasing thermal efficiency will be one of the solutions [5].

The improvement of thermal efficiency of power plant may be realized by increasing its operating temperature. This condition may lead to more excessive corrosion, especially in boiler components which exposed to high pressure and high temperature steam. Ferritic steel is a common material that used as boiler tubes in superheater and reheater area. Many researchers have reported that oxidation rate of this steel in steam condition was faster that dry atmosphere [6-8]. This is due to retardation of chromium formation and or evaporation of chromia protective layer [9, 10]. Therefore, improvement of corrosion resistant of this steel is important in order to use this material at higher operating temperature.

In this work, chromium enrichment on ferritic steel is studied by using pack cementation process. This chromium addition is expected to become a reservoir for the formation of chromia protective layer when the steel exposed to high temperature oxidation. This work was objected to study the relation between cementation temperature and homogeneity of chromium diffusion.

2 Experimental procedure

2.1 Sample Preparation

The sample used for this study is a P11 steel. The samples were cut into small pieces of specimens with the dimensions 10x10x2 mm^3 as shown in Figure 1. The surface of specimens were manually polished to a 2000grit-grade finish using SiC abrasive papers and washed with acetone in an ultrasonic bath to remove its surface rust and greasy dirt.

Figure 1: Sample of P11 steel

[a] Corresponding author: farahalia23@gmail.com

For the pack cementation process, chromium powder as the masteralloy, alumina powder as inert filler and NH_4Cl as the activator, were mixed homogenously inside ball milling for 30 minutes, according to a predetermined ratio shown in Table 1. The samples were embedded in the pack mixtures inside alumina crucible and then covered by alumina lid and glued by alumina based cement. The crucible then put into horizontal tube furnace, and loaded with argon gas. The pack cementation was conducted at 850°C, 950°C, and 1050°C for two hours.

Table 1: composition of pack mixtures (wt%)

Cr	NH_4Cl	Al_2O_3	Temperature, °C	Time, h
10	5	85	850	2h
10	5	85	950	2h
10	5	85	1050	2h

2.2 Characterization method

After the diffusion coating, surface of the samples was investigated with x-ray diffraction (Rigaku Miniflex X-Ray Diffractometer) with Cu-Kα radiation in angular range of 10-90° in 0.010° steps with 2s per step. In addition, the cross section of the samples was investigated by using SEM with EDX (Tabletop SEM TM3030Plus). Before SEM investigation, sample was mounted into epoxy resin and polished by SiC abrasive paper until 2000-grit continued with diamond paste 0.3 micron, and then cleaned with acetone.

Use A4 paper size (210 x 297 mm) and adjust the margins to those shown in Table 1. The final printed area will be 172 x 252 mm.

3 Results and discussions

Pack cementation of chromium was carried out on P11 at the temperature 850°C, 950°C and 1050°C to find the suitable condition for the formation of continuous chromium layer. Figure 2 show XRD pattern of each cementation temperature. The star (*) indicator shows the impurities that affected from the carbon that were placed at the holder plate of the samples during the XRD process was running. The peak in the energy spectrum line for element Cr is clearly different in the three layers. The energy spectrum line peak for element Cr in the temperature 950°C and 1050°C has a higher intensity as compared to 850°C, which may indicate that chromium content or thickness was higher. This result is strengthened by the cross sectional image of the steel from SEM investigation in Figure 3 until 5.

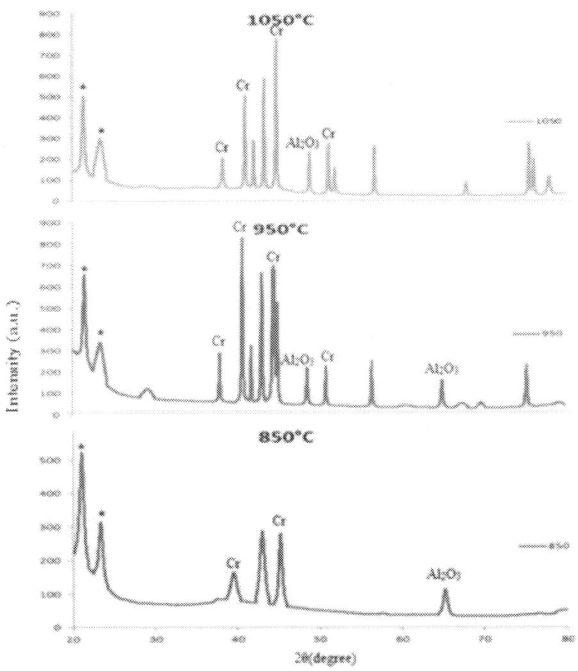

Figure 2: The XRD spectrum for different temperature for pack cementation coating process.

Figure 3 shows the cross sectional image of pack cemented steel at 850°C. The thickness and formation of the Cr is not continuous and non-uniform. The inhomogeneity in the oxide will cause the non-uniform distribution of the grain size of Cr[7]. This first approach using 850°C for the deposition of Cr using pack cementation process show a thin layer of Cr on top of the steel. Chromizing is generally thin in coating thickness because the pack continues to undergo depletion [11]. We can see from the figures, the thicknesses are very thin, which is less than 1 μm.

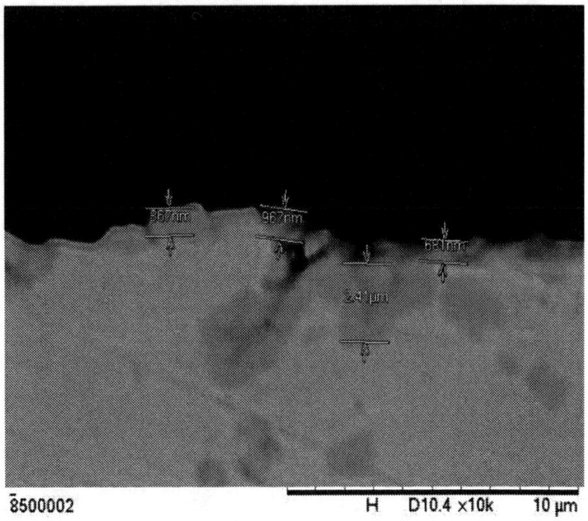

Figure 3: The image of Cr coating at temperature 850°C under BSE-SEM image

The cross sectional image of pack cemented steel at 950°C is shown in Figure 4. The Cr at this temperature starting to be more continuous and homogenous compare to Figure 3. The thickness of the Cr is increasing as the increasing in the temperature. The SEM-BSE image shown there are more pores are visible in the diffusion zone. The high activity of activator, chloride at the metal surface the iron in the substrate will react with the halide and cause the evaporation from the substrate. The influence of the pores to the in the diffusion zone soon will be investigated.

The result of pack cementation at temperature 1050°C is more likely the same with 950°C, as it shown in Figure 5. The chromium thickness at this temperature was slightly thicker due to the more Cr diffused into the steel.

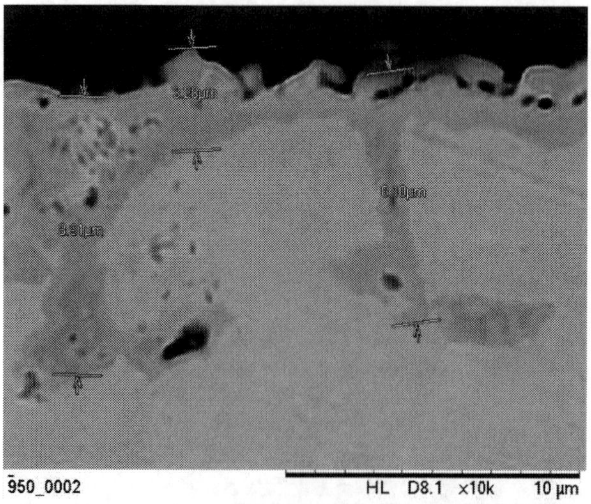

Figure 4: The image of Cr coating temperature 950°C under BSE-SEM image

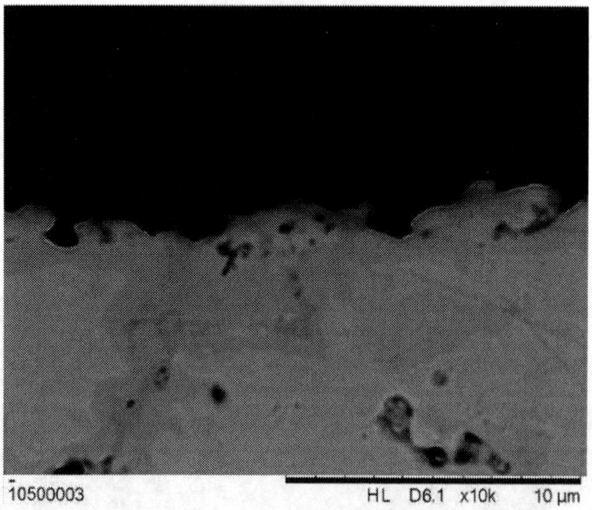

Figure 5: The image of Cr temperature 1050°C under BSE-SEM image

4 Conclusions

Enrichment of chromium on the surface of P11 ferritic steel was conducted by pack cementation process at different temperature, starting from 850°C until 1050°C. From the investigation of cross sectional image by SEM, it was clarified that continuous layer of chromium formed when the enrichment process conducted above 950°C. This continuous chromium layer is expected to become the reservoir for the formation of chromia protective layer when the steel exposed to boiler environment.

Acknowledgement

The authors would like to be obliged to Universiti Malaysia Pahang for financial assistance under project RDU140110.

References

1. van der Hoeven, M., *World Energy Outlook 2013.* (2013).
2. Jupesta, J. and A. Suwa, *Sustainable Energy Policy in Japan, Post Fukushima.* Int Assoc Energy Econ Fourth Quart Newsl, (2011). p. 23-26.
3. Viswanathan, R., K. Coleman, and U. Rao, *Materials for ultra-supercritical coal-fired power plant boilers.* International Journal of Pressure Vessels and Piping, 2006. **83**(11): p. 778-783.
4. Holcomb, G.R., et al., *Ultra supercritical turbines--steam oxidation.* 2004, Albany Research Center (ARC), Albany, OR.
5. Masuyama, F., *History of power plants and progress in heat resistant steels.* ISIJ international, (2001). **41**(6): p. 612-625.
6. Nakai, M., et al., *Correlation of high-temperature steam oxidation with hydrogen dissolution in pure iron and ternary high-chromium ferritic steel.* ISIJ international, (2005). **45**(7): p. 1066-1072.
7. Agüero, A., et al., *Oxidation under pure steam: Cr based protective oxides and coatings.* Surface and Coatings Technology, (2013.) **237**: p. 30-38.
8. Jacob, Y., et al., *The effect of gas composition on the isothermal oxidation behaviour of PM chromium.* Corrosion Science, (2002). **44**(9): p. 2027-2039.
9. Yang, Z., et al., *Effects of water vapor on oxidation behavior of ferritic stainless steels under solid oxide fuel cell interconnect exposure conditions.* Solid State Ionics, (2005). **176**(17): p. 1495-1503.
10. Ehlers, J., et al., *Enhanced oxidation of the 9% Cr steel P91 in water vapour containing environments.* Corrosion Science,(2006). **48**(11): p. 3428-3454.
11. Bose, S., *High Temperature Coatings.* (2007). **1**: p. 73-154.

MATEC Web of Conferences 74, 00037 (2016) DOI: 10.1051/matecconf/20167400037
ICMER 2015

Numerical modeling on homogeneous charge compression ignition combustion engine fueled by diesel-ethanol blends

H.Hanafi[1,a], M.M.Hasan[1], M.M.Rahman[1,2], M.M.Noor[1,2], K.Kadirgama[1,2] and D. Ramasamy[1,2]

[1] *Automotive Engineering Research Group, Faculty of Mechanical Engineering, Universiti Malaysia Pahang, 26600 Pekan, Pahang, Malaysia*
[2] *Automotive Engineering Centre, Universiti Malaysia Pahang, 26600 Pekan, Pahang, Malaysia*

Abstract. This paper investigates the performance and emission characteristics of HCCI engines fueled with oxygenated fuels (ethanol blend). A modeling study was conducted to investigate the impact of ethanol addition on the performance, combustion and emission characteristics of a Homogeneous Charge Compression Ignition (HCCI) engine fueled by diesel. One dimensional simulation was conducted using the renowned commercial software for diesel and its blend fuels with 5% (E5) and 10% ethanol (E10) (in vol.) under full load condition at variable engine speed ranging from 1000 to 2750 rpm with 250 rpm increment. The model was then validated with other researcher's experimental result. Model consists of intake and exhaust systems, cylinder, head, valves and port geometries. Performance tests were conducted for volumetric efficiency, brake engine torque, brake power, brake mean effective pressure, brake specific fuel consumption, and brake thermal efficiency, while exhaust emissions were analyzed for carbon monoxide (CO) and unburned hydrocarbons (HC. The results showed that blending diesel with ethanol increases the volumetric efficiency, brake specific fuel consumption and brake thermal efficiency, while it decreases brake engine torque, brake power and brake mean effective pressure. In term of emission characteristics, the CO emissions concentrations in the engine exhaust decrease significantly with ethanol as additive. But for HC emission, its concentration increase when apply in high engine speed. In conclusion, using Ethanol as fuel additive blend with Diesel operating in HCCI shows a good result in term of performance and emission in low speed but not recommended to use in high speed engine. Ethanol-diesel blends need to researched more to make it commercially useable.

1 Introduction

In the past decades a diesel engine has turned out to be the most standout fuel effective energy. Diesel engines have many advantages, for example simplicity, high performance, less maintenance, low fuel costs, low fuel consumption, high compression ratio, high power/weight ratio and durability [1]. But, its disadvantages are the exhaust of diesel engines are one of the major contributions to the air pollution problem, such as particulate matter, hydrocarbons, and carbon monoxide emissions [2]. The high flame temperatures that occur at the close-to-stoichiometric auto-ignition regions are the main contributor of hazardous emissions while the mixing-controlled burning of the bulk fuel contributes to the formation of soot. A reduction in NO_x emissions can be achieved by decreasing the combustion temperatures to below 1800 K [3] while the soot formation can be decreased by improving the homogeneity of the air–fuel mixture.

In order to solve the diesel engine disadvantages, researchers are now concentrating on new mode of combustion in internal combustion engine. HCCI is an alternative operating mode for an internal combustion engine and is considered as one potential solution to solve this problem [4, 5] besides that HCCI also can operated on wide range of fuel [3]. HCCI engines operates at higher thermal efficiency as high as 50% [6] than gasoline and diesel engines [7, 8] with similar displacement volume, while it also emits ultra-low particle matter (PM) and nitrogen oxides (NOx) emissions [9]. These emission reductions are achieved by physically separating the injection events from the onset of combustion, therefore makes ignition delayed long enough to ensure a lean and nearly-homogenous air–fuel mixture.

Using HCCI combustion, a wide range of fuel can be used. Ethanol is one of the next generation renewable biofuel that can be used in HCCI. It is biodegradable, low in toxicity and less polluted compared to fossil fuels [10]. Ethanol are selected for this study since they are promising alternatives in order to substitute conventional petroleum fuels in next-generation of internal combustion engines (ICEs) [11]. Ethanol can be produced from different types of sources, mainly from carbon based feed stocks. Therefore usage of alcohols in engines is another way to reduce energy resource depletion by using fossil fuels, and improve the implementation of HCCI combustion mode [12]. It also another significant research purposes to combine the advantages of HCCI combustion mode using alternative renewable oxygenated fuels. Tongroon et al. [12] found that methanol showed fastest and earliest auto-ignition, followed by ethanol using optical methods investigated in HCCI combustion mode .Christensen et al. [13] used a lean air–fuel mixture for HCCI engine that operated unthrottled, producing an ultra-low NO_x emission result.

Later on, Mack et al. [14] used wet ethanol in HCCI combustion mode and reported that increasing water concentration in ethanol reduced the in-cylinder maximum pressure rise and improved energy balance, but increased the emission of HC and CO. Viggiano and

[a] Corresponding author: hasrulazmi.h@gmail.com

Magi [15] modeled a multidimensional numerical model with a kinetic reaction mechanism for ethanol oxidation and NO_x formation, and found that CO emissions are strongly related to inhomogeneities in the near walls of the combustion chamber . Another investigations was done and found out that the tendency of alcohols to auto-ignite relatively earlier cause lower emission [16, 17]. Only few literature results are done on the modeling approach of ethanol fueled engines. Sjoberg and Dec [18] develop a multi-zone model ethanol-chemical kinetics mechanism in a single cylinder HCCI engine fueled using pure ethanol (100%). Splitter et al. [19] develop computational fluid dynamics (CFD) modeling on premixed charge compression ignition (PCCI) engine fueled by early injected pure ethanol and on dual-fuel (ethanol–diesel fuel) engine .

The above review shows not much research has been done from a numerical point of view on using alcohol fuels in HCCI engine mode. More studies need to be done to provide deeper insight and greater understanding on HCCI combustion and emission characteristics using alcohol fuels. The objective of this paper is therefore to numerically investigate performance, combustion and emission characteristics in a HCCI engine using diesel-ethanol blend.

2 Numerical modeling

In this paper, a one-dimensional (1-D) simulation was done for one cylinder, four strokes and direct ignition (DI) engine. The details of the engine parameters are shown in Table 1. The modeling was performed using the 1-D engine in simulation software. A one cylinder, four stroke, direct ignition (DI) engine was modeled to simulate the engine performance and emissions. The numerical model consists of intake and exhaust systems, cylinder, head, valve and port geometries i.e. crank angle degree (°CA). The complete model is shown in Figure 1. Operational parameters such as cam phasing, combustion phasing and duration, and air-fuel ratio were determined by built-in look-up table maps based on load and speed of the simulation case. The simulation was done using different engine speed (rpm); 1000, 1250, 1500, 1750, 2000, 2250, 2500 and 2750 with two types of diesel-ethanol blends; E5 and E10. The properties of fuels used in this research are shown in Table 2.

In order to distinguish between Compression Ignition (CI) and Spark Ignition (SI) engine model, this modeling was added with chemical reaction obtain from Lawrence Livermore National Laboratory website. A chemical reaction consists of diesel and ethanol was added before simulation was done in order to get the results. In order to make sure the modeling are correct, this model was verified with other researcher's experimental result which will be discussed later in this paper.

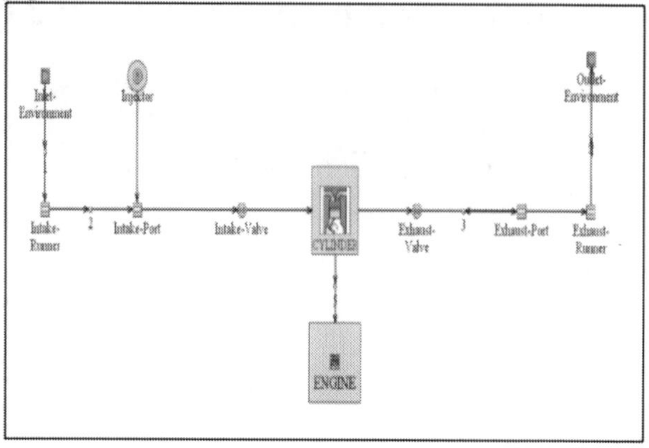

Figure 1. Numerical model setup

Table 1. Engine model specifications

Engine parameters	Value
Type	1-cylinder, air cooled, overhead valve, direct injection
Bore (mm)	87.5
Stroke (mm)	110
Displacement (*L*)	1.583
No. of cylinders	1
Compression ratio	17.5
Connecting rod length (mm)	150
Piston pin offset (mm)	1
Intake Valve Open (°CA)	351
Intake Valve Close (°CA)	-96
Exhaust Valve Open (°CA)	125
Exhaust Valve Close (°CA)	398

3 Governing equations and flow models

One dimensional flow equation is solved with and across the pipes. A pipe can be further discretized into many nodes, and the flow equations are solved at node level. Since it is one dimensional flow, variables are uniform at pipe cross section. Each scalar variable is assumed to be uniform over each node, and each vector variable is calculated at each boundary. Accordingly, the equations that involve in this modeling are:

Table 2. Fuel properties of diesel and ethanol [20]

Fuel properties	Diesel	Ethanol	E5	E10
Chemical formula	$C_{12}H_{26}$–$C_{14}H_{30}$	C_2H_5OH	-	-
Density at 20 °C (kg/*L*)	0.8485	0.7893	0.8385	0.8308
Low calorific value (kJ/kg)	44514.6	26,800	43631.8	43192.5
Viscosity at 40 °C (mm^2/s)	5.6114	1.2	5.5343	5.4564
Cetane number	50	~8	-	44.2
Flash point (°C)	53	14	24	25
Boiling point (°C)	~180-360	78	-	-
Latent heat of evaporation (kJ/kg)	250	840	-	43.4
Stoichiometric air–fuel ratio	15	9	-	-
Oxygen (% by weight)	0	34.8	-	-
Self-ignition temperature (°C)	254	363	-	-

Continuity :

$$\frac{dm}{dt} = \sum_{boundaries} \rho A v \tag{1}$$

Where, ρ = density
v = speed
A = area

Energy:

$$\frac{d(me)}{dt} = p\frac{dV}{dt} + \sum_{boundaries}(\dot{m} \text{ x } H) - h_g A(T_{gas} - T_{wall}) \tag{2}$$

Where, $\dot{m}$ = mass flow rate
p = pressure
A = area
H = height
h_g = head loss
T_g = temperature of gas
T_{wall} = Temperature of wall

Momentum:

$$\frac{d(\rho A v)}{dt} = \frac{dpA + \sum_{boundaries}(\rho A v \text{ x } v) - C_f \frac{\rho v^2}{2}\frac{dxA}{D} - C_p\left(\frac{1}{2}\rho v^2\right)A}{dx} \tag{3}$$

Where, A = area
v = velocity
C_f = coefficient of friction
C_p = specific heat of fuel at constant pressure
D = diameter of pipe

In this numerical modeling, friction across the pipe is considered. This friction is already set in the software used (built-in) to model the HCCI engine and not negligible. The friction loss factor is based on the Reynolds number and the surface roughness of the walls. In Laminar region when Reynolds number is less than 2000, $C_f = 16/Re_D$. In turbulent region, $C_f = 0.08/Re_D^{0.25}$. When the surface is rough and the flow is not laminar, the friction coefficient is calculated below:

$$C_{f\,rough} = \frac{0.25}{(2log_{10}\left(\frac{D}{2H}\right) + 1.74^2)} \tag{4}$$

Where, C_f = Coefficient of friction
D = Diameter of pipe
H = Height

Volumetric Efficiency: defined as the ratio between the air mass flow into the cylinders from the intake manifold and the air mass theoretically inside the cylinders at the manifold temperature. As such, it measure the effectiveness of the air pumping system composed by the intake manifold, the inlet port and valve, and the cylinders [21] :

$$\eta_v = \frac{\dot{m}_a}{\rho_a V_{dN/2}} \tag{5}$$

Where, ρ_a = the inlet air density
$\dot{m}_a$ = the steady-state flow of air into the engine
Vd = displacement volume

Brake Engine Torque: The measure of the engine's ability to apply power generation is called torque [22]. Torque is defined as force acting at a moment distance, their unit is N-m or lbf-ft. The equation is expressed as below [23, 24]:

$$2\pi\tau = W_b = (bmep)V_d/n \tag{6}$$

Where, W_b = brake work of one revolution
Vd = displacement volume
n = number of revolutions per cycle

For a four-stroke cycle engine that takes two revolutions per cycle,

$$\tau = (bmep)V_d/4\pi \tag{7}$$

[a] Corresponding author: hasrulazmi.h@gmail.com

Brake Power: Power is defined as the rate of work of the engine. Brake power refers to the amount of usable power delivered by the engine to the crankshaft [23, 25]. The equation is expressed as below:

$$\dot{W} = WN/n \tag{8}$$
$$\dot{W} = 2\pi N\tau \tag{9}$$
$$\dot{W} = (1/2n)(mep)A_p\bar{U}_p \tag{10}$$
$$\dot{W} = (mep)A_p\bar{U}_p/4 \tag{11}$$

Where, W = work per cycle
A_p = piston face area of all pistons
$\bar{U}_p$ = average piston speed
n = number of revolutions per cycle
N = engine speed

Brake Thermal Efficiency: The brake thermal efficiency of an engine is defined as the ratio of brake output power to input power and the brake power produced by an engine with respect to the energy contributed by the fuel [26]. The equation is expressed as below:

$$\eta_{bth} = \frac{\text{bp}}{\text{mass of fuel x calorific value of fuel}} \tag{12}$$

Brake Mean Effective Pressure: The brake mean effective pressure [27] is defined to measure engine performance and indicates an engine's capacity to produce power output over the full engine speed range. The equation is expressed as below:

$$bmep = w_b/\Delta\, v \tag{13}$$
$$bmep = 2\pi n\tau/V_d \tag{14}$$

Where, $\Delta\, v = v_{bdc} - v_{tdc}$

Brake Specific Fuel Consumption: The fuel consumption characteristics of an engine are generally expressed in terms of specific fuel consumption in kilograms of fuel per kilowatt-hour. In engine tests, the fuel consumption is measured as a flow rate-mass flow per unit time. It measures how efficiently an engine uses the fuel supplied to produce work [28]. The equation is expressed as below:

$$bsfc = \dot{m}_f/\dot{W}_b \tag{15}$$

Where, $\dot{m}_f$ = rate of fuel flow into engine

4 Model validations

To further validate the applicability of the developed reaction mechanism for engine simulations, numerical simulations were performed and compared against the diesel engine experiments. Figure 2 (a) plots the in-cylinder pressure and Figure 2 (b) plots the in-cylinder temperature obtained from experiments done by [29] and our simulations, at a fixed engine speed of 1500 rpm under full load conditions. As can be seen, the peak cylinder pressure and peak temperature timing are adequately reproduced, indicating that the important reaction pathways are very well represented. The predicted in-cylinder temperature are slightly higher and narrower than those calculated from the experimental cylinder pressure curves, suggesting that slightly higher burning rate is predicted by the developed reaction mechanism during the initial premixed burning phase.

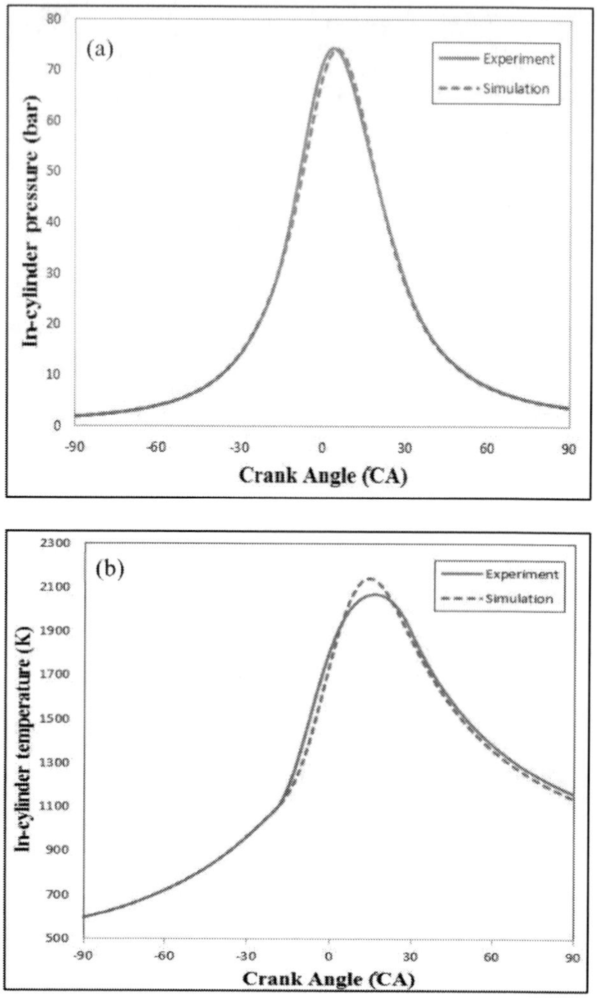

Figure 2. A comparison of simulated and experimental in-cylinder pressure (a) and temperature (b) under full load at 1500 rpm engine speed.

4 Results and discussion

4.1 Engine Performance

The numerical results are presented for diesel, diesel-ethanol blends fueled in HCCI combustion at different engine speeds. Performance and emission characteristics are analyzed in each figure.

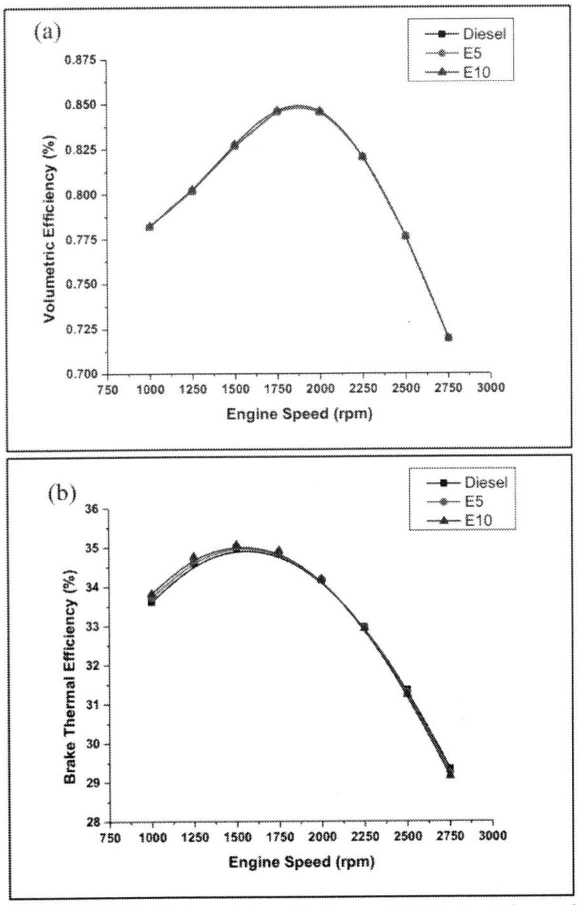

Figure 3. (a) Volumetric efficiency (b) Brake Thermal Efficiency against different engine speed

Figure 3(a) shows the result of volumetric efficiency at different speeds using 3 different types of fuels in HCCI engine. From Figure 3 (a), it can be seen that only slightest differences occur for each type of fuels. The highest volumetric efficiency happens at 1875 rpm for each test fuel. As the engine speeds increase, the volumetric efficiency decreases dramatically. This is because when the engine speed increase, the air have less time to goes inside the cylinder [30]. Hence, the amount of the air that fills the cylinder decrease and cause an incomplete combustion. It has been researched that when the ethanol amount use increases, the volatility and the latent heat of the fuel also increases [31], hence the volumetric efficiency also increased.

Figure 3 (b) shows the brake thermal efficiency (BTE) result. It show a decrease of BTE with engine speed as the amount of the ethanol amount in the blends increases. It can be seen from Figure 3 (b) that the E10 fuel gives highest efficiencies at 1500 rpm. The diesel fuel produced the lowest BTE during the slow engine speed. But as the speed goes higher, diesel fuel overtake the diesel-ethanol blends. The main reason for decrease of thermal efficiency with increase in blend ratio is shorter ignition delay which results in earlier start of combustion than for diesel. This increases the compression work as well as heat loss [32, 33] and thus reduces the efficiency [34, 35] of the engine.

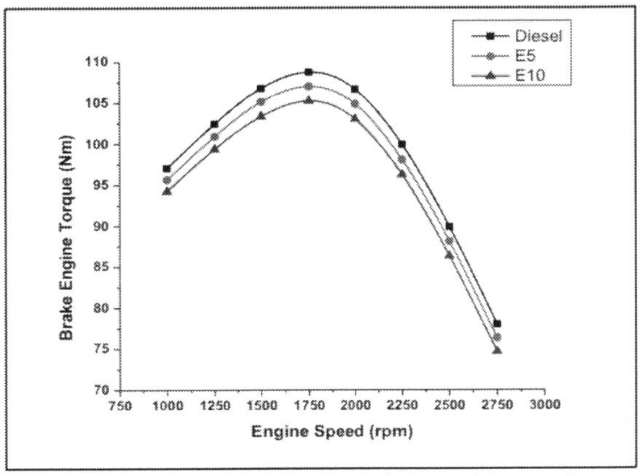

Figure 4. Brake engine torque against engine speed

Figure 4 shows the effect of engine speed and different blends of fuel on brake engine torque (BET) at different engine speeds. It is observed that at higher engine speed, BET decreasing. Diesel attained highest BET compared to E5 and E10 fuel. At higher engine speed, auto-ignition temperature is attained relatively earlier. BET obtained is significantly lower at fastest engine speeds condition. The possible explanation for this observation is as follows. At very fast engine speed (2750 rpm), the temperature in the combustion chamber is low, and the combustion efficiency is also low. As a consequence, large quantities of unburned hydrocarbons are emitted and lower BET is obtained. It is observed that BET is at peak at the 1750 rpm for all 3 types of fuel used. It is also noticed that after 1750 rpm, BET starts decreasing. At fastest engine speed, knocking combustion starts, which cause increased heat transfer to the piston and cylinder walls [36], resulting in lower BET value.

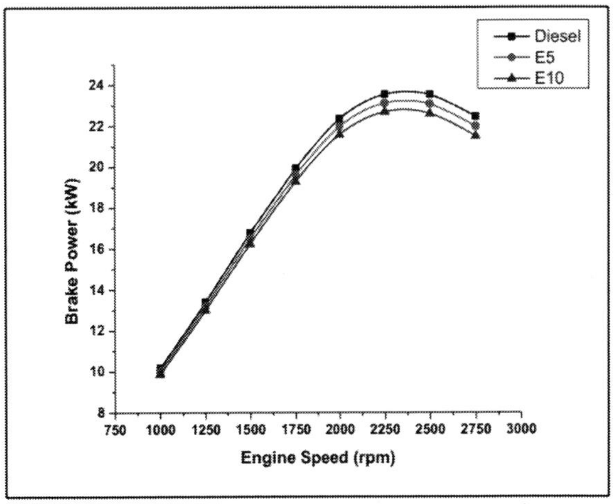

Figure 5. Brake power result against engine speed

Figure 5 shows the variation of brake power with increasing speed on the engine. It can be seen that diesel fuel alone still producing the highest brake power compared to the E5 and E10 fuel. Diesel-ethanol blends reduces engine power output as the portion of oxygenated compounds (ethanol) in the blends increases [37]. This is due to the low cetane number and calorific value and higher ignition delay of the blends, compared to diesel fuel [38]. Other researchers also found out that approximately 4.4–8.7% reduction in maximum power output by using diesel-ethanol blends compared to fossil diesel fuel [39]. Thus using these blends without any additives only will reduce engine power and torque output. These reduced torque and power can be improved and the combustion characteristics can also be optimized by using additives with these blends [40]. If no additives are used, then the portion of the ethanol/bioethanol should be kept as low as possible.

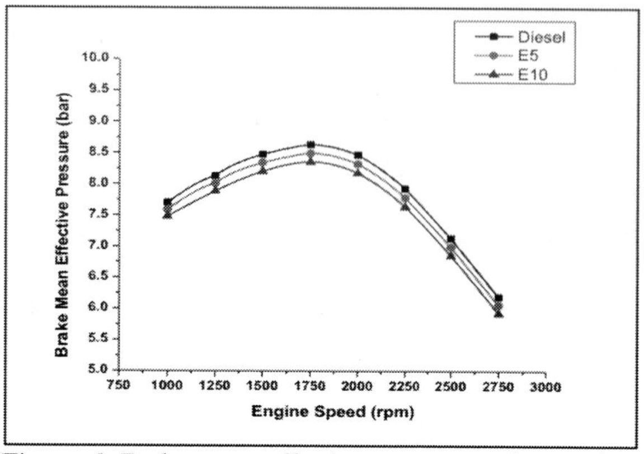

Figure 6. Brake mean effective pressure against engine speed

The result of brake mean effective pressure (BMEP) for 3 types of fuels is presented in Figure 6. As shown in Figure 6, the BMEP decrease using ethanol as additives with diesel fuels. E10 shows the lowest BMEP value throughout all engine speeds with diesel producing highest BMEP result. Higher maximum BMEP means higher stresses and temperatures in the engine hence shorter engine life or bulkier engine. As ethanol has lower cetane value, longer ignition delay and high auto ignition temperature, the BMEP decrease along with increase of engine speeds. This leads to reduced ignition delay. The lower boiling point of ethanol makes it to evaporate as the diesel gets ignited. As a result of the vaporized ethanol, the mixture burns more rapidly than diesel. This reaction results in better engine performance. The longer ignition delay of ethanol results in diffusive end burning and drops in BMEP value.

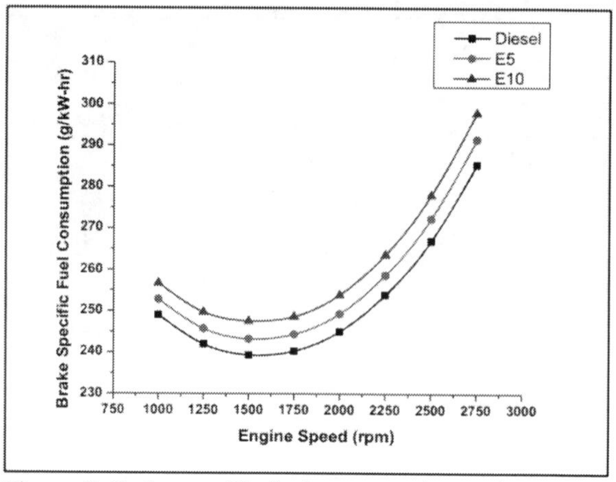

Figure 7. Brake specific fuel consumption against engine speed

The relationship between engine speed and brake specific fuel consumption (BSFC) for different diesel/ethanol blends are shown in Figure 7. The figure shows that the BSFC trend for diesel and diesel-ethanol blends are similar in nature. In all cases, diesel fuel shows the lowest BSFC as compared to diesel-ethanol blends because diesel has higher heating value and requires a lower mass than the other fuels to extract the same engine output. In general, the BSFC increased with increasing ethanol content in blend fuel. This is due to the fact that the low heat value of ethanol is about 2/3 of that of diesel [41]. The other reason is the incomplete combustion due to the ignition delay of ethanol–diesel blend fuel. Overall, ethanol addition in diesel does not create a significant change in BSFC.

4.1 Exhaust Emissions

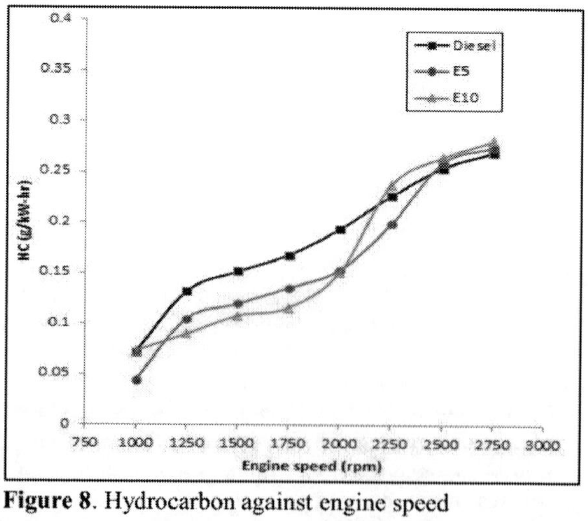

Figure 8. Hydrocarbon against engine speed

Figure 8 shows the results of hydrocarbon (HC) emission with different engine speed using 3 different types of fuels. This higher HC might be caused because of higher amounts of single hydrogen radicals in alcohol–diesel fuel in-cylinder charge. It is confirmed by investigation done by others. High content of alcohol in diesel–alcohol blends contributes to increase in HC [42]. One of the most significant reasons for the increase of HC emissions is richer charge mixture. As the engine operates with the richer mixture the whole fuel cannot be oxidized [43]. Unburned HC emissions are formed as a result of incomplete combustion. Because ethanol has a lower cetane number and high heat of vaporization as explain in Table 1, it takes longer to vaporize which leads to longer ignition delay and short duration of combustion [44]. In addition, the engine needs more air in order to complete combustion in the HCCI combustion mode. Consequently, incomplete combustion occurs and HC emissions are formed.

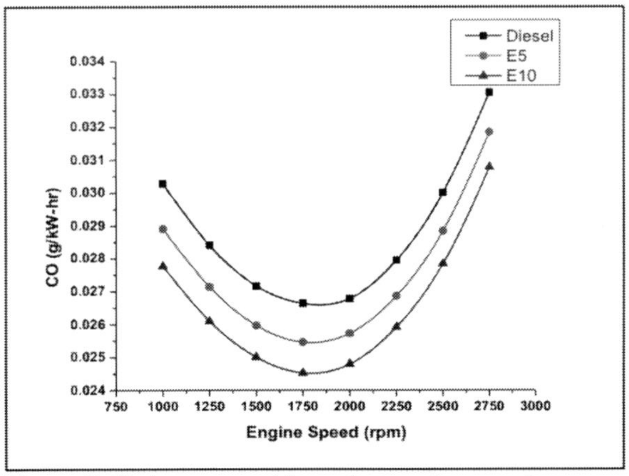

Figure 9. Carbon monoxide against engine speed

Carbon monoxide (CO) emission result is shown in Figure 9. The emission becomes higher with increased speed on the engine. The lowest CO valued achieved at1750 rpm for 3 types of fuels. From the figure it can be seen that as the percentage of ethanol in the blends increased the percentage of CO emission reduced. CO is an incomplete combustion product because of insufficient oxygen and temperature in the combustion chamber [45]. The emission reduced by dramatically as compared to diesel alone. This trend is due to the fact that ethanol has less carbon than diesel. Also, given the same fuel dispersion pattern as for diesel, the oxygen content in the blended fuels would help to increase the oxygen-to-fuel ratio in the fuel-rich regions. This resulting in more complete combustion leads to reduce of CO content in the exhaust smoke.

References

[1] Pidol L, Lecointe B, Starck L, Jeuland N. Ethanol–biodiesel–Diesel fuel blends: Performances and emissions in conventional Diesel and advanced Low Temperature Combustions. Fuel. 2012;93:329-38.

[2] Shahir SA, Masjuki HH, Kalam MA, Imran A, Ashraful AM. Performance and emission assessment of diesel–biodiesel–ethanol/bioethanol blend as a fuel in diesel engines: A review. Renewable and Sustainable Energy Reviews. 2015;48:62-78.

[3] Bendu H, Murugan S. Homogeneous charge compression ignition (HCCI) combustion: Mixture preparation and control strategies in diesel engines. Renewable and Sustainable Energy Reviews. 2014;38:732-46.

[4] Mohanamurugan S, Sendilvelan S. Emission and combustion characteristics of different fuel In A HCCI engine. International Journal of Automotive and Mechanical Engineering. 2011;3:279-92.

[5] Hairuddin A, Wandel A, Yusaf T. An introduction to a homogeneous charge compression ignition engine. Journal of Mechanical Engineering and Sciences. 2014;7:1042-52.

[6] Saxena S, Shah N, Bedoya I, Phadke A. Understanding optimal engine operating strategies for gasoline-fueled HCCI engines using crank-angle resolved exergy analysis. Applied Energy. 2014;114:155-63.

[7] Abdullah NR, Shahruddin NS, Mamat R, Ihsan Mamat A, Zulkifli A. Effects of air intake pressure on the engine performance, fuel economy and exhaust emissions of a small gasoline engine. Journal of Mechanical Engineering and Sciences. 2014;6:949-58.

[8] Yashwanth MS, Venugopal T, Ramesh A. Experimental and simulation studies to determine the effective octane number in an engine fuelled with ethanol and gasoline. International Journal of Automotive & Mechanical Engineering. 2014;10.

[9] Stanglmaier RH, Roberts CE. Homogeneous charge compression ignition (HCCI): benefits, compromises, and future engine applications. SAE Technical Paper; 1999.

[10] Najafi G, Ghobadian B, Tavakoli T, Buttsworth D, Yusaf T, Faizollahnejad M. Performance and exhaust emissions of a gasoline engine with ethanol blended gasoline fuels using artificial neural network. Applied Energy. 2009;86:630-9.

[11] Yap D, Megaritis A, Wyszynski M. An investigation into bioethanol homogeneous charge compression ignition (HCCI) engine operation with residual gas trapping. Energy & fuels. 2004;18:1315-23.

[12] Tongroon M, Zhao H. Combustion characteristics of CAI combustion with alcohol fuels. 2010.
[13] Christensen M, Johansson B, Einewall P. Homogeneous charge compression ignition (HCCI) using isooctane, ethanol and natural gas-a comparison with spark ignition operation. SAE Technical Paper; 1997.
[14] Mack JH, Aceves SM, Dibble RW. Demonstrating direct use of wet ethanol in a homogeneous charge compression ignition (HCCI) engine. Energy. 2009;34:782-7.
[15] Viggiano A, Magi V. A comprehensive investigation on the emissions of ethanol HCCI engines. Applied Energy. 2012;93:277-87.
[16] Xie H, Wei Z, He B, Zhao H. Comparison of HCCI combustion respectively fueled with gasoline, ethanol and methanol through the trapped residual gas strategy. SAE Technical Paper; 2006.
[17] Li Y, Zhao H, Brouzos N. CAI combustion with methanol and ethanol in an air-assisted direct injection SI engine. 2008.
[18] Sjoberg M, Dec JE. Ethanol autoignition characteristics and HCCI performance for wide ranges of engine speed, load and boost. SAE Technical Paper; 2010.
[19] Splitter DA, Hanson RM, Reitz RD, Manente V, Johansson B. Modeling charge preparation and combustion in diesel fuel, ethanol, and dual-fuel PCCI engines. Atomization and Sprays. 2011;21.
[20] Hasan M, Rahman M, Kadirgama K. A review on homogeneous charge compression ignition engine performance using biodiesel-diesel blend as a fuel. International Journal of Automotive & Mechanical Engineering. 2015;11.
[21] De Nicolao G, Scattolini R, Siviero C. Modelling the volumetric efficiency of IC engines: parametric, non-parametric and neural techniques. Control Engineering Practice. 1996;4:1405-15.
[22] Saber HA, Al-Barwari RRI, Talabany ZJ. Effect of ambient air temperature on specific fuel consumption of naturally aspirated diesel engine. Journal of science and engineering: volume 1. 2013;1:1-7.
[23] Rajput R. A text book of Automobile Engineering: Firewall Media; 2007.
[24] Stotsky AA. Automotive engines: control, estimation, statistical detection: Springer Science & Business Media; 2009.
[25] Hillier VAW, Coombes P. Hillier's fundamentals of motor vehicle technology: Nelson Thornes; 2004.
[26] Hairuddin AA, Yusaf T, Wandel AP. A review of hydrogen and natural gas addition in diesel HCCI engines. Renewable and Sustainable Energy Reviews. 2014;32:739-61.
[27] Guzzella L, Onder C. Introduction to modeling and control of internal combustion engine systems: Springer Science & Business Media; 2009.
[28] Van Basshuysen R, Schäfer F. Internal combustion engine handbook-basics, components, systems and perspectives2004.
[29] Ganesh D, Nagarajan G. Homogeneous charge compression ignition (HCCI) combustion of diesel fuel with external mixture formation. Energy. 2010;35:148-57.
[30] Ebrahimi R, Ghanbarian D, Tadayon MR. Performance of an Otto engine with volumetric efficiency. J Am Sci. 2010;6:27-31.
[31] Menrad H, Nierhauve B. Engine and vehicle concepts for methanol-gasoline blends. SAE Technical Paper; 1983.
[32] Behçet R. Performance and emission study of waste anchovy fish biodiesel in a diesel engine. Fuel Processing Technology. 2011;92:1187-94.
[33] Lakshmi Narayana Rao G, Durga Prasad B, Sampath S, Rajagopal K. Combustion analysis of diesel engine fueled with jatropha oil methyl ester-diesel blends. International Journal of Green Energy. 2007;4:645-58.
[34] Puhan S, Vedaraman N, Sankaranarayanan G, Ram BVB. Performance and emission study of Mahua oil (Madhuca indica oil) ethyl ester in a 4-stroke natural aspirated direct injection diesel engine. Renewable Energy. 2005;30:1269-78.
[35] Swaminathan C, Sarangan J. Performance and exhaust emission characteristics of a CI engine fueled with biodiesel (fish oil) with DEE as additive. biomass and bioenergy. 2012;39:168-74.
[36] Heywood JB. Internal combustion engine fundamentals: Mcgraw-hill New York; 1988.
[37] rahimi H, Ghobadian B, Yusaf T, Najafi G, Khatamifar M. Diesterol: An environment-

friendly IC engine fuel. Renewable Energy. 2009;34:335-42.

[38] Gumus M, Sayin C, Canakci M. The impact of fuel injection pressure on the exhaust emissions of a direct injection diesel engine fueled with biodiesel–diesel fuel blends. Fuel. 2012;95:486-94.

[39] Gowthami K, Murali S, Reddy VN. Performance and analysis of super charging and blends of rice bran oil biodiesel as an additive in diesel-ethanol for diesel engines. 2014.

[40] Li L, Wang J, Wang Z, Xiao J. Combustion and emission characteristics of diesel engine fueled with diesel/biodiesel/pentanol fuel blends. Fuel. 2015;156:211-8.

[41] Li D-g, Zhen H, Xingcai L, Wu-gao Z, Jian-Guang Y. Physico-chemical properties of ethanol–diesel blend fuel and its effect on performance and emissions of diesel engines. Renewable energy. 2005;30:967-76.

[42] Lujaji F, Kristóf L, Bereczky A, Mbarawa M. Experimental investigation of fuel properties, engine performance, combustion and emissions of blends containing croton oil, butanol, and diesel on a CI engine. Fuel. 2011;90:505-10.

[43] Cinar C, Uyumaz A, Solmaz H, Sahin F, Polat S, Yilmaz E. Effects of intake air temperature on combustion, performance and emission characteristics of a HCCI engine fueled with the blends of 20% n-heptane and 80% isooctane fuels. Fuel Processing Technology. 2015;130:275-81.

[44] Yilmaz N, Vigil FM, Benalil K, Davis SM, Calva A. Effect of biodiesel–butanol fuel blends on emissions and performance characteristics of a diesel engine. Fuel. 2014;135:46-50.

[45] Cinar C, Uyumaz A, Solmaz H, Topgul T. Effects of valve lift on the combustion and emissions of a HCCI gasoline engine. Energy Conversion and Management. 2015;94:159-68.

MATEC Web of Conferences **74**, 00038 (2016) DOI: 10.1051/matecconf/20167400038
ICMER 2015

Wettability study of lead free solder paste and its effect towards multiple reflow

Siti Rabiatull Aisha Idris[1a], Siti Zuleikha[1] and Zetty Akhtar Abd Malek[1]

[1]*Faculty of Mechanical Engineering, Universiti Malaysia Pahang, 26600 Pekan, Pahang, Malaysia*

Abstract. Nowadays, wafer bumping using solder paste has come into focus as it provides a low cost method. However, since the industries are moving towards lead-free electronic packaging, a new type of no-clean flux was produced specifically for lead-free solder paste. Therefore, this study is used to evaluate the wettability of two different types of no-clean flux onto copper substrate. Besides, its effect towards multiple reflow was also studied. Reflow soldering was conducted for both types of solder paste that contained different type of no-clean flux for up to double reflow. Two different reflow profile was used. The results showed that the Flux A exhibit better soldering performance after first and second reflow soldering. In addition, type of intermetallic compound (IMC) found after first reflow remain the same even after second reflow which was Cu-Sn based. This is shows that Flux A manage to control the diffusion process which will finally leads to a better solder joint performance. Nevertheless, mechanical testing should be carried out in order to evaluate the solder joint strength.

1 Introduction

As the electronic industry is shifting towards using lead-free solders, solder paste has become popular choice since it can form reliable solder joint in printed circuit board (PCB) through reflow soldering [1-3]. Solder paste is a homogeneous mixture of flux and also highly-graded solder alloy in powder form particles and has a great effect on solder joint quality [1]. However, to give a reliable solder joint, the flux itself must have an excellent chemical activity to remove the oxide on the surface of the substrate as well as on the surface of the solder particles. With that, it manage to promote good spreading and finally giving reliable bonding with the substrate [2]. Other than that, it also must easily be activated thermally but at the same time must not easily decomposed. Last but not least, it should be able to form benign residues that can be easily removed by washing [1, 4, 5].

Various lead-free solder alloys have been studied and used for electronics components and assemblies. Based on the results of many investigations, 96.5Sn-3.0Ag-0.5Cu solder alloy had been selected since it possesses properties which could replace lead solder [6-8]. Though, this type of solder alloy has some disadvantages such as higher melting point (216-218°C) and poor wettability as compared to tin-lead solder. Thus, the best alternative way is yet to be achieved since a lot of studies for lead-free solders are still undergoing the researches.

Apart from that, one of the crucial factors in Sn-contained solder joints reliability is the formation of a Cu-Sn intermetallic compound (IMC) between the solder alloy and copper substrate. The formation of this IMC starts with diffusion of Sn toward copper substrate after reflow soldering. Then a layer of Cu-Sn IMC starts to form which normally known as Cu_6Sn_5. This IMC will continuously grow when exposed to subsequent reflow. In some cases, Cu_3Sn was found between Cu_6Sn_5 and copper substrate. In general, IMC served as a bonding material for the solder joint, but it also one of the main factors that affect the reliability of the solder joint due to its brittle nature and thermal mismatch with the solder and printed circuit board (PCB). Therefore, excessive IMC formed at the solder/Cu interface of the solder joint will finally weaken the joint strength and finally may cause failure [9-11].

In this study, wettability of the Sn-3.0Ag-0.5Cu solder paste with different type of no-clean flux, namely Flux A and Flux B was evaluated. Besides, its effect towards multiple reflow was also studied. This study also aims at investigating the growth behaviour of IMC at the solder joint as well as the mean thickness of IMC formed at the solder/Cu interface.

2 Experimental Set Up

The samples were received from Electronic Packaging Research Society-Universiti Kebangsaan Malaysia (EPRS-UKM) where two PCB with immersion tin

[a] Corresponding author: rabiatull@ump.edu.my

 DOI: 10.1051/matecconf/20167400038

(ImSn) surface finish underwent reflow soldering at 240°C for 10 minutes. These PCBs were using Sn-3.0Ag-0.5Cu solder paste with different type of no-clean flux. Since the exact ingredient of the no-clean flux remain confidential by the Redring Solder Sdn Bhd, they were labelled as Flux A and Flux B. However, it is known that the flux is an epoxy based. Next, the samples proceeded with another reflow process at 250°C for 25 minutes in a furnace. This process was done at Universiti Malaysia Pahang (UMP). Figure 1 shows the reflow profiles used in this study. The samples were then undergoing metallography analysis in order to characterise the IMC formation. Several techniques were used to characterize the samples which include optical microscopy, scanning electron microscopy-energy dispersive xray (SEM-EDX) and imageJ software.

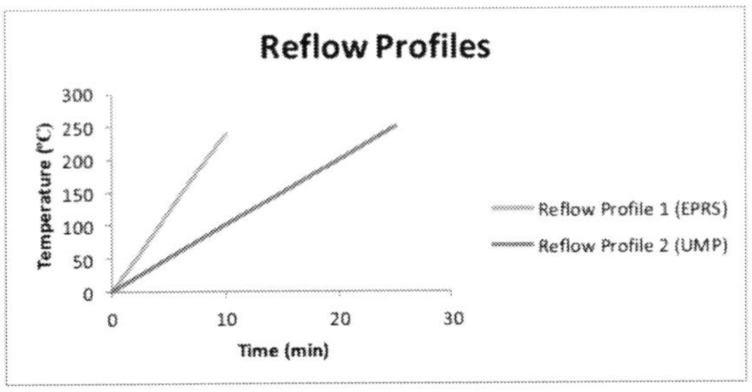

Figure 1. Reflow soldering profiles of printed circuit boards.

3 Results and Discussion

According to Table 1, results showed that Flux A had better soldering performance than Flux B. This indicated that Sn-3.0Ag-0.5Cu solder paste exhibit better wettability with Flux A on copper substrate. It is known that an optimum wetting angle of the solder joint should be less than 90°. Eventhough the composition of the flux was not mentioned by the supplier, it is believed that chemical composition inside the flux can improve wettability of the activators. These combination decomposed into acid and alkali and reacted with oxides on both substrate and solder surface during reflow soldering. Thus, enhancing the soldering capability.

Table 1. Wetting angles for intermetallic compound using Flux A and Flux B

Flux	Average Wetting Angle (θ)	Cross Sectional view of the sample
A	28.95	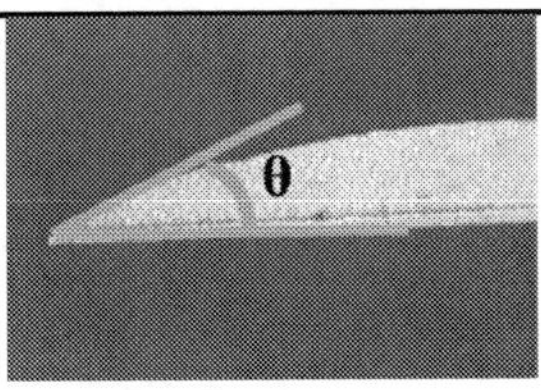
B	41.20	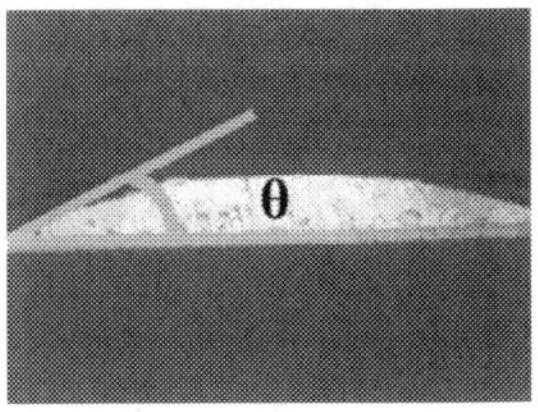

Other than that, in the soldering process, IMC formation can be affected due to many factors which include reflow temperature or profile. The effect of reflow profile for lead solder joint has fully discovered unlike the lead-free solder joint [12, 13]. In determining the IMC thickness, the reflow peak temperature and time are the critical variables. The liquidus temperature of Sn-3.0Ag-0.5Cu are between 217 and 218 °C, which is 34°C higher than eutectic SnPb solder. This is means the liquidus temperatures of lead-free alloys are higher than SnPb alloy. A thicker IMC layer is expected to form at a higher reflow temperature since the rate of dissolution of the base material in molten alloy is faster as temperature increases. However, higher lead-free solder temperature did not produce thicker IMC layer between the solder and copper substrate of the printed circuit board (PCB). The IMC layer thickness did not change significantly with different reflow profile tested in this study. Nevertheless, it was increasing when exposed to second reflow soldering as shown in Figure 2. In fact, it was found that IMC thickness for Flux A after second reflow higher than Flux B. It is believed that this is due to lower wetting angle produced after first reflow allows the diffusion rate of Sn element from the solder becoming faster. Hence, thicker IMC produced.

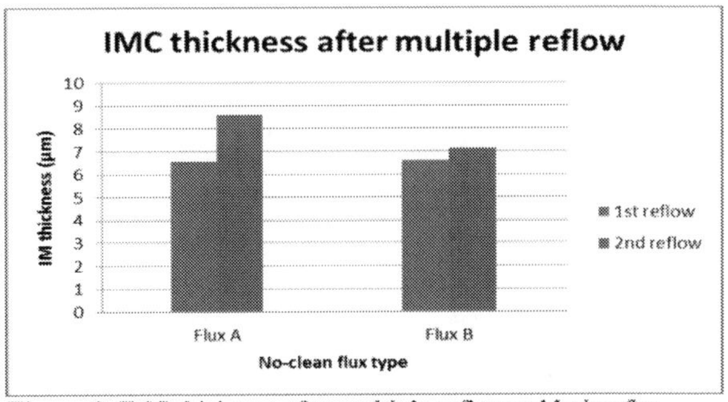

Figure 2. IMC thickness after multiple reflow soldering for both type of no-clean flux.

Moreover, after first reflow only Cu_6Sn_5 IMC was detected at solder/Cu interface (Figure 3a and Figure 3b). This was proven by energy dispersive xray (EDX) stated in Figure 4a and Figure 4b. While after second reflow, a very thin layer of Cu_3Sn started to form between Cu_6Sn_5 and copper substrate due to lack of Sn element diffused towards copper substrate. It was detected by EDX analysis shown in Figure 4c and Figure 4d. In addition, a thin layer of Cu_3Sn will have no effect on solder joint strength, but if the thickness is equal or more than those produced after isothermal aging, this flux cannot be used for long term usage.

DOI: 10.1051/matecconf/20167400038

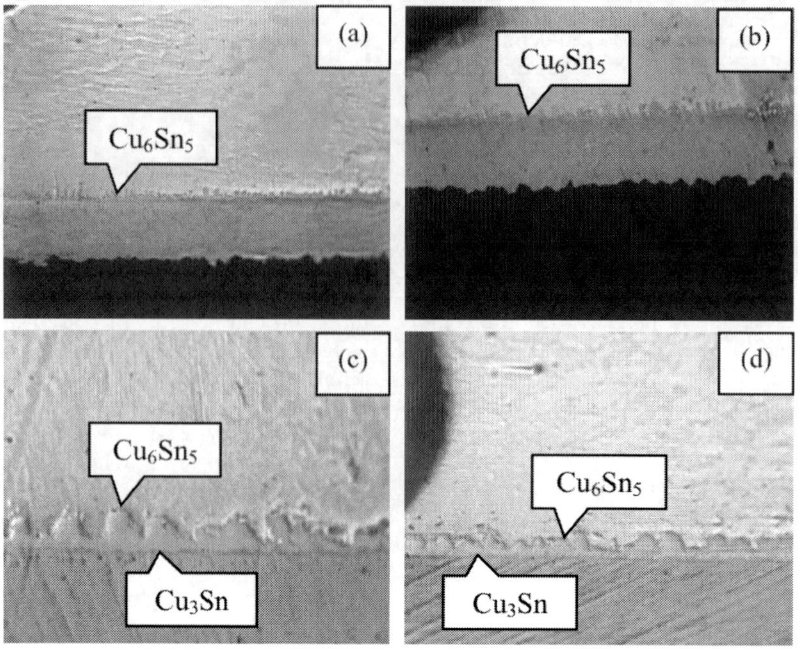

Figure 3. IMC morphology for both type of no-clean flux; (a) first reflow using Flux A, (b) first reflow using Flux B, (c) second reflow using Flux A, and (d) second reflow using Flux B.

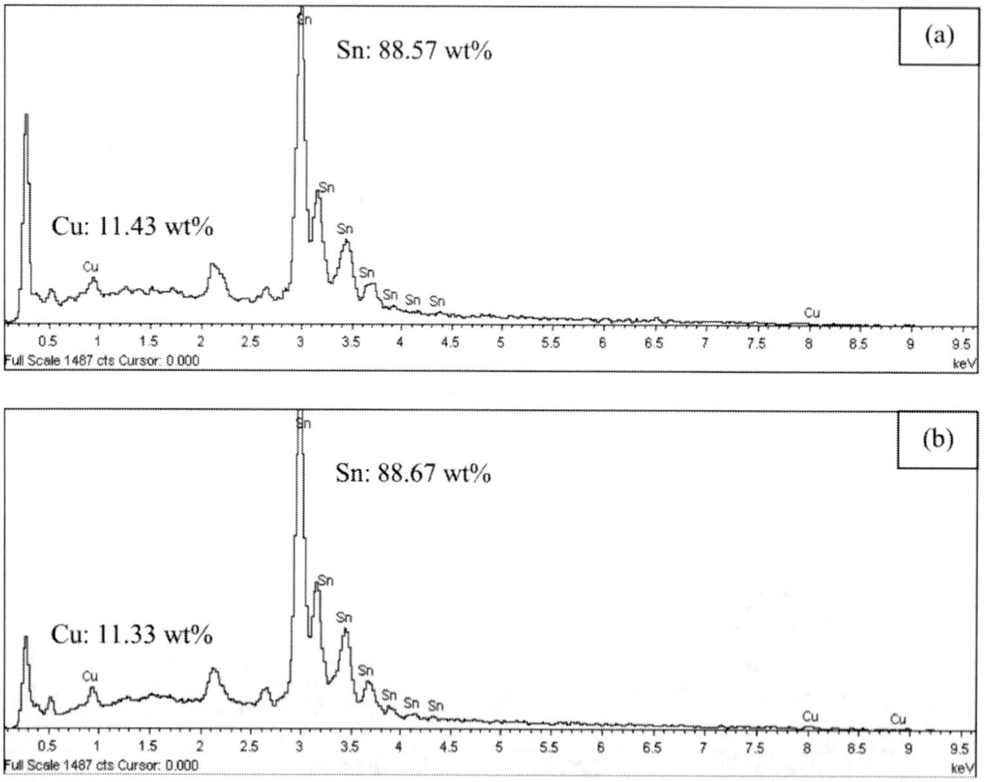

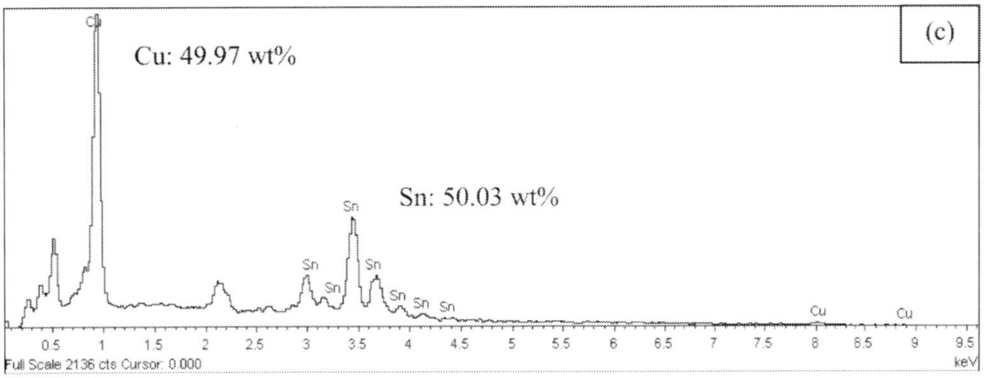

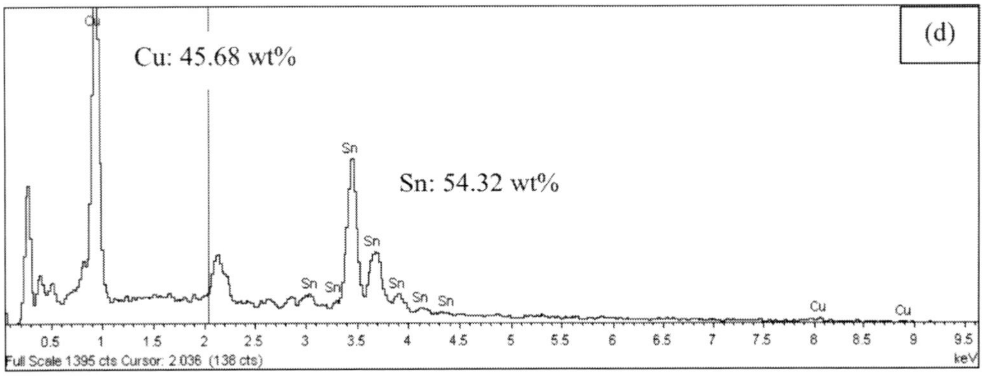

Figure 4. EDX analysis for (a) first reflow using Flux A, (b) first reflow using Flux B, (c) second reflow using Flux A, and (d) second reflow using Flux

4 Conclusion

The purpose of the current study was to evaluate wettability of the Sn-3.0Ag-0.5Cu solder paste with different type of no-clean flux, namely Flux A and Flux B. Besides, its effect towards multiple reflow was also studied as well as IMC thickness formed at the solder/Cu interface was measured. These findings suggest that Flux A is better in terms of soldering performance as compared to Flux B due to the reasons that it manage to produce lower wetting angle as well as suppress IMC thickness after first reflow soldering. However, the solder joint strength need to be evaluated for a further confirmation.

Acknowledgements

The authors would like to be obliged to Universiti Malaysia Pahang for providing laboratory facilities and financial assistance under project no. RDU130107.

References

1. X. Dongxia, L. Xusheng, W. Caiqin, and X. Bin. *Study on wettability and corrosivity of a new no-clean flux for lead-free solder paste in electronic packaging technology*. in *Mechanic Automation and Control Engineering (MACE), 2011 Second International Conference on*.(2011)
2. K. Vijay. *Miniaturization — Solder paste attributes for maximizing the print & reflow manufacturing process window*. in *Electronic System-Integration Technology Conference (ESTC), 2012 4th*.(2012)
3. R. Lathrop. *Solder paste printing and stencil design considerations for wafer bumping*. in *Electronics Manufacturing Technology Symposium, 2004. IEEE/CPMT/SEMI 29th International*.(2004)
4. I. Plotog, P. Svasta, T. Cucu, A. Vasile, and A. Marin. *Solder paste shelf life extending approach for prototyping and small series activities*. in *Electronics Technology, 2009. ISSE 2009. 32nd International Spring Seminar on*.(2009)
5. W. Schmitt. *Resin / Rosin Free Solder Pastes and Fluxes*. in *Electronics Systemintegration Technology Conference, 2006. 1st*.(2006)
6. M. Abtew and G. Selvaduray, *Lead-free Solders in Microelectronics*. Materials Science and Engineering: R: Reports, **27**(5–6): p. 95-141.(2000)
7. M. Alajoki, L. Nguyen, and J. Kivilahti. *Drop test reliability of wafer level chip scale packages*. in *Electronic Components and*

Technology Conference, 2005. Proceedings. 55th.(2005)

8. T. Gorewoda, J. Anyszkiewicz, Z. Mzyk, and Ł. Buzek, *The development of correct methods for Sn–Pb solder analysis by wavelength dispersion X-ray fluorescence spectrometry*. Spectrochimica Acta Part B: Atomic Spectroscopy, **92**(0): p. 29-33.(2014)
9. K.S. Kim, S.H. Huh, and K. Suganuma, *Effects of intermetallic compounds on properties of Sn–Ag–Cu lead-free soldered joints*. Journal of Alloys and Compounds, **352**(1–2): p. 226-236.(2003)
10. H. Nishikawa and N. Iwata, *Formation and growth of intermetallic compound layers at the interface during laser soldering using Sn–Ag Cu solder on a Cu Pad*. Journal of Materials Processing Technology, **215**(0): p. 6-11.(2015)
11. O. Saliza Azlina, A. Ourdjini, and I. Siti Rabiatull Aisha, *Effect of nickel doping on interfacial reaction between lead-free solder and Ni-P substrate*. p. 1375-1379(2012)
12. P. Liu, P. Yao, and J. Liu, *Effects of multiple reflows on interfacial reaction and shear strength of SnAgCu and SnPb solder joints with different PCB surface finishes*. Journal of Alloys and Compounds, **470**(1–2): p. 188-194.(2009)
13. I.S.R. Aisha, A. Ourdjini, M.A. Hanim, and O.S. Azlina, *Effect of Reflow Profile on Intermetallic Compound Formation*. IOP Conference Series: Materials Science and Engineering, **46**(1): p. 012037.(2013)

Experimental Investigation on Shear and Hardness of Abaca based Hybrid Composites

B Vijaya Ramnath [1*], C Elanchezhian [1], V.M Manickavasagam [2], S Gowri Prasad [3], S Arvindh Swamy [3] and R Keshav Raj [3]

[1] *Professor, Department of MechanicalEngineering, Sri Sai Ram Engineering College, Chennai - 600 044, Tamil Nadu, India.*

[2] *Research scholar, Department of Mechanical Engineering, Vels University, Chennai- 600 117, Tamil Nadu, India.*

[3] *Department of Mechanical Engineering, Sri Sai Ram Engineering College, Chennai - 600 044, Tamil Nadu, India.*

ABSTRACT. Present technology development in the area of materials replaces the conventional materials used in automobile and aerospace sector by composite materials due their less weight and bio degradability. This paper aims to fabricate and investigate the mechanical properties of abaca-raffia hybrid composite fabricated by hand layup process. Since, abaca fiber has more strength than other fibers like kenaf, banana and sisal, the composite with this fiber can be suitable replacement material for automotive applications.The properties like double shear and hardnessare evaluated andthe result shows that the double shear properties and hardness of the hybrid composites [GFRP +Abaca+Raffia] is higher than other two combinations. The internal microstructure of the hybrid composites were analysed using Scanning electron microscope (SEM).

1. INTRODUCTION

Nowadays, natural fibre composites are replacing conventional materials due to increase in environmental awareness. Beckermann and Pickering[1] studied the properties of the hemp reinforced withMaleic Anhydride-Grafted Polypropylene (MAPP) by injection moulding process and concluded that injection moulded hemp reinforced with MAPP has more tensile strength and Young's modulus compared to the hemp without reinforcement.Hautala et al[2] found a new method of making mats from hemp fiber and found that method for making mats from spring harvesting is more suitable for industrial purposes and it is also a cost efficient method. Pickering et al[3] studied the properties of alkali treated fibers and concluded that alkali treated fiber has higher crystallinity index than the untreated fiber and further concluded that by alkali treatment, the fiber strength is improved. Marianne Le Troedec et al [4]studied the treatment of the hemp fiber with Sodium Hydroxide[NaOH] and found that treatment with 6% of NaOH clean fibers by removing the amorphous compounds and it also increases the crystallinity index of the fiber . Vincent Placet [5] studied the thermal-mechanical behaviour of the hemp fiber and concluded that the rigidity of the fibers increasesduring the initial stresses and also found that the rigidity and endurance decreases due to the thermal degradation.Le [6] investigated the mechanical properties of hemp and starch composites and found that agro-composite material is used as a filling material in construction and is deformable and cannot be crushed under loading which will fail due to great displacement.Kabir [7] investigated the tensile properties of chemically treated hemp reinforced composites and concluded that chemically-treated fibres possess high tensile strength than that of untreated fibres.Adel RamezaniKakroodi et al[8] investigated the mechanical, morphological and water absorption properties of maleated polyethylene/hemp composites and concluded that tensile and flexural properties of themaleic anhydride polyethylene[MAPE] matrix,water absorption and impact strength (up to 50%)increases with hemp content. Yan et al [9] studied the reinforcement of hemp with polypropylene (PP), and concluded that the tensile, flexural and impact strengths of PP/NHF(Noil hemp fibre) and PP/SHF(scotched hemp fibre) increases with increasing amount of maleic anhydride polypropylene(MAPP).Enrico Sassoni [10] studied the application of novel sustainable hemp-based composites in the building industry. The panels exhibited promising physical, thermal and mechanical properties while the low density panels exhibited a good thermal conductivity and the medium density panels exhibited a higher thermal conductivity.Shinji Ochi [11] studied the cultivation of kenaf and its application to biodegradable composite materials and it was concluded that the highest temperature that does not affect the fibre strength is 160° C. Yousif et al[12] investigated the flexural properties of the treated and untreated kenaf fibre reinforced epoxy (KFRE) composites and their result revealed that flexural strength of the treated kenaf composite was superior when compared to the untreated composites. Vijaya Ramnath et al [13,14,15] investigated mechanical properties of hybrid composites made up of Jute with Abaca, banana fibers and concluded that hybrid composites has high strength than mono fiber composites. Ghani et al [16] studied the mechanical properties of Kenaf/Fiber glass polyester hybrid composite and concluded that, as the moisture penetration into the

vijayaramnath.mech@sairam.edu.in

composites increase, the mechanical properties of the kenaf fiber deteriorated. Nor Amalina Nordin et al [17] studied the wear rate of kenaf fiber composites. Long kenaf fibers were used as the raw material and it was concluded that sliding distance plays an important role in specific wear rate between the composites. Vijaya Ramnath et al. [18, 19, 21 ,22] evaluated the mechanical properties of Abaca and pineapple hybrid composite and concluded that natural fiber composites are found to have good mechanical properties and suitable for electrical and automotive applications. They also found mechanical properties of twisted kenaf fibers and concluded that fiber twisting has effect on mechanical strength of composites. Anuar al [20]studied the mechanical properties of reinforced thermoplastic elastomer composite with kenaf bast fibre and concluded that kenaf fibre played an important role in increasing the impact strength of the composites.

2. MATERIALS USED

In this work Abaca fiber which is also known as Manila hemp and Raffia which is a palm of Madagascar are used to fabricate the composites. Epoxy resin (Araldite LY 556) along with hardener is used to fabricate the composite. Glass fiber laminates are used as top and bottom most layer while the natural fibres are used as reinforcements between them.

3. FABRICATION PROCEDURE

The composites are fabricated by hand layup process. The GFRP is kept as top and bottom most layer of the composite. The first type composite is fabricated using GFRP with Abaca (composite I),while the second type composite is fabricated using GFRP and Raffia(composite II) and the third typecontains Abaca, Raffia and GFRP(composite III).Epoxy resin is used and the entrapped air is removed from the surface of the fiber by the use of roller.A load of 10-12kgs is applied for the curing time of15 hours to get required composite laminate.

4. TESTING OF MATERIALS

4.1DOUBLE SHEAR TEST

The shear test attachment is placed on the lower table, which contains a cutter. Now, the Load is applied such that the specimen breaks in two or three pieces. If the specimen breaks in two pieces it is single shear and if it breaks in three pieces then it is double shear.During test at a particular stage, straight fracture occurs, breaking the test piece into three pieces. Maximum load recorded by the pointer is noted down.Fig 1 shows double shear test specimen as per ASTM.

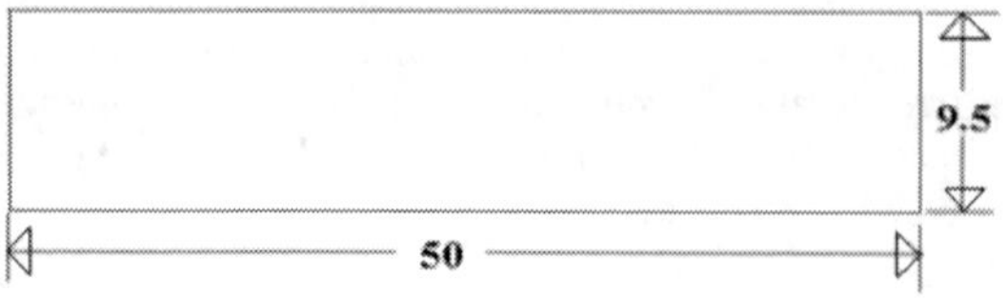

Figure 1: Double shear test specimen [ASTM: D5379]

4.2 HARDNESS TEST

Hardness is a measure of resistance offered by a material for indentation in compressive load. In this work, Rockwell hardness test was conducted by applying 60Kgf load with hardened steel ball indenter.

5. RESULTS AND DISCUSSIONS

5.1 DOUBLE SHEAR TEST

The result of double shear test of composite is shown in table 1.Double shear test was conducted on the composite specimens namely composite I, composite II and composite III and the load Vs. displacement graphs are plotted as shown in fig 2, 3, 4.

Table 1: Results of double shear test

Composites	Break Load (KN)	Displacement at F_{MAX} (mm)	Maximum Displacement (mm)	Ultimate Stress (kN/mm^2)
GFRP + Abaca	1.935	1.600	3.900	0.037
GFRP + Raffia	0.425	1.500	1.700	0.019
GFRP + Abaca + Raffia	2.290	2.100	4.700	0.044

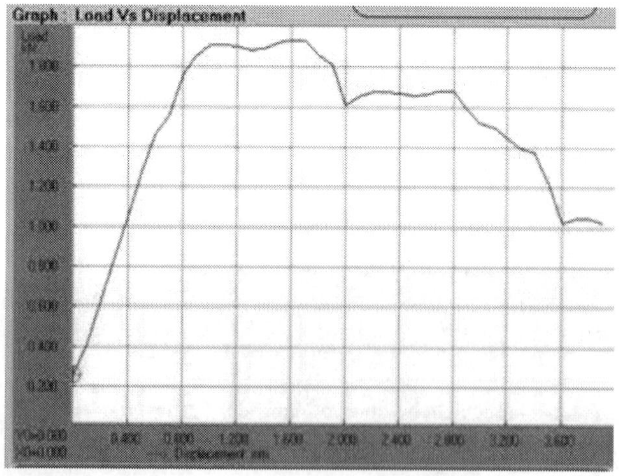

Figure 2. Double Shear Test Graph- (Composite I)

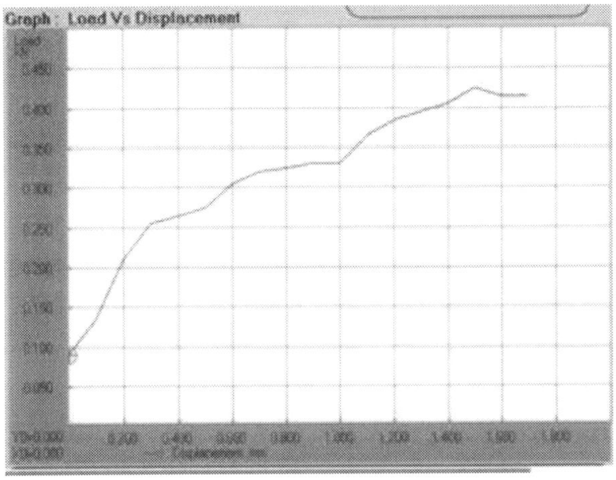

Figure 3. Double Shear Test Graph – (Composite II)

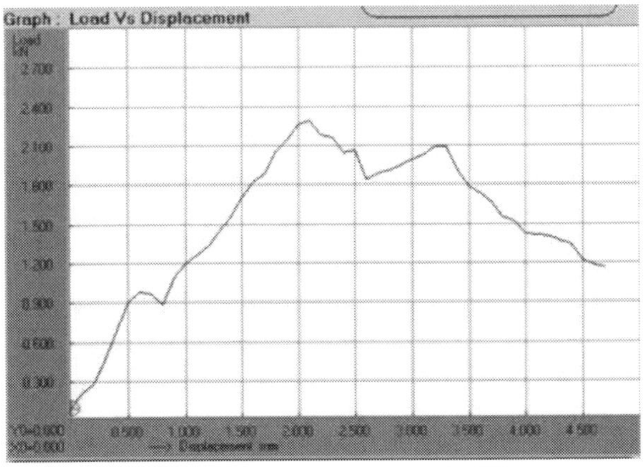

Figure 4. Double shear test graph (composite III)

From the figure 2, 3, and 4 it can be seen that the double shear properties of the hybrid composite III (GFRP+Abaca + Raffia) are higher than other two.

5.2 HARDNESS TEST

The result of hardness test of composites is shown in table 2 from which it is observed that composite III which consists of GFRP + Abaca + Raffia has more hardness than others. This is due to the presence of Raffia fiber along with Abaca which makes it harder to resist the indentation.

Table 2: Result of Hardness test

Sl.No	Composite	Hardness (HRC)
Composite I	GFRP+ Raffia	96
Composite II	GFRP+Abaca	99.5
Composite III	GFRP+Abaca+Raffia	109.5

6. MORPHOLOGICAL ANALYSIS

Morphological analysis is carried outto study the internal surface characteristics of the tested specimens.Scanning electron microscope(SEM) appliesbombardment of electrons on the fractured surface of the composite to form the image of the internal structures.

6.1SEM ANALYSIS OF COMPOSITES

Before carryingout SEM analysis, the tested samples are dried and coated with 15–20 nm thick gold layers using an Ion - Sputter coater device. The figure 5 shows the SEM image of the double shear tested specimen which indicates the fiber pullout and fiber breakage. This is due to inadequate time for curing of composite. Also, it is observed that dry surface in the inner layer of laminates which is the result of improper application of resin.

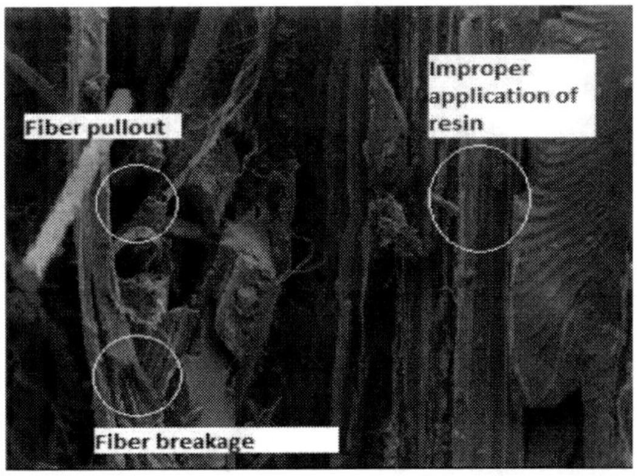

Figure 5. Double shear test SEM image

7. CONCLUSION

In this work, Abaca, Raffia based fiber composites are fabricated using hand layup method. The mechanical properties like double shear and hardness are found. The test result shows that among three composites the composite III (Abaca+Raffia+GFRP), shows better properties than others. Based on the results, following conclusions were inferred;

1.The ultimate strength in the double shear test of the composite III(GFRP + Abaca + Raffia) composite is 0.044 kN/mm^{2}which is higher than that of the GFRP + Abaca composite with 0.037 kN/mm^2 and GFRP + Raffia composite with a value of 0.019 kN/mm^2.

2. Also, GFRP+Abaca+Raffia composite exhibits high hardness of 109.5 HRC.

REFERENCES

[1]G.W. Beckermann, K.L. Pickering. Engineering and evaluation of hemp fibre reinforced polypropylene composites: Fibre treatment and matrix modification. *Composite Part A*: 39 (2008) 979–988.

[2]M. Hautala, A. Pasila, J. Pirila. Use of hemp and flax in composite manufacture: a search for new production methods. *Composite Part A* 35 (2004) 11–16.

[3]K.L. Pickering, G.W. Beckermann, S.N. Alam, N.J. Foreman. Optimising industrial hemp fibre for composites. *Composite Part* A 38 (2007) 461–468.

[4]Marianne Le Troedec, David Sedan, Claire Peyratout, Jean Pierre Bonnet, Agnes Smith, Rene Guinebretiere, Vincent Gloaguen, Pierre Krausz. Influence of various chemical treatments on the composition and structure of hemp fibres. *CompositePart A* 39 (2008) 514–522.

[5]Vincent Placet Characterization of the thermo-mechanical behaviour of Hemp fibres intendedfor the manufacturing of high performance composites.*Composite Part A* 40 (2009) 1111–1118 .

[6]A.T. Le , A. Gacoin , A. Li , T.H. Mai , M. Rebay , Y. Delmas. Experimental investigation on the mechanical performanceof starch–hemp composite materials. *Construction and Building Materials*61 (2014) 106–113.

[7]M.M. Kabir , H. Wang , K.T. Lau ,F. Cardona. Tensile properties of chemically treated hemp fibres as reinforcement for composites. *Composite Part B 53* (2013) 362–368.

[8]Adel Ramezani Kakroodi, Yasamin Kazemi, Denis Rodrigue. Mechanical, rheological, morphological and water absorption properties of maleated polyethylene/hemp composites: Effect of ground tire rubberaddition *CompositePart B* 51 (2013) 337–344.

[9]Z.L. Yan , H. Wanga, K.T. Lau , S. Pather, J.C. Zhang , G. Lin , Y. Ding. Reinforcement of polypropylene with hemp fibres.*CompositePart B* 46 (2013) 221–226.

[10]Enrico Sassoni, Stefania Manzia, Antonio Motoria, Matteo Montecchid, Max Cantie. Novel sustainable hemp-based composites for application in thebuilding industry: Physical, thermal and mechanical characterization.Energy and Buildings77 (2014) 219–226.

[11]Shinji Ochi. Mechanical properties of kenaf fibers and kenaf/PLA composites. Mechanics of Materials 40 (2008) 446–452

[12]B.F. Yousif, A. Shalwan, C.W. Chin , K.C. Ming. Flexural properties of treated and untreated kenaf/epoxy composites.Materials and Design 40 (2012) 378–385

[13] B Vijaya Ramnath, V.M.Manickavasagam, C Elanchezhian, C Vinodh Krishna, S Karthik, K Saravanan. Determination of mechanical properties of intra-layer abaca–jute–glass fiber reinforced composite, *Materials and Design*, Volume 60(2014), Pages 643-652.

[14] B.V ijaya Ramnath, V. Santhosh Kumar, P.V. Nirmal, S.Karthick, G. Prem Kumar, C. Elanchezhian , S.Rajesh , K. Suresh. Experimental Investigation of Mechanical behavior of Jute-Flax based Glass Fiber reinforced composite", *Fibers and polymers, Volume 15 (2014)*, Pages 1251-1262.

[15] B.Vijaya Ramnath, R. Sharavanan, M. Chandrasekaran, C. Elanchezhian, R. Sathyanarayanan, R. Niranjan Raja, and S. Junaid Kokan, "Experimental Determination of Mechanical Properties of Banana Jute Hybrid Composite", *Fibers and polymers, Volume 16*, Pages 164-172, 2015.

[16]M.A.A Ghani, Z. Salleh, Koay Mei Hyie, M.N.Berhan, Y.M.D. Taib, M.A.I. Bakri., Mechanical Properties of Kenaf/Fiberglass Polyester Hybrid Composite. *Procedia Engineering* 41 (2012) 1654 – 1659

[17]Nor Amalina Nordin, Fauziah MdYussof, SalmiahKasolang, Zuraidah Salleh and Mohamad Ali Ahmad.Wear Rate of Natural Fibre: Long KenafComposite. *Procedia Engineering* 68 (2013) 145 – 151

[18] B.VijayaRamnath, C. Vinodh Krishna, S. Karthik, K. Saravanan, V.M. Manickavasagam, C. Elanchezhian. Evaluation of the Flexural Properties of Pineapple Reinforced Polymer Compositefor Automotive and Electrical Applications, *Advanced Materials Research*. 893, (2014) 271-274.

[19] B.Vijaya Ramnath, R. Niranjan raja, S. Junaid Kokan, R. Sathya Narayanan, S. Rajesh, and V.M. Manickavasagam. Fabrication and testing of Abaca fiber reinforced Epoxy composites for automotive applications *Advanced Material Research*, Vol.718-720,2013, pp 63-68.

[20] H. Anuar , A. Zuraida. Improvement in mechanical properties of reinforced thermoplastic elastomer composite with kenaf bast fibre. *Composites Part B 42* (2011) 462–465.

[21] B.Vijaya Ramnath, S. Rajesh, C. Elanchezhian, A. Santosh Shankar, S. Pithchai Pandian, S. Vickneshwaran, and R. Sundar Rajan, Investigation on Mechanical Behaviour of Twisted Natural Fiber Hybrid Composite Fabricated by Vacuum Assisted Compression Molding Technique, Fibers and polymers, Vol.17,2016, pp 80.

[22] B.Vijaya Ramnath, R. Niranjan Raja, S. Junaid Kokan, R. Sathya Narayanan, A.R.Rajendra Prasad, C.Elanchezhian and V.M.Manickavasagam, "Evaluation of mechanical properties of abaca–jute–glass fibre reinforced epoxy composite", Materials and design, Vol. 51(2013), Pages 357-366.

AUTHOR INDEX

AUTHOR INDEX